Services and metropolitan development

T0264818

The dynamics of national and international urban systems, as well as individual metropolitan areas, are now more closely connected than ever with the decisions and actions of firms and institutions in the service sector. *Services and Metropolitan Development* explores the processes guiding both the development and the spatial impacts of services on the urban system and individual urban areas. The book describes the symbiotic relationship between the internationalisation of services and the effects of this restructuring on urban systems.

The multidisciplinary nature of the subject and its global development are reflected by the international range of contributors; specialists in geography, business management, economics and public administration. The book analyses the theoretical, conceptual and measurement issues confronting research on the development of services in North America, Western Europe and Australia.

Services and Metropolitan Development will be of interest to students and academics in urban studies, geography, economics, planning and business studies.

P.W. Daniels is Professor of Geography at Portsmouth Polytechnic and Director of the Service Industries Research Centre.

Services and metropolitan development

International perspectives

Edited by
P.W. Daniels

Routledge
Taylor & Francis Group

LONDON AND NEW YORK

First published 1991 by Routledge

2 Park Square, Milton Park, Abingdon, Oxon OX14 4RN
711 Third Avenue, New York, NY 10017, USA

Routledge is an imprint of the Taylor & Francis Group, an informa business

First issued in paperback 2016

Transferred to Digital Printing 2005

Copyright © 1991 P.W. Daniels
Typeset by Leaper & Gard Ltd, Bristol, England

British Library Cataloguing in Publication Data

Services and metropolitan development.
    1. Urban regions. Economic aspects
    I. Daniels, P. W. (Peter Walters) *1945–*
330.91732

    ISBN 978-0-415-00852-5 (hbk)
    ISBN 978-1-138-98168-3 (pbk)

Library of Congress Cataloging in Publication Data

Services and metropolitan development / edited by P.W. Daniels.
        p.   cm.
    Includes bibliographical references and index.
    ISBN 0-415-00852-2
    1. Service industries   2. Service industries workers–Supply and demand.
   3. Economic development.   4. Metropolitan areas. 5. Urbanization.
   l. Daniels, P.W.
   HD9980.5.S43   1991
   338.4–dc20   90-47660
                                                          CIP

Contents

Tables

Figures

Contributors

Antoine Bailly is Professor of Geography at the University of Geneva and past president of the Association de Science Regionale de Langue Française. He is the author of many books in urban, behavioural and medical geography, and with Professor D. Maillat author of the book *Le Secteur Tertiaire en Question* (1987 and 1989). He has undertaken researches and publications on the development of service activities in Europe and Canada.

William B. Beyers ia Professor of Geography at the University of Washington. He is author of 'Speed, information exchange, and spatial structure' in *Information Society and Spatial Structure*, and 'Structural change in interregional input-output models: form and regional development implications' in *Frontiers of Input-Output Analysis*. He has also recently completed a major study of US producer services for the US Economic Development Administration entitled 'The producer services and economic development in the United States: the last decade'.

Joanne Brion is a graduate student in the Urban Planning Program at New York University where she is completing the Master of Urban Planning degree.

M.T. Daly is McCaughey Professor of Geography and Director of the Research Institute for Asia and the Pacific, University of Sydney. He is author of *The Brittle Rim: Business and Finance in the Pacific Region* (1989) and *Sydney Boom, Sydney Bust* (1982). His most recent work has concentrated on the relationship of finance to development in the Asian nations of the Pacific Rim. His current research is focused on the investment impacts on Australia of the European Single Market.

P.W. Daniels is Professor of Geography and Director of the Service Industries Research Centre, Portsmouth Polytechnic. He is the author of *Office Location: An Urban and Regional Study* (1975), *Service Industries: A Geographical Appraisal* (1985) and editor of *Spatial Patterns of Office*

Growth and Location (1979). He has undertaken research and publication on aspects of metropolitan office location and on the development of producer services in the United States, Australia and Europe. He is currently directing a two-year ESRC-funded project on office development, information technology and planning in the City of London.

Gregor Dürrenberger has a PhD in geography and is researcher in the human ecology group of the Geography Department at the Swiss Federal Institute of Technology (ETH) in Zurich. He is co-author of *Telearbeit-von der Fiktion zur Innovation* (1987). He is undertaking research and publication in the field of new technologies and metropolitan development.

David W. Edgington is Assistant Professor at the Department of Geography, University of British Columbia. At the time of contributing to this volume he was Co-ordinator, Economic Development Unit, Ministry for Planning and Environment, State of Victoria. He has undertaken research and publications on aspects of urban and regional change in Australia and countries of the Pacific Rim.

Michael A. Goldberg is the Herbert R. Fullerton Professor of Urban Land Policy in the Faculty of Commerce and Business Administration, University of British Columbia. His most recent research focuses on questions relating to international financial and business issues. His book *The Chinese Connection: Getting Plugged in to Pacific Rim Real Estate, Trade and Capital Markets* (1985) explores the real estate investment behaviour of southeast Asian ethnic Chinese. He is also conducting extensive background research for the Province of British Columbia on the feasibility of establishing Vancouver as an international financial centre. He is currently on leave from UBC to serve as Executive Director, IFC Vancouver, a provincially sponsored non-profit society dedicated to promoting Vancouver as an international financial centre. He is also Chairman of the Real Estate Foundation, on the Board of Directors of the Hong Kong Canada Business Association in Vancouver, and a member of the Vancouver Board of Trade.

Robert W. Helsley is the Real Estate Foundation Junior Faculty Professor of Urban Land Economics at the Faculty of Commerce and Business Administration, University of British Columbia. He holds a PhD in Economics from Princeton University, and has been at UBC since 1984. His research interests lie in analytical urban economics and public finance. His recent research examines land pricing under growth and uncertainty, the microfoundations of agglomeration economies, urban capital markets, and equilibria in systems of cities.

Thomas A. Hutton is Assistant Manager (Policy) in the Economic

Development Office, Finance Department, City of Vancouver. He has also held teaching positions since 1982, and is currently Adjunct Professor, in the School of Community and Regional Planning, University of British Columbia. He has published widely on office location, urban service sector development, and metropolitan planning policy. At present, he is engaged in research on urban development within the Pacific Rim.

Carlo Jaeger, born in Switzerland, works with the human geography ecology group at the Swiss Federal Institute of Technology (ETH) in Zurich. He obtained his masters degree in Sociology at the University of Berne and his PhD in Economics at Johann Wolfgang Goethe University in Frankfurt. He has worked at length on the topics of industrial relations and technological change. He is co-editor of *Information Society and Spatial Structure* (1989). His current research activities are focused on the development of interdisciplinary theory in the social sciences.

P-Y. Leo is a senior research fellow in economics, University of Aix-Marseille III. He is the author of numerous articles on regional development and small industrial firms. He is co-author of *PMI: Strategies Internationales* (1989). He is completing research on producer services economic profiles and on international relations of rank B cities.

Maurice D. Levi is Bank of Montreal Professor of International Finance at the Faculty of Commerce and Business Administration, University of British Columbia. His papers in macroeconomics and international finance have appeared in most leading economics journals including *The American Economic Review, Journal of Political Economy, The Journal of Finance, Econometrica, Journal of Monetary Economics, Journal of International Economics* and *Journal of Econometrics.* He is author of several books in finance and economics inlcuding *International Finance* (1983) and *Economics Deciphered* (1981).

David Ley is Professor of Geography at the University of British Columbia, Vancouver. He is author of *A Social Geography of the City* (1983), co-author of *Neighbourhood Organizations and the Welfare State* (forthcoming) and co-editor of *The Social Geography of Canadian Cities* (1990). His research is concerned with downtown and inner city issues, including relations between downtown labour markets and inner city housing markets. He is presently drawing this work together in a book, *Remaking the Canadian City.*

Denis Maillat is Professor of Economics at the University of Neuchatel and Director of the Institut de Recherches Economiques et Regionales de Neuchatel. He is author of many books on regional development in Switzerland and wrote with Professor A. Bailly a book *Le Secteur Tertiaire* *en Question.* He has undertaken researches and publications on production systems and labours markets in Europe.

Mitchell L. Moss is Professor of Public Administration and Planning and Director of the Urban Research Center, New York University. He has conducted research on the relationship of communications technology to urban development and has served as a consultant on telecommunications policy to public and private organizations. Professor Moss's articles have appeared in *Urban Studies, Journal of Communications, Urban Affairs Quarterly,* and *The New York Times.*

Thierry Noyelle is Senior Research Scholar and Deputy Director, Conservation of Human Resources, Columbia University. He is the author, co-author, and editor of ten books, including *The Economic Transformation of American Cities* (1983) co-authored with Thomas M. Stanback, Jr. His current research focuses on international trade in services. Also, with Penny Peace, he is currently managing the development of a directory and database of the world's 500 largest service corporations on behalf of the United Nations Centre on Transnational Corporations.

Kevin O'Connor is Senior Lecturer in the Department of Geography and Environmental Science, Monash University, Melbourne. His research centres on the patterns of development of metropolitan areas, both at the inter-urban and intra-urban scale. In recent years he was seconded to the State Government Ministry for Planning and Environment, and has been involved in the development of planning policy for Melbourne. He is still engaged as a consultant in that capacity. Current research includes an Australian Research Council funded project on recent economic change in Australia.

Penny Peace is researcher at Conservation of Human Resources, Columbia University. She is co-author of *The Information Industries: New York's New Export Base* and *Computer Software and Data Processing in New York City.*

J. Philippe is a Senior Research Fellow in Economics, University of Aix-Marseille III. He is the author of various articles in collective books on service industries and on African development. He is co-author of *PMI: Strategies Internationales* (1989) and editor of *Nigeria: un pouvoir en puissance* (1988). He is currently undertaking research on information technology and its impact on services and on the quality of producer services.

Peter J. Rimmer is Senior Fellow, Department of Human Geography, Research School of Pacific Studies, The Australian National University,

Canberra. In the producer services field, he is the author of two monographs: *Consultancy Services: Supply to Southeast Asia from Australia* (1984) and *Australian and Japanese International Consultancy Services* (1986). Stemming from his original interest in transport and communications, he has undertaken research and publication on the development of engineering consultancies, particularly within the Western Pacific Rim. In turn, this has led to research into construction contracting in Japan and the role of agents involved in translating capital into the built environment.

Pieter P. Tordoir is Senior Researcher at the Institute of Spatial Organization of the Netherlands Organization of Applied Science Research (INRO-TNO). He has an MA in economic geography, University of Amsterdam and a PhD from Johns Hopkins University, Baltimore. Mr Tordoir has conducted various research projects concerning the economics and geography of services for government agencies and private enterprises. His recent research deals with future developments in high-value intermediate services and viable development strategies for private enterprise and public agencies.

Barney Warf (PhD 1985, University of Washington) is an Assistant Professor of Geography at Kent State University, Ohio, USA. During the mid-1980s he was employed as an Economic Analyst for the Port Authority of New York and New Jersey. His research interests include political economy, international trade, services in telecommunications, and regional development.

Preface

The dynamics of national and international urban systems, as well as individual metropolitan areas, are now more closely connected than ever with the decisions and actions of firms and institutions in the service sector. One of the clearest implications of this new phenomenon is the way in which trade in services (much of which has its origins and destinations in large cities around the world) has recently become a prominent and contentious issue in the Uruguay Round of Negotiations for the General Agreement on Tariffs and Trade (GATT).

It therefore seemed timely to bring together a collection of papers which reflects this additional dimension of urban and economic analysis. Since the processes involved are truly manifest at the international scale it has been considered important to assemble a group of contributors from around the world. A second objective has been a search for the breadth conferred by a multi-disciplinary approach which emphasizes the most recently completed, as well as on-going, research. Geographers, business management specialists, economists, and public administration specialists from Australia, Canada, United States, France, Switzerland, the Netherlands and the UK have been asked to explore the processes guiding both the development and the spatial impacts of services on the urban system (national and/or international) and, where appropriate, in individual urban areas. All the contributors are leading researchers in their specialist subject; each has been encouraged to include material on the theoretical, conceptual, and measurement issues confronting research on the development of services, notably producer services. Most have chosen to present some of the results of research recently completed or in progress on topics which have international, national or local ramifications. Wherever possible all the contributors have been encouraged to evaluate the relevance of their research findings for the formulation of public policies towards services in an urban development context. While the interpretation of the brief for the volume has inevitably varied according to research priorities and disciplines there is evidence for a good deal of agreement about the symbiosis between the continuing internationalization and restructuring of services and their effects on the global urban system. There is greater variation at the lower level of national urban systems. Although the evidence is still rather limited there are indications that the impact of highly dynamic service growth on individual large metropolitan areas is characterized by convergence with respect to social, economic and land use effects.

P.W.D.
Service Industries Research Centre
Portsmouth Polytechnic, UK

Acknowledgements

I am grateful to all the contributors for quietly enduring numerous requests for revisions, additional information or in some instances, reminders that a manuscript was overdue! The enthusiasm shown by them all for this project has been encouraging and it is hoped that the finished product does justice to their support. The Cartographic Unit in the Department of Geography, Portsmouth Polytechnic re-drew some of the figures and Roger Homer, the Department's photolithographer produced the bromides. I am particularly appreciative of Carol Smith who carefully re-typed all the tables in the manuscript and typed numerous revised versions of Chapter 1 most patiently.

Editing books, in common with most other scholarly pursuits, requires peace and quiet as well as the occasional diversion to remind one that there is a world beyond producer services and cities. My wife, Carole, and our children, Paul and Charlotte, have always been a crucial part of this process. I dedicate this book to them.

1 Service sector restructuring and metropolitan development: processes and prospects

P.W. Daniels

ECONOMIC TRANSFORMATION AND SERVICES

The economic and urban landscape is increasingly shaped by forces operating at a global rather than a national scale (Cooke 1986; Dicken 1986; Knox and Agnew 1989). Although the transition from pre-industrial, through industrial to post-industrial economic structures is variable in time and space, particularly between core and periphery, it is clear that an increase in the 'openness' of individual national economies is a key characteristic of the transition. As the Industrial Revolution unfolded long-distance economic interaction became a necessity because of the growing discontinuity between economic and political institutions (earlier economic landscapes had been dominated by single major players such as China) (Knox and Agnew 1989). Such fragmentation encouraged capitalist modes of production and more diverse division of labour. But some nations within this more fragmented economic system responded more quickly than others to the opportunities for trade or market control, thus promoting unevenness in rates of development and therefore in the spatial effects of capitalist production.

The distinction between dominant cores, specializing in particular types of manufacturing industry and in international trade, and peripheries dominated by cash-crop production on large estates or plantations was already part of the pre-industrial foundation for the subsequent development of global economic landscapes (Knox and Agnew 1989). The way in which the world economy was beginning to function was accompanied by significant impacts on settlement and urbanization (Johnston 1980). The hierarchical systems of settlement both within countries and accompanying colonial expansion into overseas territories were further strengthened. Mercantile capitalism was the basis for these developments which were subsequently carried over to industrial capitalism with some additional dimensions. These included agglomeration of manufacturing activity, the introduction into the urban hierarchy of specialized towns (in heavy manufacturing, transport, resource extraction) and a continuation, often with an accelerating trajectory, of the expansion of the primate cities already extant during the pre-industrial/mercantile capitalist period. There was also an important extension of a world-economy, initially largely centred in Europe, to a truly global scale with increasing interdependence between core and periphery and further developments in the spatial division of labour (see Wallerstein 1979).

According to Lash and Urry (1987) industrial capitalism makes a further transition to organized capitalism (see also Martin 1988). Among the principal features are the increasing separation of ownership from control of economic organizations (including a related development of managerial hierarchies), and the dominance of particular regions by large metropolitan areas. This is a necessary prerequisite for the emergence of advanced capitalism. It also signifies that global interdependence is not a feature of economic organization that has emerged only recently. Indeed the 'ascent of the industrial core regions could not have taken place without the foodstuffs, raw materials and markets provided by the rest of the world' (Knox and Agnew 1989: 169); the events that have unfolded during advanced capitalism are to a considerable extent a 'fine tuning' or 'restructuring' of an existing system of metropolitan hierarchies and organization.

It will also be apparent that the foundation of industrial capitalism and its global expression was provided by manufacturing production and the harnessing of human and physical energy. Service industries were essentially cast in a supporting role. The harnessing of knowledge or information for the economic development process was in its infancy during organized capitalism but has since, arguably, become more important than physical energy in the development process. Advanced capitalism is founded on knowledge and information as capital rather than on human energy and mechanical energy. One indicator is the way in which the manufacturing component of advanced economies has been contracting (Table 1.1): employment in transformative activities (construction, manufacturing, other) decreased from 28.2 per cent to 23.1 per cent of all employment in the United States between 1973 and 1984, from 42.3 per cent to 33.9 per cent in Japan (1971–84) and 38.7 per cent to 32.2 per cent in France (1971–84) (Elfring 1988; UNCTAD 1989; see also Riddle 1986; Cuadrado and Rio 1989). The share of services in the gross national product of fifty-eight countries (out of a total of eighty-three providing data to the World Bank) has increased over the last twenty years from an average of 55 per cent in 1965 to 61 per cent in 1985 for developed market economies, and from 42 per cent to 47 per cent respectively for developing countries (Dunning 1989). Dunning lists six conditions as determinants of the share of services in a country's GNP: the level and pattern of demand for consumer services; the extent to which services enter into the exchange economy; the complexity of the production process and the role of services within it; the organization of service production; the economic structure of the country; and the state of technology in supplying services or goods-embodied services.

Table 1.1 Employment by industry (%) in United States, Japan and France

Industry	United States 1973	United States 1984	Japan 1971	Japan 1984	France 1971	France 1984
Extractive[1]	6.3	5.6	16.1	10.2	11.6	9.5
Manufacturing	21.9	17.5	26.2	23.7	27.1	22.7
Services	71.8	76.9	57.9	66.0	60.7	68.0
Construction	5.4	5.1	8.6	9.0	9.3	7.4
Transport/Comm.	5.1	4.6	6.3	5.9	5.8	6.5
Wholesale	5.0	5.3	6.4	7.0	4.4	4.6
Retail	11.4	11.5	10.7	12.1	8.5	8.9
Financial	4.8	5.8	3.1	3.9	2.7	3.4
Business	3.9	6.5	3.4	5.1	3.3	4.5
Domestic/Personal	10.9	12.2	8.9	10.1	7.5	7.7
Social	16.2	17.9	6.9	9.5	11.6	16.1
Government	9.1	8.0	3.6	3.4	7.6	8.9
Total	100.0	100.0	100.0	100.0	100.0	100.0
Average annual growth 1979–85	1.4		−0.3		1.0	

Note: 1. Includes utilities, agriculture, mining
Source: UNCTAD (1989), Tables 22 and 23

This is not to say that manufacturing industries do not use knowledge and information. Quite the contrary, these have become essential inputs to the production process. The decisions made by the manufacturing sector on how to use or to acquire these services for product innovation and development has been one of the factors contributing to the growth of certain service industries (see for example MacPherson 1989). Such is the specialization of the inputs represented by some services, however, that it is sometimes easier to purchase them from specialist suppliers than to produce them in-house. The process of externalization has therefore stimulated the growth of service producers, although some would argue that it is simply a redistribution of work which essentially owes its origins to the manufacturing sector. It reflects a shift in how production is taking place as well as in what is being produced (Noyelle and Stanback 1988). Thus, between 1975 and 1981 expenditure on producer services in manufacturing as a proportion of output increased from 13.7 per cent to 16.1 per cent in Italy, 7.7 per cent to 10.1 per cent in the UK, and from 12.8 per cent to 14.1 per cent in West Germany (Green 1985).

Advanced capitalism has been accompanied by very large increases in the geographical extent of markets, especially in relation to services. This is intimately connected with developments in technology which have revolutionized both the quantity and types of information which can be handled using electronic methods. In addition, of course, technology has transformed the time/distance environment within which information (services) is exchanged; satellite technology and glass fibre optics, for example, enable almost instant communication between locations scattered around the globe. The interaction between the flexible geography of markets and the enabling function of technology has provided the spur for the growth of large companies which can afford the costs of operating in extensive markets (including the investment in technology) and which are best able to protect their market share by diversifying and/or controlling as much of the supply side as possible. Large manufacturing corporations emerged during organized capitalism; large service corporations have become prominent players in those economies that have moved towards advanced capitalism.

The appearance of completely new markets for services or new services for markets has not only contributed to the expansion of service organizations; governments have become increasingly involved in creating more deregulated, yet properly monitored and controlled, environments for business. The latter is seen as essential if national economies are to retain or even to increase their share of advanced services which in certain respects, have more flexibility in their choice of location than ever before. But deregulation of financial markets or telecommunications services also requires mechanisms and institutions for ensuring that the more 'open' environment operates fairly and with due regard for the public interest. New service-type jobs therefore result. In addition, the rising incomes that have accompanied post-war growth of advanced economies have encouraged the demand for public non-profit services such as health and education.

There have therefore been some selective trends in the growth of service employment. While the growth arising from final demand (consumer services) levelled off during the 1980s, it has been rapid in public sector (non-market) services provided collectively to consumers. The highest rate of employment growth, however, has taken place among services associated with intermediate demand (producer services) from other industries (see for example Stanback 1981; Gershuny and Miles 1983; Marshall *et al.* 1988; Elfring 1988, 1989).

EXPLANATIONS AND SYMPTOMS OF THE SHIFT TO SERVICES

It is important to stress that the factors already cited as contributing to the growth of service activity have actually only entered the equation comparatively recently. There are a number of more established explanations. First, that the shift to services is related to the increasing demand generated by final consumers with greater disposable income as societies become more prosperous (the 'income elasticity' explanation). Second, the growth of services is related to the increasing demand for producer services (from other sectors in the economy). These explanations were originally advanced by Clarke (1940), Fisher (1935) and later by Greenfield (1966). Third, the lower rate of increase in productivity of services compared with manufacturing has ensured a steady transfer of labour as demand for service output has grown. It has been difficult for these 'conventional' explanations to attain credibility largely because of the problems of measurement. The productivity of services is notoriously difficult to measure (see for example Gershuny and Miles 1983; Noyelle and Stanback 1988; Elfring 1989); the output is unstandardized, quality varies widely, price indexes are not available and, for the public sector services, the prices do not represent the full costs of production because of the use of subsidies. Noyelle and Stanback (1988: 22) conclude that 'for at least a number of the services, productivity is not only unmeasurable ... but may, indeed, be an inappropriate concept to evaluate the contribution of services to the workings of advanced economies'. In relation to the income elasticity of demand explanation, Gershuny and Miles (1983) suggest that it is the price elasticity of demand for services that is more significant since the rising relative price of services will cause users to find innovative ways of providing service functions. This is the key, they suggest, to their more optimistic view of the prospects for continuing expansion of service sector employment during the 1990s (see also Gershuny 1978). Elfring (1989) also provides data showing that price increases in services have been above average between 1960 and 1984.

While the arguments continue about how best to measure and explain the emergence of services in all economies, there is probably more agreement about their impact on the labour force. After acknowledging the even greater problems of assembling suitably comparable occupation data, Gershuny and Miles (1983: 58) conclude that 'the patterns of change over time, are remarkably similar between countries', that the '"tertiarization" of the economy is more a consequence of changing occupational profiles within industrial sectors than of changing patterns of demand between them' (p. 65). On the basis of data for France, Ireland, Italy and the UK they show that manual occupations declined everywhere, administrative, technical and clerical staff increased, as did clerical workers although the increases for the latter slowed down during the 1970s. Transport workers in services declined in the 1970s but then numbers increased in manufacturing (as large food producers, for example, adopted greater vertical integration to include distribution functions (see for example, McKinnon 1989)). Put another way, white collar occupations have been increasing while blue collar occupations have been declining, in all four countries. This change is not unrelated to innovations in production in both manufacturing and service industries; these require greater knowledge and information inputs than physical inputs.

But the significance of the shift to white collar occupations (in relation to the theme of this volume) becomes clearer when some related changes in the workforce are also outlined. The first major change has been the increasing participation of women in the white collar labour force, many of them in part-time employment. In the US for example, 14.5 million jobs were added to the economy in the 1980s; some 25 per cent were part-time and almost 66 per cent have been filled by women (Christopherson 1989). Approximately 89 per cent of the part-time labour force is employed in services with a large proportion concentrated in retail and in wholesale trade. But the fastest growing group is temporary workers most of whom are women who are mainly employed by large firms in both consumer and producer services (Christopherson 1989). It will be apparent that these changes in the service industry labour force represent a more flexible approach to production necessitated by the preference of women for part-time work or the needs of firms wishing to respond quickly to business cycles which are much shorter. Rapidly fluctuating demand for a wider range of more specialized service products also creates variable labour demand. The use of temporary workers allows them to get around social security and related employment regulations which inhibit rapid adjustment of labour inputs in very competitive markets.

A second change that has accompanied the shift to services and white-collar occupations has been a rise in demand for better-qualified labour (Bertrand and Noyelle 1986). An expansion of administrative, professional and technical occupations has accompanied the above-average growth of producer services. As knowledge and information-intensive activities they require a high proportion of employees with specialized knowledge or professional qualifications in areas such as computer programming, design, marketing, securities dealing, risk assessment, actuarial work, or futures trading, for example. The combination of high growth in producer services, globalization of large service organizations and expanding international trade in services means that the availability of highly trained, well educated labour is a crucial discriminant for the locational choice of such services (Bertrand and Noyelle 1986; Dunning 1989). Differences between national educational systems produce contrasting labour markets at inter-national level while there will also be wide variations within countries such as France (Pumain and Saint-Julian 1986). But there is also a simultaneous process of polarization in services employment; low-skill, low wage occupations have also experienced significant growth, especially in consumer services and to a lesser degree in non-profit, public services.

LOCATION AND INTERNATIONAL TRADE IN SERVICES

It has been necessary to outline the relationship between the transfor-mation of economies and services because it ultimately translates into distinctive spatial outcomes at the national and international level. This is most evident for producer services which are the focus of attention in most of the contributions to this volume. Before turning to the locational outcomes that have accompanied the expansion of producer services it is, however, necessary to consider one other development which helps to understand these patterns; trade in services.

A good deal of the international provision of services such as shipping, transport and other travel is trade-orientated while banking, finance services and telecommunications are foreign direct investment (FDI) orientated. Thus, the services share of FDI outflows has risen from 41 per cent at the beginning of the 1970s to 49 per cent in 1978–80 for the UK and from 20 per cent to 67 per cent in 1984 for Japan (United Nations 1983). Such trade had been taking place between countries such as the UK and its trading partners since the nineteenth century but until recently it had comprised but a minor share of total world trade. Information technology has, however, transformed the conduct of transactions in banking, accountancy, insurance, advertising, computer and information technology services and some government and education services. One indicator of the scale of the recent increase in services trade is that it has, for the first time, been included on the agenda for the Uruguay round of negotiations for the General Agreement on Tariffs and Trade (GATT).

An immediate problem here is the definition of services trade; what, for example, is the distinction between trade, on the one hand, and income derived from property and labour, on the other (they are distinguished differently in different national fiscal systems). National statistics are not at all comparable so that measurements of the true scale and significance of trade in services remain speculative. It is agreed, however, that it is increasing if only because the contribution of goods to world current accounts has declined from 73 per cent in the early 1960s to 68 per cent in 1986 (UNCTAD 1989). The implied increase in services trade must nevertheless be interpreted cautiously; it seems that only trade in 'other services' (banking, business, computer, management, research and development services, etc.) has actually increased in relative terms even though invisible (service) transactions were valued at more than $1 trillion in 1986. Large increases in interest payments on international bank loans account for a large proportion of this (approximately 35 per cent) while the importance of other invisible items such as savings from labour and property, transportation and tourism has decreased in relative terms. Nevertheless, the relative increase in trade in 'other services' (14 per cent of invisible trade by value in 1986) is significant (in the context of the theme of this volume) since the majority is generated by producer services which have revealed particular organizational, operational and technological tendencies which are significant for location.

More than 78 per cent of world receipts and 73 per cent of world payments arising from international trade in services involve developed market-economy countries (principally between members of the Organization for Economic Co-operation and Development (OECD)) (UNCTAD 1989). Of particular interest here is the variable role of 'other services' in national current account profiles. They offset deficits in merchandise trade for the UK, with significant revenues from financial services (banking, insurance and brokerage in particular), consultancy services and technical co-operation. For France, other services also make a strong contribution, especially in technical co-operation and construction engineering, together with tourism. In the case of the United States property revenue and interest payments make a more significant contribution to the current account than 'other' services. Traditionally an area of strength, the US has recently showed its first deficit in services trade for 30 years (*The Times* 1989). The problem could continue as more funds need to be transferred overseas to service US debt in foreign hands. The UK's seasonally adjusted surplus in invisibles also began to fall in 1988, underlining the need for countries not to be complacent about their ability to compensate for declines in merchandise trade with increases in services trade. The other major positive contribution to US services trade comes from intra-corporate trade in services, i.e. other private services to/from affiliates. This is precisely the kind of trade which may be causing underestimation, worldwide, of the value and volume of trade in services.

By contrast some of the developed countries, such as Australia are stronger on the export of goods and raw materials which compensate for deficits in most services. Such profiles are more like those of developing countries such as Brazil and Indonesia which have strong relative positions in goods but very negative positions with respect to interest payments. A small number of developing countries, Republic of Korea and Singapore for example, have economies that have been growing rapidly on the basis of a strong position in services which offsets deficits in goods trade (UNCTAD 1989). But these countries have also been improving their level of merchandise exports in response to demand from developed countries and the growing demand for services from the developing countries. By and large, however, developing countries derive most positive returns from services trade which involves travel: the movement of persons inwards (tourism) or the movement of persons outwards as providers of services in other countries. UNCTAD (1989) observes that the developing economies showing the most rapid growth of manufacturing exports are those which have always had a strong relative position in services trade.

The locational significance of trade in services can be related to two factors: the importance of comparative advantage and the characteristics of internationalization (Hindley and Smith 1984; Deordorff 1985; Ochel and Wegner 1987). For the latter particular attention must be given to the actions necessary for a transaction to take place. At least one, or combinations, of the following must cross national frontiers: goods, capital, persons or information (UNCTAD 1989) with the movement of persons covering both consumers (e.g. tourists) and labour (e.g. construction workers). Each of these can cross a border to provide or to receive a service; many of these actions require the movement of persons either to provide a service directly (managing an engineering project, arrangements for trade financing) or indirectly (by operating equipment or manning a ship).

Whether or not these transactions are identified as 'trade' depends on whether they are undertaken within companies or between them. Legal requirements will ensure that the latter is more accurately recorded than the former. The likelihood of being able to trace either has actually been reduced by information technology; in most cases the transactions using this mode of delivery do not (cannot) be recorded by some official agency outside the country of origin. Therefore, it is likely that there is a substantial underestimation of international trade in services using this medium. But whether delivery of services at an international level involves people or information technology it will be necessary to utilize routes or channels; international airline services or international satellite/land line services, for example, and if this is the case then the location of service users and producers will inevitably be circumscribed (Hepworth 1987; Moss 1989). Most affected will be producer services or activities included in the 'other' services group which have made the most positive contribution to increases in services trade during the last decade.

The second factor with consequences for the location of growth associated with the recent international expansion of services is comparative advantage (see for example Nusbaumer 1987). While there is no agreement that the law of comparative advantage as it relates to trade in goods can also be applied to all services, it is reasonable to hypothesize that those countries (or locations within them) which have the most appropriate infrastructure for the delivery of services internationally will be able to retain and to generate a larger share of service trade flows than those countries which do not have such facilities. This not only represents a barrier to trade but also a barrier to entry, an effect compounded by the probability that service companies in locations with favourable infrastructure (often accompanied by a regulatory environment conducive to growth) will be more innovative and therefore more competitive. The quality of services will also likely be higher so that reputation may also enhance the competitive edge of certain countries.

THE INTERNATIONAL CORPORATE DIMENSION

Throughout the 1980s it has become clear that the restructuring of national economies has been increasingly dominated by the decisions and actions of large corporate organizations. These corporations have been engaged in both vertical and horizontal acquisition and merger activity designed, for example, to increase their representation in world markets, to protect or preferably to increase the market shares for their output, or as a response to the changing scale and complexity of international transactions and projects. There is now a substantial literature devoted to the analysis and interpretation of these activities and their significance for metropolitan and regional development or for national economies as a whole (Clairmonte and Cavanagh 1984; Dunning 1985; Enderwick 1989). The location and investment decisions by multinational enterprises (MNEs) contribute to the participation of individual countries in international trade in goods or information services and in the circulation of financial services. The ability of regions within countries to participate fully in economic restructuring is also influenced by MNE behaviour at the international level (Cooke 1986). Such enterprises have therefore become important targets for public policy initiatives devised to influence their investment and location decision making. Until recently almost all the research and publication about MNEs has, not unreasonably, focused on the manufacturing sector. Many of the multinationals that emerged during the 1960s and 1970s were in the food, electrical, engineering or petroleum industries, for example. However, during the 1980s corporate internationalization has diversified to embrace service organizations. The rationale has to some extent been similar to that for manufacturing MNEs but information technology and telecommunications, which have allowed effective expansion of service enterprises outside their national boundaries while concurrently increasing the potential for competition within domestic markets, have rapidly enhanced the opportunities for certain types of location. Most notable are large urban areas where many of the service organizations at the forefront of internationalization are grouped in a way which allows them to fully capitalize on comparative advantage (Dunning and Norman 1987). In an effort to maintain their position or to divert growth, national, state or institutional jurisdictions have deregulated certain service operations and markets. They have concentrated on those most vulnerable to the revolution in the distance, volume and speed of data and information transfer made possible by telecommunications technology, i.e. producer (or intermediate) services such as corporate banking, finance, legal, accountancy, advertising or marketing and the whole spectrum of management, computing, engineering and related consultancies. Even if the internationalization phenomenon is discounted there is abundant evident pointing to a metropolitan orientation in the growth and location of these services (although there are exceptions, see for example Illeris 1989).

METROPOLITAN AREAS AND THE LOCATION OF PRODUCER SERVICES

Sectoral shift, internationalization of service corporations, trade in services, advances in information technology and the theory and practice of comparative advantage have ultimately affected the location patterns of producer services. There is no doubt that the pivotal locations are metropolitan areas which have attracted a disproportionate share of producer service growth (see for example Wheeler 1986; Thrift 1987;

Beyers 1989). Such centralization allows service firms to reduce uncertainty in a highly volatile trading environment. Some of the reasons for the prominence of metropolitan areas can be deduced from the preceding discussion; labour market diversity, access to information technology services, comparative advantage, availability of vital information about local borrowers, market conditions or client behaviour and priorities. They have helped to maintain metropolitan dominance and have contributed to the emergence of 'world cities' (Friedmann 1986; Lambooy 1988). At the same time, the relationship between the growth of services and the development of domestic metropolitan relationships should not be overlooked (see for example Wheeler 1986; O'hUallacháin 1988; Stanback and Noyelle 1981 for the United States; Leyshon *et al.* 1988 for the UK). Finally, the metropolitanization of producer service growth in the 1980s has profoundly changed both the direction and the consequences of social and economic processes within some of the key cities (see for example Leyshon *et al.* 1987; Warf 1987; Brake 1988).

The 'world cities' such as London, New York and Tokyo rest at the top of the global urban hierarchy and tend to dominate lower order metropolises in peripheral locations. This dominance can be measured by the concentration of control as indicated by the number of headquarter offices (of both manufacturing and service companies) concentrated in them or, conversely, by examining the number of branches in lower-order metropolises controlled from the dominant cities. Thus, cities in the core areas of North America and Europe are the headquarters locations for 78 per cent of the international banks with branch offices in South America (Meyer 1986). New York, London and Tokyo are the control points for 40 per cent of the international banks headquartered in the sixteen largest cities (thirty-eight of ninety-five banks). The concentration of control exhibited by international banks at the global scale is equally evident for legal firms (Moss 1987), accountancy and management consultancy firms (Leyshon *et al.* 1987) and engineering consultancies (Rimmer 1988). Another measure of the dominance of the world cities is the distribution of banks and of traders engaged in foreign exchange dealing (Levich and Walter 1989). New York totally dominates the North American distribution of banks and traders in 1983 and 1985 with some 70 per cent of the total but Western Europe has the largest share of the 907 banks and 5,410 traders listed in 1985. However, London does not dominate in proportional terms (46 per cent in 1988) to the same degree as New York in North America even though it had twice as many banks and traders in 1985 and a daily trading volume in 1986 of $90 billion compared to $50 billion for New York and $48 billion for Tokyo (Levich and Walter 1989). This probably reflects the effects of national sovereignty for the countries within Western Europe as well as cultural and language differences which necessitate traders operating from several other cities such as Zurich or Paris which are smaller than Chicago or Toronto in North America but still have far more traders. In Asia and the Middle East Tokyo's position is much weaker relative to other regional competitors such as Singapore and Hong Kong which rank close to many of the West European centres outside London. Tokyo also has a minor standing relative to London or New York. The difficulty of operating within its highly regulated and protected financial markets (modifications are taking place but only very slowly) explains Tokyo's poor position in relation to this index.

The mechanisms responsible for the concentration of the services in dominant cities are technology, historical inertia, cumulative causation and agglomeration economies. After all information technology, especially telecommunications, has been hailed as the great opportunity to substitute time for distance as the determinant of transactional flows (Nilles 1975; Bell 1979; Mandeville 1983; Kellerman 1984; Hepworth 1987); dispersal of all economic activity (not just manufacturing) is possible because telecommunications allow contact to occur in real time almost irrespective of the distance separating the communicators. Certainly there is evidence showing some regional and even national redistribution of services within countries for reasons related to telecommunications flexibility (Moss and Dunau 1986; Leyshon *et al.* 1989; Noyelle *et al.* 1989). But at the international level the economics of telecommunications investment as well as the economies of scale arising from high levels of utilization of trunk routes have restricted the locational alternatives. This effect has also been encouraged by deregulation of the telecommunications industry (Moss 1987); this has caused suppliers of such services to identify the safest and most profitable investment opportunities. Some business users, especially producer services, are the biggest purchasers of advanced international telecommunications services. This results in an in-built tendency to look towards existing sources of business for any additional investment. This further reinforces the comparative advantage of the world cities (for examples of the kind of infrastructure investment which has been taking place in recent years, see Moss 1987). It is also the case that all the world cities have been (or still are) major centres for manufacturing and/or transportation (particularly as ports) (Hall 1984). Links with international markets have long been present and often supported by services such as shipping agents, trade financiers and insurance companies.

Just as these cities have been able to adjust to the changes engendered by the shift from industrial to organized capitalism, so they have been the areas most prepared to adjust their economic base to the needs of advanced capitalism. Which leads us to cumulative causation. The acquisition of one round of investment in infrastructure or a location choice by an international investment bank increases the chance that the conditions have been made even more attractive for yet another round of investment or location. The disconcerting fact for smaller, lower order, cities is that cumulative causation seems to be retaining its momentum with respect to producer services even though there are noticeable diseconomies, such as escalating property rents or high labour costs, accompanying their concentration in the world cities. A simple example of cumulative causation is the purchase of external services by foreign banks in London and New York (Daniels 1987). This not only revealed a trend towards increased externalization of expenditures on producer services but showed that more than 70 per cent of transactions were with suppliers located in downtown Manhattan or in the City of London. The presence of a range of specialized services further reinforces the attraction of these two locations for other actual or potential user firms. Kindleberger (1974) also suggests that the identity of world cities as international financial centres depends on the international prominence of the national currency and the international investment position of the nation. Since both of these are subject to change over time it is unlikely that one city will become totally dominant. The US for example has relinquished its position as the main creditor country of the world, Japan has become the largest creditor nation and Switzerland has strongly resisted an international role for its currency (Levich and Walter 1989).

O'hUallacháin (1988) argues that if we are to understand the locational preferences of producer services for large metropolitan areas or specialized cities it is necessary to establish whether urbanization or localization economies are most significant. The former arise from the benefits of access to the diverse business and infrastructure of large cities and the latter from the lower costs per unit of output in large markets where firms can specialize, acquire the market and other information required to perform adequately and obtain the necessary skilled labour. O'hUallachain produces a four-fold classification of the locational determinants of employment and establishment growth. First, there is a large group of services (e.g. business services, legal services, wholesale distribution, engineering and architectural services) that are primarily attracted to cities providing good intra-industry externalities. A second group (e.g. research and development, personnel supply services, management and public relations) has similar requirements but also needs good connections through the urban hierarchy. The communications industry is the only member of a third group in which establishment growth through intra-industry externalities is prominent and urbanization economies are important for employment growth. The fourth group (e.g. advertising, health services) is dominated by the influence of urbanization economies on establishment and job growth and position in the urban hierarchy is therefore very significant. This analysis also serves to underline an important distinction between services and manufacturing which makes the application of classical industrial location theory difficult, namely that search costs (for information, human resources, etc.) may well be far more significant for location than wages, transportation or capital. Perhaps, as O'hUallacháin (1988) stresses, this goes a long way towards explaining why New York, London and Tokyo on the world scale are sound choices for international producer service control and related functions.

The resulting network of 'world cities' linked by the electronic transactions and business travel generated by multinational corporations, international financial, business and professional services is organized into three tiers (Friedmann and Wolff 1982). The dominant cities have already been mentioned (London, New York, Tokyo). The sub-dominant or major world cities are Los Angeles, Chicago, Sao Paulo, Singapore, Paris, Frankfurt, Brussels, Zurich. Forming the third tier are the secondary world cities, i.e. Sydney, Johannesburg, Buenos Aires, Rio de Janeiro, Caracas, Mexico City, Miami, Houston, San Francisco, Toronto, Madrid, Milan, Vienna, Rotterdam, Bombay, Bangkok, Manila, Hong Kong, Taipai, Osaka and Seoul. There are two notable features; the presence of three regional sub-systems (Asia centred on Tokyo; Western Europe centred on London; and North America centred on New York) and the dominance of cities in the core over those in the periphery with Sao Paulo and Singapore the only major or dominant world cities in the periphery. The size of metropolitan area is clearly not important here, nor is output and employment. All are significant because they contain a large number (and diversity) of control functions extending to other cities and regions within the sub-systems of which they are a part and to other places within the global urban system. There are analogies with the 'hub-and-spoke' structure of airline networks in the US. Although the strength of the linkages as measured by flows of people, trade, information, etc. in this global metropolitan system is related to the size of the origin and destination cities, the differentiation between them is not strictly determined by distance decay effects. Historically, this has indeed been the case but information technology now allows the major linkages to take place between the cities furthest apart; i.e. London, Tokyo and New York.

What are the prospects for metropolitan areas not yet considered to be part of the world system of cities? A perspective on this has been provided by Polese (1983) in his assessment of Montreal's attempts to become an international business centre. Since it is already overshadowed by Toronto as the dominant Canadian location for international head-office functions and international financial services Montreal's position might be perceived as impossible, yet it has a labour force which is as well educated and as cosmopolitan as Toronto, it has an attractive setting with a good quality of life and a bilingual cultural environment giving businesses access to French- and English-speaking economies. The latter has ensured that Montreal has business services that adequately meet the needs of its regional market (mainly within Quebec) but it has not (along with the other conditions) proved effective in attracting international business functions. Indeed Polese (1983) implies that Montreal's very diversity might be a positive disadvantage; most established international financial centres are 'homogenous and unilingual in character compared to many other places around the globe ... *and it* ... is important to distinguish between cities which have built up a *generalized* comparative advantage for international business functions, based on many decades of office agglomeration ... and those cities which are able to build up *specialized* comparative advantages for certain functions or markets, based on a particular locational advantage, on specific legislation or distinct linguistic characteristics relevant to certain markets' (Polèse 1983: 18–19). In order for cities like Montreal to become part of the world city systems they must, it would appear, make the transition from specialized to generalized comparative advantage.

The concentration of international producer service functions at the global scale is complemented by some evidence for simultaneous centralization and decentralization at the level of national urban systems (Kindleberger 1974; Kirn 1987; O'Connor 1987; O'hUallacháin 1988; Beyers 1989; Thrift *et al.* 1989). Kindleberger lists nine factors that influence the emergence of a city as the premier national financial centre: central bank culture, tradition, economies of scale, central location, administrative capital, transport, national and local policies, corporate headquarters, but stresses that the connection between these factors and the location of the premier financial centre is not clearly defined. This has to be seen within the context of the transformation of metropolitan economies in line with the shift to advanced capitalism. Noyelle and Stanback (1981) first drew attention to this in the United States. They identified two key changes in the US urban system: the transformation of cities that were formerly major manufacturing centres to functional structures dominated by services (especially producer services); and a geographical shift in above average service industry growth to metropolitan areas in the south and west of the country. These cities did not have a traditional manufacturing base although they were targets for employers moving their production facilities from the north east as well as for population in-migration (Kirn 1987).

After analysing the employment and functional structures of 140 US cities, Noyelle and Stanback (1981) suggest that a three tier system has developed. At the top level are national nodal centres (New York, Los Angeles, Chicago and San Francisco), regional nodal centres (Philadelphia, Seattle, Boston, Minneapolis for example), and sub-regional nodal centres (examples include Oklahoma City, Charlotte). All have significant concentrations of national and regional headquarters of manufacturing and service corporations, a diverse set of producer service firms, major medical, educational and public infrastructures and are often foci for wholesale distribution functions. It can also be observed that the national nodal centres are all part of the world city system, thus demonstrating the dual role of successful, functionally dominant cities. Unless cities like Montreal become national nodal cities their chances of operating as gateways at the interface between national and international urban systems will be negligible. The range of service activities in second level cities is more restricted and more likely to be linked with the specific needs of the industries associated with particular cities. Research and development, technically-orientated functions as well as headquarters offices comprise these functional nodal centres. Examples are Pittsburgh, Cleveland and Detroit and most are in the 'manufacturing belt' of the north east. An additional sub-group of second level cities are metropolitan areas specializing in government/education functions such as Raliegh-Durham, Austin or Washington DC. Service functions are least significant in the profiles of third level cities which Noyelle and Stanback sub-divide into four groups, namely manufacturing centres (Gary, Buffalo), industrial/military centres (Norfolk, San Diego), resort/retirement/residential centres (Las Vegas, Orlando, Santa Barbara), and mining/industrial centres (Bakersfield, Duluth). The key control points in this service-orientated hierarchy of US cities are those at levels one and two and the cities with the most diversified service structure are most able to strengthen their influence on the system as a whole. They are also least vulnerable to fluctuations in the business cycle, for example, which can easily adversely affect the more specialized, mainly smaller cities at level three.

This does of course assume that the relationships between the levels in the urban hierarchy remain constant. Recent evidence compiled by Kirn (1987) and by Beyers (1989) indicates that this is not necessarily the case. Using data for the period 1958–77, Kirn (1987) concludes that the service structures at different levels of the US urban hierarchy appear to be converging, a process also found for the US divided into four major regions. This is attributed to significant downward filtering of certain individual activities in financial, business and professional services. But certain services showed little change and some such as banking, became more concentrated in large metropolitan areas. The decentralization revealed in Kirn's data is explained by increased demand at lower levels in the urban hierarchy consequent upon the long-established movement of population and manufacturing industry in the same direction. As yet there is no substantial evidence that economic activities in lower level centres do trigger sufficient demand for locally-provided producer services; import substitution is possible but it is equally likely that cities which are regional dominants will be the principal beneficiaries from such demand since, as we have seen, many producer services can be provided at a distance.

Although using different statistical units, Beyers (1989) analysed employment data for 1974–85 and found that more than nine out of ten producer service jobs were in metropolitan territories in 1985 compared with five out of six of all jobs. In common with Kirn he also shows that producer service employment is strongly concentrated in larger metropolitan areas; only about one in six of the BEA economic areas used in his study had producer service location quotients in 1974 and 1985 above 1.0. Again, however, smaller metropolitan areas seemed to be performing better than larger centres and those in the south and west better than those in the north central and north east regions. Beyers (1989) stresses, however, that although first- and second-level metropolitan areas reveal the largest negative net shifts in producer services employment (using shift-share analysis) they all still gained jobs. New York had the largest negative net-shift but still gained 550,000 producer service jobs.

INTERNATIONALIZATION AND LOCAL METROPOLITAN EFFECTS

The processes that are causing differential growth among the nodes that comprise the international urban system also have a significant impact at the local scale. Levich and Walter (1989: 67) emphasize the benefits of being a leading national (and international) financial centre as a function of 'the value-added by the factors of production (labour and capital) employed by the financial institutions'. The scale of value-added will depend upon the level of skills required and on the volume of linkages with other economic activities in the same trade. Value-added is greatest in market-making, industry mergers and acquisitions, product development or risk management, for example. The local benefits for cities with a large concentration of such activities, e.g. London or New York will be far greater than in Singapore which is principally an offshore depositing centre with an emerging financial futures exchange. Leyshon *et al.* (1987) have demonstrated the connection between the rapid expansion of producer service jobs in and around the City of London between 1976 and 1986 and residential and property values, the characteristics of housing demand, rental values for housing and office space or levels of investment in information technology and its consequences for the life expectancy of office buildings. Such effects are not confined to areas immediately adjacent to the City but also extend into the furthest reaches of the metropolitan region, especially for housing (Leyshon *et al.* 1987). The Regional Plan Association (RPA) (1987) has also examined the role of New York in the global economy and its implications for the region. Several of the contributors to the RPA project have since continued systematically to evaluate the local impact of internationalization of services on the New York region (Moss and Dunau 1986; Warf 1987; Brake 1988; Noyelle 1989; Noyelle *et al.* 1989).

The three principal global financial centres have certain unique characteristics. Tokyo is the world's largest source of international capital although it is not the world's largest capital market. It also has a much larger export sector in goods and related services than either London or New York. London is the leading centre for international finance in terms of the number of international banks represented there, its diverse connection with markets around the world and its superior time zone position, which allows it to handle transactions in New York and Tokyo during the same (extended) working day. New York's distinctiveness arises from its prominent position in the world's largest market economy, the innovativeness of its financial and other institutions, its role as the centre for the world's most widely used currency (the dollar) and its willingness to replace redundant building structures with new ones rather than try to conserve architectural legacies while at the same time remaining competitive (as in the City of London).

Yet all three cities have been undergoing change and transition along similar lines. Office space and administrative, technical and professional occupations now comprise the dominant share of activity in downtown which has the highest absolute concentration of decision-making function in each metropolitan area. This is assisted by the growing separation between control/decision-making functions in downtown and 'back-offices' that are moved to lower-cost fringe locations or even further into the surrounding metropolitan regions. Evidence from other cities such as Toronto (Code 1983; Gad 1985) or Pittsburgh (Kutay 1986) also demonstrates that these processes are not just taking place in the largest cities. But there is less agreement about the relative importance of centralization and decentralization. The latter is encouraged by high levels of office construction in locations around the edges of the metropolitan regions such as Stamford (New York) or Reading (London). In the residential market, especially of New York and London, rehabilitation of properties much nearer to downtown (gentrification) is well developed; the Docklands area in London or Manhattan's East Side have been transformed from blighted to highly desirable locations. This process is almost exclusively driven by demand from high-income workers in downtown service and corporate headquarters; there is only limited construction or rented accommodation for lower-income households. This is not only leading to greater social polarization but is also responsible for changing the structure of retailing, for example, in a way which further disadvantages the less affluent.

In the case of New York, service sector growth in general, and producer services in particular, have been responsible for an economic renaissance (Warf 1987; Brake 1988). During the 1970s New York lost more than one million people, corporate headquarters moved out in a steady stream, residential and office property markets collapsed, infrastructure deteriorated and per capita income fell. But the crisis 'laid the preconditions for a startling recovery, and ushered in a powerful restructuring process currently reshaping every facet of the region's economic and social landscape' (Warf 1987: 1). The reasons are given by Warf as follows. First, the internationalization of the US economy in terms of the shift of US production to developing countries and the attraction of foreign investment to the US. Second, a deregulated financial environment which has (as in London) triggered extensive merger, takeover and leveraged buyouts among service and manufacturing corporations. Third, a growing division between skilled and less skilled occupations but with a noticeable growth of demand for well qualified labour at all levels. Fourth, extensive changes in the application of technology to office and related work and New York's success in attracting private investment in advanced telecommunications.

Finally, the transformation in the use of space in the city as represented by gentrification and mass displacement of low-income residents on the social scale, and functional restructuring of the central city (in favour of financial and business services) and the decentralization of back offices. By 1983 finance and business services generated some 47 per cent of the income of the New York region economy with all services becoming more export orientated between 1958–83 (Drennan 1985). The relationship between global and local restructuring in which services have played a prominent part is therefore very clear and leads Warf (1987: 29) to observe that this 'perspective effectively bridges the artificial and analytically disastrous chasm between the office and industrial location literature on the one hand and issues concerned with housing on the other'. Perhaps most important for metropolitan areas in general is that although it has solved one set of problems, New York has generated others in the process. A recent example is the success achieved in stimulating the emergence of many new computer software and data processing firms during the 1980s (Noyelle *et al.* 1989) which now looks like presenting the city with the problem of how to keep them. The strain on the labour supply (100–150,000 employed in the sector in 1988) and the high costs of being in New York means that three-quarters of the firms in Noyelle *et al.*'s study said they would consider moving out of Manhattan. This must also be seen in the context of continuing outmigration of computer software and data processing facilities to locations outside the city. Similar trends have also been evident in London during recent years (Edward Erdman Research 1989).

It will be evident that at all three levels of analysis the dynamics of the metropolitan system involves a good deal of rivalry between cities (Lambooy 1988). The successful record of New York following its serious fiscal crisis during the mid-1970s is no guarantee that its success during the 1980s will be continued through the 1990s. Following 'Big Bang' in London, which among other things opened up the Stock Exchange firms to foreign ownership and participation and removed fixed scale commissions, there was a massive expansion of employment in financial and business services and in ancillary services. The crash which started on the New York Stock Exchange in October 1987 and was almost simultaneously transmitted to Tokyo and London (a less desirable feature of telecommunications technology) has since generated a more pessimistic view of the City of London's standing as a global financial centre. Should the position deteriorate it will create ripple effects in the local metropolitan economy; several thousand jobs have already been shed in financial services, depressing the market for residential property in the adjacent Docklands.

While the City's growth has depended a good deal on the relatively liberal approach of the UK authorities to regulation, tax and the efficiency of markets, other centres are deregulating in order to attract international business back to their markets. Thus, the loosening of US registration requirements whereby Eurobonds cannot be sold into the US until the end of a 90-day period, for example, could attract business back to New York. Similar effects could follow if the Glass–Steagal legislation is further relaxed to allow US commercial banks to enter the securities business in the domestic market. Business will likely to be transferred from London to Tokyo as it introduces interest rate deregulation and therefore becomes a more efficient banking centre. Paris is about to abolish fixed commission rates and French equity business now being transacted in London is likely to be repatriated.

Some markets are so international, e.g. foreign exchange, swaps, government bonds, commercial paper, and price more significant than commercial relationships that some big firms believe that they do not need to be run from separate offices in London, Tokyo and New York. In view of the high operating costs, relative to New York in particular, London might lose if these big firms move their control functions. As the implications of the European internal market post-1992 become clearer a trend towards insurance company mergers, already evident in France, will lead to bigger companies with more limited requirements for services such as re-insurance. This raises questions about the future of the Lloyd's market in London which generates more invisible earnings than all the banking activity combined. The point may now have been reached where, instead of encouraging competitors to give up their financial markets to international competition, the City of London should welcome the way in which Tokyo or Frankfurt are restricted by illiberal, highly regulatory national policies. Elsewhere, Hong Kong is certainly the Asia-Pacific region's leading financial centre (ahead of Sydney and Singapore) for the syndication of foreign loans, insurance, fund management and gold (Thrift 1984). In 1980 producer services accounted for 27 per cent of Hong Kong's GDP and some 5 per cent of employment. But there is now a real possibility of an exodus, including many service firms, as a result of the political uncertainty surrounding the transfer of the colony to China in 1997. Thrift (1984) suggests that the withdrawal will have three main components: regional headquarters of large foreign-based multinational corporations, relocation of some large indigenous companies (especially those which have internationalized their activities) and withdrawal of assets of Chinese-owned firms overseas. Related to these three events will be the subsequent withdrawal of producer service firms linked to these activities. If this happens Hong Kong's role as a regional financial centre will be greatly undermined, allowing its rivals in the region to enhance their status.

CONCLUSION

The world of services is expanding as the global economy is shrinking. Producer services, in particular, are no longer largely confined to local or national markets. The way in which they have come to exert an influence on the international economic and urban landscape is manifest most clearly in those countries that have moved towards advanced capitalism. But those countries that have not yet achieved this level in the development sequence are not immune from the organizational behaviour of service firms based primarily in advanced economies but increasingly involved in global distribution and supply of services. The purpose of this chapter has been to draw attention to some of the key explanations for the growing inter-nationalization of services, especially information and knowledge-intensive activities, and to consider some of the assumptions. It has been emphasized that information technology and the goals and objectives of service organizations, often acting in concert, have enabled services to influence the development of the global urban system as well as individual urban areas. It is important not to emphasize the global scale of the changes at the expense of understanding impacts on the urban system at the national level. Such is the locational specificity of international service expansion that the local effects within individual metropolitan areas are also worth exploring. The con-tributors to this volume are better able to substantiate some of the processes and relationships sketched in this chapter.

To summarize, the contributions to this book demonstrate:

1. The interrelatedness of changes in corporate structure and the re-inforcement of pivotal cities in the national and international urban system (Rimmer, Ch. 4; Tordoir, Ch. 10; Leo and Philippe, Ch. 14).
2. The importance of telecommunications and information technology for overcoming the tyranny of distance and 'channelling' a good deal of transactional activity. This again operates in the interests of the major, established corporate complexes (Daly, Ch. 2; Rimmer, Ch. 4; Moss and Brion, Ch. 12).
3. The dynamism of the international urban system within the context of economic restructuring towards services. Established dominance cannot be assumed to continue, the differentials between cities fluc-tuating according to political circumstances, the quality of labour provided by national education systems, the ability of cities to adapt their economic and physical infrastructure to meet the needs of international business (Daly, Ch. 2; Goldberg, Helsley and Levi, Ch. 3; Jaeger and Dürrenberger, Ch. 5; Bailly and Maillat, Ch. 6).
4. The replication of hierarchical concepts and service development at international level to relationships between cities within national urban systems (Beyers, Ch. 7; Ley and Hutton, Ch. 8; O'Connor and Edgington, Ch. 9; Tordoir, Ch. 10).
5. The connection between internationalization and changes/impacts at the local metropolitan level (Warf, Ch. 11; Moss and Brion, Ch. 12; Noyelle and Peace, Ch. 13).

In order to understand the changes taking place inside contemporary metropolitan areas we must untangle the influence of factors operating at a much larger scale.

REFERENCES

Bell, D. (1979) 'The social framework of the information society', in M.L. Dertouzas and J. Moses (eds) *The Computer Age: A Twenty-Year View*. Boston, Mass.: MIT Press, 163–211.

Bertrand, O. and Noyelle, T.J. (1986) 'Changing technology, skills and skill formation in French, Japanese, Swedish and US financial service firms: preliminary findings'. A report to the Center for Educational Research and Innovation, OECD, Paris and New York.

Beyers, W.B. (1989) 'The producer services and economic development in the United States: the last decade'. Final report prepared for US Department of Commerce Economic Development Administration Technical Assistance and Research Division. Seattle: University of Washington.

Brake, K. (1988) *Phoenix in der Asche: New York verandent seine Stadtstruktur*. Oldenburg: University of Oldenburg.

Christopherson, S. (1989) 'Flexibility in the US service economy and the emerging spatial division of labour'. *Transactions, Institute of British Geographers*, NS14, 131–43.

Clairmonte, E. and Cavanagh, J. (1984) 'Transnational corporations and services: the final frontier', *Trade and Development*, 5, 215–73.

Clarke, C.A. (1940) *The Conditions of Economic Progress*. London: Macmillan.

Code, W.R. (1983) 'The strength of the centre: downtown offices and metropolitan decentralization policy in Toronto', *Environment and Planning A*, 15, 1361–80.

Cooke, P. (1986) *Global Restructuring, Local Response*. London: Economic and Social Research Council.

Cuadrado, J.R. and del Rio, C. (1989) 'Structural change and evolution of the services sector of the OECD', *The Service Industries Journal*, 9, 439–68.

Daniels, P.W. (1987) 'Foreign banks and metropolitan development: a comparison of London and New York', *Tijdschrift voor Economische en Sociale Geografie*, 78, 269–87.

Deordorff, A. (1985) 'Comparative advantage and international trade and investment in services', in R.B. Stern (ed.) *Trade and Investment in Services: Canada/US Perspectives*. Toronto: Ontario Economic Council, 39–71.

Dicken, P. (1986) *Global Shift*. London: Harper and Row.

Drennan, M. (1985) *Modelling Metropolitan Economies for Forecasting and Policy Analysis*. New York: New York University Press.

Dunning, J.H. (1985) *Multinational Enterprises, Economic Structure and International Competitiveness*. Chichester: John Wiley.

Dunning, J.H. (1989) 'Multinational enterprises and the growth of services: some conceptual and theoretical issues', *The Service Industries Journal*, 9, 5–39.

Dunning, J.H. and Norman, G. (1987) 'The location choice of offices of international companies', *Environment and Planning A*, 19, 613–31.

Edward Erdman Research (1989) 'Overseas head office location', *Property Now*, Summer, 21–2.

Elfring, T. (1988) *Service Employment in Advanced Economies*. London: Gower.

Elfring, T. (1989) 'The main features and underlying causes of the shift to services', *The Service Industries Journal*, 9, 337–56.

Enderwick, P. (1989) (ed) *Multinational Service Firms*. London: Routledge.

Fisher, A.G.B. (1935) *The Clash of Progress and Security*. London: Macmillan.

Friedmann, J. (1986) 'The world city hypothesis', *Development and Change*, 17, 69–83.

Friedmann, J. and Wolff, G. (1982) 'World city formation: an agenda for research and action', *International Journal of Urban and Regional Research*, 6, 309–44.

Gad, G. (1985) 'Office location dynamics in Toronto: suburbanization and central district specialization', *Urban Geography*, 6, 331–51.

Gershuny, J.I. (1978) *After Industrial Society?* London: Macmillan.

Gershuny, J.I. and Miles, I. (1983) *The New Service Economy*. London: Frances Pinter.

Green, M. (1985) 'The development of market services in the European Community, the United States and Japan', *European Community*, September.

Greenfield, H.I. (1966) *Manpower and the Growth of Producer Services*. New York: Columbia University Press.

Hepworth, M.E. (1987) 'Information technology as spatial systems', *Progress in Human Geography*, 11, 157–80.

Hindley, B. and Smith, A. (1984) 'Comparative advantage and trade in services', *The World Economy*, 7, 369–90.

Illeris, S. (1989) *Services and Regions in Europe*. Aldershot: Avebury.

Johnston, R.J. (1980) *City and Society*. Harmondsworth: Penguin.

Kellerman, A. (1984) 'Telecommunications and the geography of metropolitan areas', *Progress in Human Geography*, 5, 222–46.

Kindleberger, C.P. (1974) *The Formation of Financial Centres*. Princeton: Princeton Studies in International Finance.

Kirn, T.J. (1987) 'Growth and change in the service sector of the US: a spatial perspective', *Annals, Association of American Geographers*, 77, 353–72.

Knox, P. and Agnew, J. (1989) *The Geography of the World Economy*. London: Edward Arnold.

Kutay, A. (1986) 'Effects of telecommunications technology on office location', *Urban Geography*, 7, 243–57.

Lambooy, J.G. (1988) 'Global cities and the world economic system: rivalry and decision-making', in Dben, L.; Heinemeijer, W.; van der Vaart, D. (eds) *Capital Cities as Achievement*. Amsterdam: Centre for Metropolitan Research, University of Amsterdam.

Lash, S. and Urry, J. (1987) *The End of Organized Capitalism*. Cambridge: Polity Press.

Levich, R.M. and Walter, I. (1989) 'The regulation of global financial markets', in Noyelle, T.J. (ed) *New York's Financial Markets: The Challenges of Globalization*. Boulder: Westview Press, 51–89.

Leyshon, A., Daniels, P.W. and Thrift, N.J. (1987) 'Internationalization of professional producer services: the case of large accountancy firms'. Working Papers on Producer Services No. 3. Liverpool and Bristol: University of Bristol and University of Liverpool.

Leyshon, A., Thrift, N.J. and Daniels, P.W. (1987) 'The urban and regional consequences of the restructuring of world financial markets: the case of the City of London'. Working Papers on Producer Services No. 4. Liverpool and Bristol: University of Bristol and University of Liverpool.

Leyshon, A., Thrift, N. and Tommey, C. (1988) 'South goes North: the role of the British financial centre'. Bristol and Portsmouth: Working Papers on Producer Services No. 9.

McKinnon, A. (1989) *Physical Distribution Systems*. London: Routledge.

MacPherson, A. (1989) 'Factors associated with export success among small manufacturers'. SUNY, Buffalo: Canada-United States Trade Centre Occasional Paper No. 3.

Mandeville, T. (1983) 'The spatial effects of information technology', *Futures*, 15, 65–72.

Marshall, J.N., Wood, P., Daniels, P.W., McKinnon, A., Bachtler, J., Damesick, P., Thrift, N., Gillespie, A., Green, A. and Leyshon, A. (1988) *Services and Uneven Development*. Oxford: Oxford University Press.

Martin, R.L. (1988) 'Industrial capitalism in transition: the contemporary reorganization of the British space-economy', in D. Massey and J. Allen (eds) *Uneven Re-Development: Cities and Regions in Transition.* Milton Keynes: Open University Press, 202–311.

Meyer, D.R. (1986) 'The world system of cities: relations between international financial metropolises and South American cities', *Social Forces,* 64, 553–81.

Moss, M.L. (1987) 'Telecommunications, world cities and urban policy', *Urban Studies,* 24, 534–46.

Moss, M.L. (1989) 'The information city in the global economy'. Paper presented at Third International Workshop on Innovation, Technological Change and Spatial Impacts, Cambridge (UK), September (mimeo).

Moss, M.L. and Dunau, A. (1986) *The Location of Back Offices: Emerging Trends and Development Patterns.* New York: Sylvan Lawrence Research and Data Centre.

Nilles, J.M. (1975) 'Telecommunications and organizational decentralization', *IEEE Transactions on Telecommunication,* Com-23, 1142–7.

Noyelle, T.J. (ed.) (1989a) *New York's Financial Markets: The Challenge of Globalization.* Boulder: Westview Press.

Noyelle, T.J. (1989b) 'New York's competitiveness', in Noyelle, T.J. (ed.) *New York's Financial Markets: The Challenge of Globalization.* Boulder: Westview Press, 91–118.

Noyelle, T.J. and Stanback, T.M. (1988) 'The post-war growth of services in developed economies'. A report to the United Nations Commission on Trade and Development, Geneva, April.

Noyelle, T.J., Peace, P. and Kahane, L. (1989) *Computer Software and Data Processing in New York City.* New York: Conservation of Human Resources, Columbia University.

Nusbaumer, J. (1989) *Services in the Global Market.* Boston: Kluwer.

O'Connor, K.J. (1987) 'The restructuring process under constraint: a study of recent economic change in Australia', *The Australian Journal of Regional Studies,* June, 23–36.

Ochel, W. and Wegner, M. (1987) *Service Economies in Europe: Opportunities for Growth.* London: Pinter.

O'hUallacháin, B. (1988) 'Agglomeration of services in American cities'. Paper presented at Annual Meeting, Association of American Geographers, Phoenix, April.

Polèse, M. (1983) 'Montreal's role as an international business centre: cultural images versus economic realities'. Paper presented at a Workshop on Non-Capital Cities, Hosts to International Organizations. City University of New York, October (mimeo).

Pumain, D. and Saint-Julien, T. (1986) 'The impact of recession on the socio-economic structure of the French urban system', in J.T. Borchert, L.S. Bourne and R. Sinclair (eds) *Urban Systems in Transition.* Amsterdam: Netherlands Geographical Studies 16, 142–52.

Regional Plan Association (1987) 'New York in the global economy: studying the facts and the issues'. New York: Regional Plan Association (Report prepared for World Association of Major Metropolises).

Riddle, D.L. (1986) *Service-Led Growth: The Role of the Service Sector in World Development.* New York: Praeger.

Rimmer, P.J. (1988) 'The internationalization of engineering consultancies: problems of breaking into the club', *Environment and Planning A,* 20, 761–88.

Stanback, T.M. (1979) *Understanding the Service Economy,* Baltimore, Md.: Johns Hopkins University Press.

Stanback, T.M. and Noyelle, T.J. (1981) *The Economic Transformation of American Cities.* New York: Conservation of Human Resources.

The Times (1989) 'First US services deficit for 30 years'. 14 September.

Thrift, N.J. (1984) 'The internationalization of producer services and the genesis of a world city property market'. Paper presented at a Symposium on Regional Development Processes/Policies and the Changing International Division of Labour, Vienna, August.

Thrift, N.J. (1987) 'The fixers: the urban geography of international commercial capital', in J. Henderson and M. Castells (eds) *Global Restructuring and Territorial Development.* Newbury Park, CA: Sage.

United Nations (1983) *Salient Trends and Features in Foreign Direct Investment.* New York: United Nations.

United Nations Committee on Trade and Development (UNCTAD) (1989) *Trade and Development Report 1988.* New York: United Nations.

Wallerstein, I. (1979) *The Capitalist World-Economy.* Cambridge: Cambridge University Press.

Warf, B. (1987) 'Service sector growth since the New York renaissance'. Paper presented at Annual Meeting, Association of American Geographers, Portland, April.

Wheeler, J.O. (1986) 'Corporate spatial links with financial institutions: the role of the metropolitan hierarchy', *Annals, Association of American Geographers,* 76, 262–74.

2 Transitional economic bases: from the mass production society to the world of finance

M.T. Daly

The world on entering the last decade of the twentieth century is poised at a significant turning point in the international economic system; and at the heart of the changes occurring is the international financial system. The characteristic of the manufacturing era that dominated most of the twentieth century was its highly concentrated nature. The financial system is even more concentrated. Decisions made on a regular basis in a handful of world cities dictate the operations of critical parts of the entire world. Not since the metropolitan dominance of the great imperial powers of the nineteenth century has such control been exercised so clearly. There are two basic differences: communications now link the world in a web of extraordinary complexity; and at the heart of the system is private enterprise rather than government.

The superstructure of cities reflect their economic bases. In the 1950s and 1960s large scale manufacturing was the distinctive driving force in sustaining economic growth. By the 1980s finance, and a set of related industries, provided the focal force of economic growth. In each era the characteristics and operational modes of the cities, and the relationships of the cities to each other, reflected the impact of these primary forces. The main focus of this chapter is the changing economic system and the forces which have created those changes. The rise of finance as the most critical of the service industries is discussed in the context of broad economic development patterns. The later part of the chapter turns to the impact of the international financial system on the network of cities which manage its operations.

Looking back at the start of each new decade commentators, in such journals as *The Economist* and *The Far Eastern Economic Review*, are given to remarking on how different the world is to that of a decade, two decades, or three decades before. In good times there is a smug self-congratulatory air about these declamations; in bad times soul-searching prevails, and certain questions are regularly debated: has the world undergone some kind of fundamental transformation?; are the structures of the past fading into uselessness?; and what will replace them?

Such questions are always worth asking for the dynamics of industrial capitalism are such that jerky growth is the norm rather than the exception (Schumpeter 1934: 223). The reasons for this lie in the process of technological innovation (and its relationship to product and trade cycles), the process of accumulating capital, and the way in which profits rise and fall in the passage from industrial infancy to industrial old-age. Organizational forms evolve which are designed to counter inefficiencies and instability in the system at large (the rise of scientific management illustrates this). The relationship between industrial and finance capital swings through processes of competitive adjustment and collusion. Governments might at various times be arbitrators, accomplices or participants in creating or destroying economic structures. Economic systems are caught in the interplay between politics, ideology and commerce that produce the disruptions of war and the threat of war.

Small wonder that commentators, as each new decade dawns, can find evidence of shifts in economic structures that make the immediate past or the immediate future of especial significance (and their comments, in turn, of particular noteworthiness).

In fact in over two hundred years of industrial capitalism there have been relatively few fundamental turning-points, when the system at large has been transformed. The Great Depression of the 1930s was seen at the time to be a crisis of such proportions that capitalism might be destroyed or, at least, utterly transformed. Time showed that period to be a serious, but temporary, interruption of industrialization which regained strength when economic recovery produced a pick-up in new investment. The turning-points of long-term significance occurred in the latter part of the eighteenth century when mechanization was applied to the textile industry; in the middle of the nineteenth century when the iron and steel industries began a process of transformation which brought a transport revolution and, through that, an urban revolution; in the early years of the twentieth century when electrification and assembly-line methods were introduced.

It is argued here that the three decades beginning in the 1960s have produced a process of economic change that has unalterably shifted the direction of growth. The 1970s was a decade of change; the 1980s a decade of challenge; and the 1990s is likely to be a period of resolution of the interacting forces that have created the shift.

Central to the new order has been the transcendence of global forces: global production, global marketing, and global finance dictate the operations of the emergent system. This world economy manifests itself in myriad ways every minute of every day, yet it remains a shadowy concept. Nigel Harris observes:

> The world economy, despite a plethora of facts, is almost as obscure a subject as was the human body in ancient times. We interpret obscure shadows on the wall. Yet despite the mystery, the world economy is decisive for all that is most important for the inhabitants of the globe.
>
> (Harris 1983: 9)

The triumph of the global economy is the setting for the fundamental shift that has taken place in the latter half of the twentieth century.

Three things are basic to understanding the nature of this shift. First, the major contribution to the growth of gross domestic product (GDP) and employment in the industrialized world has come from the service sector. The beginnings of this process could be discerned in the early 1950s when service sector employment in the USA outstripped manufacturing. By the 1980s, as is well known, service sector contributions to GDP and employment were predominant in each of the OECD nations. Trade in services also became the fastest growing, and most profitable, element of international trade. At both national and international levels the revolution wrought in financial markets was at the core of the change. It is the principal focus of this chapter.

Second, the character of manufacturing has been changed with the shift from energy-intensive to knowledge-intensive industry. Allied to this change the age of mass production–mass consumption is coming to an end. A vast restructuring of manufacturing has been under way for two decades and the reorganization of manufacturing, engineered by the century's greatest take-over and merger wave, was the highlight of the 1980s. It reorganized the ownership and control of the production system. Take-overs and mergers occurred because the transformed financial sector provided the capital, the strategies, and the instruments.

Third, the pattern of world trade has been altered. The two sets of rises in OPEC prices in the 1970s turned the terms of trade in favour of commodity producers. A massive flow of funds to OPEC resulted. They were unable to employ the bulk of such money in domestic consumption or new productive schemes, and so they directed large flows into the rich markets of the USA and Europe, and they invested up to 30 per cent of their surpluses in the international capital markets. Banks recycled OPEC funds lending unprecedented amounts to the developing world whose capacity in manufacturing and resource utilization was greatly increased as a result. When the oil cartels' dominant power collapsed in the 1980s, world markets found themselves with too much of almost everything (significantly because of the expansion of capacity in the preceding period). World trade shrank as many countries faced regimes of tight money and billowing deficits (at both trade and government levels). The great trading nations of eastern Asia, led by Japan, ran up huge trading surpluses. Japan became the principal creditor nation and the USA the world's greatest debtor. Debt (corporate, governmental, and personal) became a monster within nations; and international debt threatened to break many developing nations, and the banks of the industrialized world.

The transformation of the world system was a product of changes wrought in each of the three areas of trade, manufacturing, and services; and, most importantly, by the interactions between them. At the heart of each of these aspects of change stood finance.

In the subsequent pages the international financial system is considered from two angles:

(a) its role in helping to transform the character of the capitalist system, in the general context of the rise of service sectors; and
(b) the technological changes and organizational networks (especially at the city level) which have enabled finance to move to centre-stage in the transformed world.

STAGES OF ECONOMIC DEVELOPMENT AND THE ROLE OF FINANCE

Rybczyniski's model

Manufacturing has long been at the heart of development theory. In the great period of decolonization after the Second World War attention was directed at promoting the growth of the recently free, and poor, nations of the world. Popular attention was given early on to import substitution models of development, largely because it was thought that unfavourable terms of trade for commodity producers would persist; and that they would create an ever-widening division between rich and poor nations (Prebisch 1950). Later, the mystique of export-orientated industrialization cast its spell over the developing world as import substitution schemes failed to promote rapid enough growth (Ichiyo 1983). Throughout, however, manufacturing was deemed the critical element.

Models of economic development were cast at the level of the nation, the container of people's development hopes. From the time that Great Britain led the rest of the world into the industrial age national wealth had been largely perceived as a simple product of industrial prowess. There were only two occasions when broad attention was shifted from this preoccupation with manufacturing. In the early years of the twentieth century Hilferding (1912), Hobson (1914), and Lenin (1917) saw the great period of European imperialism as being related to the declining market opportunities in competitive Europe: monopoly capitalism was seen to replace market capitalism; bank capital merged with finance capital; and the export of capital in place of commodities became of central importance. Then war, inflation, a disjunction between the political status and economic power of the major nations, and finally the Great Depression moved the focus of analysts down different paths. Bilateralism focused attention on the nation, while the Great Depression once again switched attention to the behaviour of banks.

War soon intervened once more. Post-war, in the thirty years to 1973, the world system operated under the Bretton Woods Agreement which fixed exchange rates between nations in relation to the US dollar that was tied, in turn, to the price of gold. The idea was to tame financial instability

and uncertainty; so, in the continuing debate about how the poorer parts of
the world might share in the prosperity of the industrialized core the nature
and functioning of the capital markets received low priority. Aid, direct
investment, and savings would provide the capital base for the various
programmes of industrialization.

It was not until the Bretton Woods system began to fissure in the 1960s
and then collapsed in the early 1970s that widespread attention was given
to the capital markets. It was then not until the 1980s that the role of
financial systems, in the broad sweep of long-term economic development,
began to be appraised. Rybczyniski (1988) is one of the few to address this
question (as opposed to the very large number of scholars who have
considered the technical intricacies of finance and the equally large number
of commentators who have sought to explain how the system went wrong
in terms of generating the debt crisis).

Rybczyniski proposes that the finance industry has made three basic
contributions to the process of economic growth. The first stage was the
bank-orientated stage; the second the market-orientated stage; the third
the securitized stage.

He argues that the early stages of industrial development, the transformation of agrarian economies into industrial economies, involved the
transfer of *new* savings into industry. In this stage investment tended to be
in heavy capital-intensive manufacturing (iron and steel and shipbuilding),
natural resources (coal and iron ore), and infrastructure (transport). This is
the bank-orientated phase where the bulk of external company funds are
raised from banks in the form of loans.

The next stage is the market-orientated phase where a large proportion
of external funds raised by non-financial firms are obtained through the
capital markets. These markets become the means by which ever larger
proportions of personal savings are invested; they also become a conduit
for some of the savings collected in the non-bank sector (life assurance
offices and investment trusts). Ownership and control of productive
resources becomes separated and the markets reflect the risk/rewards
attributes of past savings. Restructuring depends on the perceptions of
individuals and institutions who determine the degree of risk that an
economy assumes, who can change the management teams, and who allow
the expansion of old firms and the entry of new in a competitive environment.

The market-orientated phase broadly covers the lead up to the Second
World War and the thirty years after it. More recently the industrialized
world has moved into the securitized phase (securities understood as any
interest-bearing piece of paper traded in the financial market). Here the
financial system becomes dynamic in changing the economic system. In
addition to wide and deep primary and secondary capital and credit
markets (covering the bulk of flows of new and old savings) the two main
and new elements of the securitized system are: the market for corporate

control and the market for venture capital and associated markets.

The market for corporate control is concerned with the phasing out and changing of past savings embodied in physical assets (and linked to it human and financial assets). The venture capital and associated markets (leveraged buy-outs) are concerned with the transfer of new savings to new industries that have different risk characteristics.

In this latter stage of financial development Rybczyniski describes the role of the markets in the following terms:

> The securitised financial system is characterised by a heavy reliance of firms on external funds raised through capital and credit markets and the important role of non-depository institutions in collecting new savings and managing old savings. It not only helps but is also *the active and indispensable ingredient of re-structuring all mature and de-industrialising economies* if all its constituent parts function effectively (author's emphasis).
>
> (Rybczyniski 1988: 10)

The outstanding feature of the latter part of the 1980s was a record wave of mergers and acquisitions centred on the manufacturing, resource, and distribution industries. In 1987 in the USA there were 2,000 mergers and acquisitions costing $150 billion. At the end of the third quarter of 1986 £11.3 billion were spent on acquisitions in Britain: a total exceeding that of any *full* year prior to 1986; three years later that total had doubled. The average level of acquisitions in the USA in the 1980s (around $170 billion a year) was three times the average yearly level from the late 1960s (Scouller 1987: 20). Record levels of mergers and take-overs were recorded in West Germany and Canada in the 1980s.

Take-overs grew larger as well as more prevalent. Between 1969 and 1980 there were only twelve acquisitions exceeding $1 billion; from 1984 to 1986 more than ninety acquisitions exceeded $1 billion. The Gulf-Socal merger involved an outlay of $13.5 billion in 1984 and remained the world's largest until the $25 billion RJR–Nabisco deal at the end of 1988. Corporate structures in all the major industrial nations were being transformed.

There were strong international forces operating as well. British companies made 1,572 separate direct investments (including 1,252 mergers) in the USA between 1976 and 1986 with a total value of $36.6 billion; the average value of transactions rose from $11 million in 1976 to $98 million in 1986 (Hamill 1988: 4). Mergers and acquisitions abroad by Japanese companies rose from 60 in 1984 to 228 in 1987 (there were only 219 domestic acquisitions in that year). In 1987 Japanese companies acquired 106 US companies for a total of $5.5 billion.

There are many criticisms of the restructuring of economies through the securitized financial system. They are concerned with the use of debt and the emphasis on short-term profits; and they invoke a range of technical, accounting and legal issues, especially the public scandals of insider trading. None the less corporate restructuring has become a central fact of industrialized societies. The mode of achieving restructuring is significantly different to the experience of the past. It has to be given a context within the broad scope of stages of economic development: this is precisely what Rybczyniski's model attempts to do.

Rybczyniski's model is incomplete in two regards. It implicitly assumes that corporate (and especially manufacturing) restructuring is a positive, natural, and inevitable development; but there is a need to establish why such a restructuring has taken place. The next section addresses this and in the process discusses the rise of service industries.

The model also gives no account of why the financial industry has moved into such a critical position in organizing and shaping modern societies. That issue is discussed in the subsequent section.

The rise of knowledge-intensive industries

One interpretation of the great reorganization of industry in the 1980s is that businesses acted out of confusion amidst a state of crisis:

> Forecasting models that in the prosperous 1950s and 1960s could predict the development of national economies within a narrow band now cannot foretell even the direction of change from one month to the next. Business executives are unsure of which products to make, which technologies to use to make them, and even how to distribute authority within their firm.
>
> (Piore and Sabel 1984: 13)

In fact what was happening in the 1980s was the end of the dominance of one industrial era (the mass production era) and the beginning of a new era (where the links between manufacturing and the service sector are much stronger and more intricate). Throughout the twentieth century modern manufacturing became synonymous with mass production methods. Mass production, in turn, built on a much longer tradition that saw the core of economic development in increases in productivity which depended on the increasingly specialized use of resources. Progressive increases in efficiency through the product-specific use of resources depended on an accompanying growth in demand. When demand was adequate growth through increasing division of labour was theoretically self-sustaining.

In the early twentieth century the assembly line and scientific management brought to full flowering the long development of mass production methods. Progress along this technological trajectory brought higher profits, higher wages, lower consumer prices, and a range of new products. The increased scale of production implied a parallel growth of a mass consumption society. Mass production, however, demanded large investments in highly specialized equipment operated by narrowly trained workers. The system at large was 'dedicated'.

Mass production therefore was profitable only with markets that were large enough to absorb an enormous output of a single, standardized commodity, and stable enough to keep the resources involved in the production of that commodity continuously employed. Markets of this kind did not occur naturally. They had to be created.

(Piore and Sabel 1984: 49)

The triumph of mass production was achieved within the capitalist system, a system that has had an inherently unstable pattern of growth, as discussed above. Mass production could only prosper, perhaps only survive, under special conditions. In the thirty years after the Second World War these conditions were met throughout the western world and mass production-mass consumption was able to support a unique period of thirty years' general prosperity.

There were three special sets of factors supporting the system at the national level. First, the organizational form of the corporation reached maturity, and solved a basic problem faced by mass production. A piece of machinery dedicated to the production of a single part cannot be turned to another use if the price falls or demand dies; and in unstable markets both things were inevitable at various times. Corporations managed the co-ordination of supply and demand in individual markets. Second, sta-bilization had to be achieved in input markets as well, and the labour market was the most important of these. Agreements on wages and conditions, and mechanisms to produce such agreements were critical. Third, economic fluctuations at the national level had to be evened out. Here the contribution of Keynesian management policies of government were an outstanding success up to the early 1970s.

Each of these sustaining conditions collapsed. A long-term decline in the rate of profit of corporations can be traced back to the 1960s (Daly and Logan 1989). In the 1970s in the seven largest OECD nations the net rate of return on corporate capital was 10 percentage points above the long-term real interest rate; by 1987-8 it was only just over 6 points above. By the early 1970s the most profitable corporations in the 'old' industries were transnationals which organized production, finance and marketing to maximize profits over the global, rather than the national, sphere. The management of production and marketing at the national level was much less significant.

The collapse of cosy arrangements between labour, employers, and the state was seen in the rise of unemployment in the 1970s. Nations reacted in different ways. The United States boosted economic growth (through inflation-creating expansion of the money supply under President Carter and through increasing debt with President Reagan). In the USA there was a great redistribution of industry to cheap labour areas (responsible in part for the sunbelt–rustbelt contrast). In Europe tough monetary policies were allied to a belief that many 'old' industries were inefficient and ought to be phased out. In the early 1980s unemployment was high in every industrialized nation except Japan (varying from 6.1 per cent in West Germany in 1982 through 9.5 per cent in the USA to 12.3 per cent in Britain). Unemployment fell in the USA through the 1980s (6.1 per cent in 1987), rose in West Germany (6.5 per cent 1987), and stayed high in Britain (10.3 per cent). In some European nations unemployment had reached extremely high levels by 1987 (Spain 20.1; Ireland 19.0 per cent). Long-term unemployment (the proportion of unemployed out of work for more than twelve months) rose throughout the 1980s in each of the major industrialized nations including Japan. Hard core unemployment affected men more than women as a result of the shift to service sector employment.

The 1970s saw a loss of faith in Keynesian management methods. Inflation, unemployment, and slow growth went hand in hand. The status of governments was also weakened by a jump in their level of debt. In the twenty-four OECD countries government spending rose from 29 per cent of GDP in 1960 to 41 per cent in 1989. Alongside this, from 1975 onwards, governments financed their extra expenditure through borrowing. Of the fourteen major western industrial nations all but two (Britain and Norway) showed persistent increases in debt:GDP ratios from 1975 to 1985; the annual average increase varying from a maximum of 6.2 per cent (Belgium) to 0.3 per cent (USA).

There was also a set of external conditions that supported mass production–mass consumption. They were: the establishment of a global trading system embodying principles of free trade under the auspices of the General Agreement on Trade and Tariffs (GATT); a system of monetary arrangements (the Bretton Woods Agreement) designed to ensure a stable international environment; and a guaranteed supply of cheap energy. 'The expansion of the mass economy was fuelled, literally and figuratively, by this fundamental fact: during the past hundred years, labour has been replaced by fossil fuels' (Hawken 1984: 17). The world price of oil (in constant dollars) in 1900 was $3.80 a barrel and in 1950 was $3.23. In 1970 it was $1.79 (Hawken 1984: 27). The constancy of the price of oil and its fall during the period of the long post-war boom was the vital secret of success of the mass economy. When OPEC forced up the price of oil (in 1974 it was $7.92; in the early 1980s it climbed as high as $35) mass production as the mainstay of the growth of industrialized economies was doomed. Free trade was then endangered by the heightened competition of the 1980s. The collapse of the Bretton Woods Agreement between 1971 and 1973 (discussed further in the next section) introduced a period of vast financial change and instability.

The critical factor has been the changing ratio between materials and information in goods and services. This is expressed in the way goods are produced and in the goods themselves. Production methods increasingly involve the use of computers (computer-aided and computer-designed manufacturing), lasers, and robotics. Manufacturing becomes more flexible and on a smaller scale. New fields emerge: biotechnology, super-computers, materials science, superconductors. Products change: there is a greater emphasis on quality and durability, and the amount of material per product declines. Products tend to be customized rather than mass produced. The life-cycle of new products is drastically reduced, compared to mass-produced goods, and R&D costs increase.

At this point it is necessary to focus the discussion squarely on the rise of services in general, and the role of finance in particular. Much of the expanding literature on services has concentrated on the meaning and consequences of the rise of the service sector (the 'post-industrial' school and its critics and derivatives); or on the definition of services and their impact on local economies (Daniels 1985). Too little of the literature has been concerned with placing the rise of services in its specific economic–historical circumstances. The first part of this section paints the broad setting; the following part concentrates on services in particular.

Services grew naturally out of the mass-production system. Scientific management permitted the growth of nationwide corporations and as organizations grew specialized tasks were assigned to individuals; then as marketing and financing tasks grew in importance more of the internal structure of manufacturing companies was concerned with service activities. Externally manufacturing was supported by a large array of services: legal, accounting, financial, transport, advertising, sub-contracting, and maintenance. Mass production succeeded because it increased productivity. Increased productivity increased wages, and a wealthy society could consume a greatly expanded volume of personal services. When firms expanded into the international arena the need for levels of intermediation between production and final consumption grew.

As mass production went into decline the giant firms it had bred faced fierce competition. They still had large assets and handled very large cash flows; a significant number returned large profits. Because growth in the 'old' industries was limited such corporations were faced with a small number of investment opportunities. They could move into the 'new' industries but this generally proved difficult (as General Motors' ill-fated purchase of Electronic Data Systems showed). They could exploit the more profitable aspects of their industries: in 1985 Chrysler purchased Finance-America (from the Bank of America) and E.F. Hutton Credit Corporation; Ford Credit bought First Nationwide (a savings and loan corporation); and General Motors bought two mortgage companies to make it the second biggest mortgage banker in the USA. And there was an even easier way: the profits that the international money-markets offered turned industrialists into financial dealers. The Japanese called it *zaiteku* or financial engineering. In Japan in 1986 the ten biggest earners from *zaiteku* obtained at least a third of their total profits from this source. For Nissan it was the difference between profit and loss for the year.

More generally the lack of opportunities caused the wave of mergers and acquisitions of the 1980s with the break-up and restructuring of corporate empires formed during the long period of growth of the mass-production system. The pace of industrial change (Compaq Computer of Houston became the fastest growing company in history: in 1982 it did not exist; in 1987 it had a turnover of $1.2 billion) called for much larger quantities of venture capital than in the past. These two factors (corporate funding and venture capital) underlie Rybczyniski's model.

The growth of knowledge-intensive industries has blurred the distinction between manufacturing and service industries. Economists from Adam Smith and Marx onwards liked to distinguish between 'useful, productive' activities (agriculture, manufacturing, some utilities) and 'semi-parasitic' activities (services). The distinction, always dubious, becomes outmoded with the rise of knowledge-intensive industries. A great deal of work has gone into defining services according to their role in the economic system with the distinction between producer and consumer services being especially important (Daniels 1985).

Services have been particularly important in setting the direction and pace of change in world trade. The Japanese *sogo shosa* (giant trading houses) played a critical role in the expansion of Japanese exports.

We are like the air, invisible but pervasive,
providing essential things to sustain life.

Thus Yohei Mimura as president of Mitsubishi Corporation (Yoshino and Lifson 1986: 1) characterized his corporation's role. Services provide the invisible link that unites the modern world of trade. Because services are heterogeneous and intangible their importance is often overlooked in assessments of international trade.

International finance

Finance is the most dominant and most critical of the international services. The rise of the finance industry to the centre-stage of world affairs was an outcome of the set of forces which were critical in causing the demise of the mass production–mass consumption system.

The Bretton Woods Agreement was built around the need to provide credit for post-war reconstruction, and to disperse the gigantic US trade surplus. It attempted to revive world trade under orderly conditions fixing the value of each nation's exchange rate to the US dollar, and the dollar to the price of gold. That system could only be strong as long as the US trade surplus was strong (giving faith in the value of the dollar) and as long as the supply of gold could match the growth of world trade. Neither thing was sustainable through the 1960s. Speculation both against currencies and gold spelt the end of that system long before President Nixon's renunciation of the link of the US dollar to gold in 1971.

The restrictions of the Bretton Woods era and the inability of the system to service the appetite of the expanding global economy led to the rise of the Euromarkets. The rate of growth of these markets through the 1960s was remarkable: averaging 30 per cent expansion per year, and in some years 50 per cent. Most importantly they allowed banks to operate largely outside the control of governments. Market forces could dictate interest rates and, after the collapse of the Smithsonian agreement in 1973, exchange rates for a number of key currencies. When the OPEC surpluses began to flood the world the international banks had the network and the means to recycle a large portion of those funds; banks moved to the centre-stage of world affairs.

THE BANKERS' WORLD: THE DOMINANT INTERNATIONAL SERVICE

Atinc *et al.* (1984: 145) describe international trade in services as occurring in any of four ways:

(a) services provided by resident firms or individuals across national boundaries to non-resident firms or individuals abroad;
(b) services provided through contractual arrangements;
(c) services provided within national boundaries to non-residents; and
(d) services provided through a foreign affiliate of a parent company.

The remarkable thing about banking is that it fits each of the categories. In addition banking is a natural partner to the diverse list of operations that now go under the heading of international services. Banking is so significant and so basic to other activities that there is a temptation to treat it as if it had a life of its own. The preceding discussion was designed to place the rise of banking in a broad context (disabusing the notion of banking existing and growing under its own momentum). This section focuses on banking itself: the characteristics of the modern system; its technological base; its worldwide networking and city-based activities; and its status at the beginning of the 1990s.

A world on the move: international capital markets

Three characteristics have distinguished modern international capital markets: the volume of capital within them; the mobility of that capital; and the volatility of the price of money (interest rates) and the price of currencies (exchange rates).

The volume of money traded in the international money markets each day is around twenty-five times as great as the daily volume associated with international trade of goods. The wide, and ever-increasing, range of financial products available is attractive in a world where investment in manufacturing and resource development during the 1980s has been sluggish. The markets are driven by a host of factors (movements of broad economic indicators in various countries, the profit and loss situation of companies, mergers and acquisitions), respond to very fine margins, and operate twenty-four hours a day. The markets provide funds for giant development projects and cover governments in debt; they also service customers with overnight cash needs. They hedge those involved with other countries' currencies. They provide means of dealing with interest rate fluctuations. They have means of handling obligations made in the past and they allow people to trade into the future.

Modern banking began its strong growth in the 1970s in a world that was flooded with money. First, the breaking of the nexus between the US dollar and gold produced a sudden flow of devalued dollars and, through a chain reaction (US dollars were the world's reserve currency) an expansion of the world's money supply. Through the 1960s this had grown at around 8 per cent per annum but in 1971, 1972, and 1973 world money supply expanded by more than 20 per cent a year (Parboni 1981). Inflation became the great problem of the 1970s and, with it, came a general expansion in the supply of money. The USA devalued on two other occasions during the 1970s exacerbating the general problem of over-expanded money supplies. The first round of OPEC price rises produced a $133 billion surplus from 1974 to 1976, and $49 billion of this was lodged directly in the international capital markets. The second round of OPEC increases produced a combined OPEC surplus one-third larger than the first, and the banks received a commensurately larger flow of funds.

The new capital markets of the 1970s and 1980s were extremely volatile. In the 1970s interest rate fluctuations averaged 32 basis points a month. In the 1950s they had averaged 6 to 8 points a month, and from 1933 to 1949 interest rates in the USA had averaged no more than 3 basis points per month. Interest rates in the USA jumped from 8.2 per cent in early 1979 to 22.0 per cent at the end of 1980, and were still fluctuating over a 7 to 8 per cent range in 1982. These fluctuations had a profound impact on the world economy and ushered in the crisis of Third World debt. As the 1980s progressed the level of change in rates declined but finer adjustments still produced sizeable effects. In 1988 US prime rates moved from 8.59 per cent in the first quarter to 10.5 per cent in mid-December; three-month Eurodollar rates moved from 6.91 per cent to 9.50 per cent over the same period.

Currency rates were the most pervasively fluctuating element in the world capital markets through the 1970s and 1980s. Long experience with managed rates had not prepared the world for the opportunities, or the problems, of fluctuating exchange rates. Banks were the medium through which rates were determined and the first to realize the enormous opportunities that exchange dealing offered. By the early 1980s currency trading was providing one-third of the profits of the largest international banks, and many learnt that their central position enabled them to turn dealing into manipulation (Hutchinson 1986). Governments eventually decided, that whatever the theoretical advantages of floating rates, intervention and management were necessary to sustain world trade and to limit the worst consequences of volatility and manipulation (on the part of both banks and governments). The Plaza Accord of September 1985 was the result; the Group of Five agreed that US–Japan trade imbalances had reached the point where a US dollar devaluation and yen revaluation were necessary. At the end of August 1985 the US dollar had been trading at 237 yen to the dollar; at the start of 1988 it was 128 yen to the dollar; and a year later was just above 121. The prime place of the dollar in world trade and the prime place of the Japanese as world traders ensured that the world currency markets would spin through a general period of instability over the latter part of the 1980s. Through 1988 yen:US dollar rates roamed between 122 and 136.

Volatility was most notoriously expressed in the late 1980s in the world's stockmarkets. The crash of 19 October 1987 punctured years of uninterrupted bullish growth. Stock trading reflected an important element of the internationalization of financial markets. Deregulation broke down the barriers that had separated banking institutions and others in the financial area, and technology combined to make world trade in stocks possible. On 20 October 1982 the Dow Jones average had been 1,013.80; on 20 October 1986 it was 1,811.02; during 1987 it peaked at 2,722.42; and on 20 October 1987 had collapsed to 1,013.80. Between 1982 and 1985 Japanese share prices had risen by more than 350 per cent and Australian prices by over 400 per cent; the Nikkei index fell to 21,910.08 (year high 26,646) and the Sydney index to 1,549.5 (year high 2,305.9) on 20 October 1987. Many feared that the gigantic markets created in the 1980s had been destroyed. The remarkable resilience of the international capital markets was shown by the climb of share prices in 1988: world average prices moved up 20.6 per cent; US 12.1; Japanese 30.8; and Australian 42.6 per cent.

There were other important sources of volatility. Commodity prices, highly variable throughout the post-war period, were a major source of economic change throughout the 1980s. The price of oil slid between a peak of $34.40 in 1981 to a low of $14.20 in 1986. Gold peaked at $458 an ounce in 1981 and fell to $318 in 1985. The index for all commodity items fell from 100 in 1980 to a low of 71.3 in 1985 but ended 1988 at 109. Metal prices soared in 1987 and 1988. From an index of 100 in 1980 they had fallen to 63.5 in 1986 but zoomed to a peak of 157.1 in June 1988, and ended that year at 149.7 (*The Far Eastern Economic Review* 5 January 1989: 29–37).

Volatile and voluminous markets were necessarily mobile markets. Deregulation, pioneered and engineered by the international banks, made mobile markets possible. The array of actors in the capital markets (commercial banks, merchant banks, security houses, government agencies, corporate treasuries) made the degree of mobility extremely high.

Stepjumps in technology: the basis of the banking revolution

The international financial revolution has been built on the back of other changes: the broad sweep of production changes outlined earlier in the chapter, and the breakthroughs in communication technology summarized here. Alexander Graham Bell's famous invention of 1876 quickly had an impact (by 1878 over 10,000 telephones had been sold in the USA and by 1892 240,000; London's first telephone exchange was opened in 1879); but the impact was confined to a few nations and a few activities for almost a century. In 1956 the USA had half the world's telephones. In 1960 only half the skilled workers of West Germany had a telephone. In 1975 in France there were only 6 million telephones in the whole country.

Computers allowed the telecommunication industry to revolutionize itself. Optical fibres enabled signals to be sent as pulses of light. Computerization of the signals turned exchanges and lines of work into digital networks. Satellites enabled signals to be transmitted instantaneously around the world. Facsimile machines permitted documents to pass from country to country very quickly. The system pushed out into new areas with the so-called VAN services. These include electronic mailboxes, remote databases, inventory and accounting control, and remote banking.

There has been a dramatic interplay between the growth of the service sector and the growth of telecommunications. They interact to speed each other's development. Information flows grew in the USA at a rate of over 15 per cent per annum in the 1980s. Demand in the USA reached the equivalent of 200,000,000,000,000 computer bits per *second* by the end of the 1980s (*The Economist* 17 October 1987: 37). Network access lines reached 180 million in western Europe and 150 million in the USA in 1989. They were forecast to grow to 220 million in Europe and 180 million in the USA by 1995. Furthermore these networks would be highly integrated. Integrated Services Digital Network (ISDN) provides technical standards of a higher quality and cheaper networks; it is seen as the major new development in global communications through the 1990s.

Of all the areas of the service sector finance has been the first user of technology in most instances, and the most effective user. Global money markets could not operate efficiently without instant communications. Information and dealing are the most obvious aspects of technological dependency but clearing operations and document exchanges have also been developed (the various generations of SWIFT and CHIPS for example); the integration which technology provides has been a dominant force in shaping the global financial system.

World cities shaped by finance

A common theme in the literature of the 1960s and 1970s was the enormous growth of the world's industrial companies. The world being ruled from the skyscrapers of New York was a common enough image. Then in 1974 New York itself was on the verge of bankruptcy: the city owed its bankers almost $14 billion. A new image began to dawn, and much more attention was given to these bankers, the roles they performed, and the locations they operated from.

The giant industrial companies remained giants. The world's top 500 industrial companies in 1984 had sales equivalent to nine-tenths of the gross national product (GNP) of the world's largest producer (Kidron and Segal, 1987: 30). There was less relationship between manufacturing and city growth or city functioning than in the past. In New York 600,000 jobs were lost from 1969 to 1977, and in that time forty-nine large companies (included in the top 500 companies) shifted their headquarters out of New York. From 1977 to 1987 New York gained 300,000 jobs. The driving force in the change of fortunes was the growth of finance and related services.

International banking is essentially a city-based activity, and there are relatively few cities that can lay claim to being international financial centres. By 1985 the total assets of the world's 500 largest commercial banks was more than two-thirds the GNP of the whole world. That power was remarkably concentrated. Of the top 500 banks 190 were from North America; 77 from Japan; 42 from West Germany; 29 from Italy; and 16 from UK. Japanese banks held 27 per cent of the top 500 banks' assets.

Large banks not only came from a small number of countries but they were clustered in a small number of cities within those countries. Banks located close to each other because they depended on information flows clustered at a few points. They relied on each other to create profitable money markets and they joined with each other in funding very large projects or in servicing government deficits. They needed to be close to trade flows and the centres of international service activities. They gathered the savings of large areas and dispersed them to clients around the world.

Banks and banking activities were at the core of financial centres but commercial banks supported and, in turn, relied upon other activities: merchant banks, insurers, stockmarkets, commodity traders, currency dealers, legal and accounting firms, international transport agencies. They also relied on high quality infrastructure, particularly communications.

Few centres were capable of offering such facilities and so, over the two decades of the 1970s and 1980s, a definite hierarchy of centres emerged. There have been several attempts to define the levels of the hierarchy (Daly, 1984) which contains four distinct levels: global, zonal, regional, and national. London, New York and Tokyo struggle for dominance at the peak of the hierarchy.

The City of London was long seen as the hub of the global system. It became the centre of the Euromarket through a mixture of historical circumstances, infrastructure, and a legal system attuned to the growth of international banking. In 1967 there were 114 foreign banks in London; in 1974 there were 336; and in 1984 there were 460 (Rybczyniski 1984: 29). By the late 1970s New York was challenging London for the title of financial capital of the world, and a decade later Tokyo loomed at the top of the ladder.

Zonal centres serve large geographical areas and make possible the twenty-four hour trading system. Singapore and Hong Kong are outstanding examples of zonal centres. The strength of zonal centres depends on more than position within time zones and high class supporting infrastructure. The size, savings and investment capabilities of surrounding nations, and their level of trade surplus (or deficit) are important. Both Hong Kong and Singapore have grown with the astounding growth of the Pacific Rim region. This region contains the world's largest surplus nation (Japan), the nation with the largest foreign reserves (Taiwan), the world's fastest growing trader (South Korea) and a diversified set of other nations that have made the region the most rapidly growing in the world. In thirty years the region has lifted its share of world GDP from 8 per cent to 18 per cent (1989).

The arena of international banking is fiercely competitive. Throughout the 1970s it was dominated by US banks but in the 1980s these gave way to Japanese banks. By 1988 they controlled 36 per cent of all cross-border banking assets. Six of the top ten banks in the world were Japanese, and they were the most profitable banks.

Financial centres competed to enter the international arena during the 1980s. Regional centres, such as Sydney, Sao Paulo and Chicago, attempted to introduce new products or change tax or other regulations to give them an advantage. National centres such as Melbourne or Miami strove to improve their competitive positions within nations as a step to entering fully the international arena.

The greatest influence on the location and practice of global banking during the 1980s was the debt crisis of the Third World. Banks, which had vigorously spread themselves into new locations and offered international loans on an unheralded scale for fifteen years to 1982, went into a full retreat once the enormity of the debt problem (around $3,000 billion in 1988) became apparent. The retreat took many back to the wealthy nations of North America, Europe and east Asia. It created a scramble to find profitable outlets for their money. It fed the merger and acquisitions boom that played such a role in restructuring western economies in the 1980s. It paid for trade and government deficits of many nations through a gigantic recycling of funds. It fed an orgy of personal and corporate debt (personal debt in the USA in 1988 was $505 billion and corporate debt around $100 billion). It transformed the status of nations and the power bases of groups within nations. It was the key to the future of the world.

REFERENCES

Atinc, T., Behnam, A., Cornford, A., Glasgow, R., Skipper, H. and Yusuf, A. (1984) 'International trade in services and economic development', *Trade and Development*, 5: 141–214.

Daly, M.T. (1984) 'The revolution in international capital markets: urban growth and Australian cities', *Environment and Planning A*, 16: 1003–20.

Daly, M.T. and Logan, M.I. (1989) *The Brittle Rim*, Melbourne: Penguin.

Daniels, P.W. (1985) *Service Industries*, London: Methuen.

Hamill, J. (1988) 'British acquisitions in the United States', *National Westminster Bank Quarterly Review*, August, 2–17.

Harris, N. (1983) *Of Bread and Guns*, Harmondsworth: Penguin.

Hawken, P. (1984) *The Next Economy*, Sydney: Angus and Robertson.

Hilferding, R. (1912) in Owen, R. and Sutcliffe, B. *Studies in the Theory of Imperialism* (1972) London: Longman.

Hobson, C.K. (1914) *The Export of Capital*, London: Constable.

Hutchinson, R.A. (1986) *Off the Books*, New York: William Morrow.

Ichiyo, M. (1983) 'The free trade zone and the mystique of export-led industrialisation', in E. Utrecht, *Transnational Corporations in South East Asia and the Pacific*, Sydney: University of Sydney, Transnational Corporations Research Project.

Kidron, M. and Segal, R. (1987) *The New State of the World Atlas*, London: Pan.

Lenin, V.I. (1917) *Imperialism: The Highest Stage of Capitalism*, Moscow: Progress.

Parboni, R. (1981) *The US Dollar and Its Rivals*, London: Verso.

Piore, M.J. and Sabel, C.F. (1984) *The Second Industrial Divide*, New York: Basic.

Prebisch, R. (1950) 'Commercial policy in the underdeveloped countries,' *American Economic Review Proceedings*, 49: 251–73.

Rybczyniski, T.M. (1984) 'The UK financial system in transition', *National Westminster Bank Quarterly Review*, November, 26–42.

Rybczyniski, T.M. (1988) 'Financial systems and industrial restructuring', *National Westminster Bank Quarterly Review*, November, 3–13.

Schumpeter, J.A. (1934) *The Theory of Economic Development*, Cambridge: Harvard University Press.

Scouller, J. (1987) 'The United Kingdom merger boom in perspective', *National Westminster Bank Quarterly Review*, May, 14–30.

Yoshino, M.Y. and Lifson, T.B. (1986) *The Invisible Link*, Cambridge: The MIT Press.

3 The growth of international financial services and the evolution of international financial centres: a regional and urban economic approach

Michael A. Goldberg, Robert W. Helsley and Maurice D. Levi

INTRODUCTION

Background issues

The world of international finance has long been viewed as the very pinnacle of high finance with all its attendant glamour and mystique. However, over the past decade the explosive growth of international trade and the rise of east Asia, most notably Japan, but also Hong Kong, Taiwan and Singapore, has led to the growth of a number of venues for international financial transactions outside the traditional bastions of Zurich, London and New York. This growth in international finance (Mayer 1985; Hamilton 1986; Kaufman 1986; Wachtel 1986) has attracted enormous attention and headlines. With it, growing interest has also been expressed in financial centres and, in particular, in the prospects for the development of second- and third-tier centres such as Vancouver, British Columbia.

Part of the reason for this interest in second- and third-tier centres is the growing realization that services, and in particular producer services such as finance, are increasingly important elements in the urban and regional economic base (Enderwick 1987). This has been illustrated in the United Kingdom by Shelp (1981), Daniels (1986), Wood (1986) and Leyshon *et al.* (1987) and for Australia by O'Connor and Edgington (1987).

The growth of international financial centres (IFCs) is not an independent process but rather needs to be seen in the context of the recent rapid growth of international financial transactions. This growth in turn has resulted not only from booming world trade but also from dramatic changes in the regulation and conduct of international financial services (Mendelsohn 1980; Ferris 1984; Bryant 1987). These services came to be provided in a relatively small number of international financial centres.

Reed (1981) sees IFCs as key components in the developing global economy, functioning as communication and management centres. State-of-the-art telecommunication systems, a dense network of airline connections to other IFCs, and management services such as accounting, marketing and advertising, law, printing, and trade are essential adjuncts of both IFC status and IFC functioning. The relative autonomy granted to managers in IFCs and their need for support services further reinforces these tendencies (Reed 1981: 69–74). However, despite the growth and importance of IFCs there is remarkably little hard information on their evolution.

Identifying and analysing empirically the forces that lead to IFC development has considerable utility to urban and regional scholars and policy makers. First, relatively little is known about these forces and since they are obviously important regional building economic elements, more needs to be known. Second, more research needs to be available to make effective and appropriate policy. Accordingly, this chapter presents the results of two studies that identify factors linked with IFC development. First we report on cross-sectional evidence from the growth of financial activity in thirty US states. Second, this evidence is strengthened with nation-based time series and cross-sectional studies.

The spatial scale: a brief digression

Ideally, we would have liked to conduct our analysis at the level of the metropolitan region. Indeed, our earlier efforts were directed at this target. However, it quickly became apparent that the difficulties involved in putting together a meaningful and consistent data set were insurmountable, at present, for any set of IFCs in a given recent year, or for one or a few IFCs through time. Thus, our focus shifted to a cross-sectional analysis of US states and to both cross-sectional and time-series analyses for nations. Given the consistency of the findings reported upon below, we do not feel that this larger spatial scale weakens our analysis or the generality of our findings.

On the contrary, there are reasons to believe that the state and national level are the appropriate scales at which to analyse the development and evolution of metropolitan-based international financial activity. First, for virtually all nations (and states) there is only one IFC so that the international financial services in that metropolitan region alone are accounted for in the national (and state) data. Second, international financial transactions are governed by national, not local, rules further reinforcing the utility of the national level of analysis. Finally, since national economic forces are important for the development of IFCs (Reed 1981; Kindleberger 1984), it is appropriate that these national factors be included explicitly to gain insights about IFCs. Thus, we infer IFC development factors from national, not urban, economic and financial variables.

EVIDENCE ON THE DEVELOPMENT OF INTERNATIONAL FINANCIAL CENTRES

We begin with evidence for US states and then proceed in subsequent sections to explore similar data at the national level.

Cross-sectional evidence for US states

We adopt several measures of the level of financial activity in each US state, the dependent variable in this phase of our work. These are:

(a) total assets and deposits in domestic banks;
(b) total assets and deposits in foreign banks; and
(c) total state employment in finance, insurance, and real estate.

No equilibrium models exist of financial centre development or of the regional location of financial activity. Indeed, the only conceptual idea that appears repeatedly in the meagre literature on financial centres is that there are economies inherent in banking and finance that encourage financial firms to concentrate geographically.[1] However, models from international finance of the size of the financial sector in a country may be combined with concepts from location theory to reveal factors likely to influence financial location among regions of a nation. Despite the national and state spatial scale used here, the work in fact draws heavily on concepts from the smaller spatial scale of urban and regional economics.

To explain the level of financial activity, country-based models suggest a number of potentially relevant variables.[2] In general, theory holds that a country will contain more banking and financial activity the greater is its level of economic development. On the supply side, savings and the supply of loanable funds tend to rise with the level of economic activity. On the demand side, the demand for goods and services, including financial services, tends to rise with the level of economic activity. We conjecture that these effects also appear at the regional or local level. We use state personal income as a proxy variable for the level of economic development.[3]

International financial theory also suggests that the amount of banking and financial activity in a country should vary positively with the volume of international trade. This occurs because international transactions utilize financial services more intensively than domestic transactions. The level of international trade is captured by including the total dollar value of imports and exports booked through the thirty states which had customs offices. The import data used in our analysis pertain to the point of unlading, rather than the point at which the merchandise first enters the United States.[4] Imports and exports are entered as separate variables, since they generate different demands for financial services, and hence have different effects on the location of international financial services. Reed's (1981) study of international financial centres suggests that a disproportionate share of banking will occur in those states in which large companies maintain their head offices. Thus, we include three measures of corporate activity in a state as explanatory variables:

(a) the number of US multinational corporation headquarters;
(b) the number of foreign corporations represented in the state; and
(c) the total number of corporations.

Finally, the average corporate state income tax rate was initially included in our analysis. However, consistent with most studies of business location, it was never statistically significant and is not included in the results reported below.

In sum, the following regression equation was estimated using ordinary least squares for the thirty states containing US customs offices and the District of Columbia for 1984 (i.e. for thirty-one entities):

$$\text{Financial activity} = \beta_0 + \beta_1 \text{Income} + \beta_2 \text{Imports} + \beta_3 \text{Exports} + \beta_4 \text{Corporations} + \mu.$$

The first explanatory variable, state personal income, controls for variations in the level of economic development. The second and third explanatory variables, the dollar value of imports and exports cleared through the customs office in a state, capture the effects of international trade. The fourth explanatory variable accounts for the impact of the number of corporations located in a state. *A priori*, we expect the coefficients on all the explanatory variables to be positive.

Results

The models account for a very high fraction of the variation in the dependent variable: the adjusted R^2 lies between 0.90 and 0.99 in all regressions (Tables 3.1–3.3). The models using employment in finance, insurance, and real estate as the dependent variable consistently have the highest R^2 and those using foreign assets or deposits the lowest. Thus, it appears that the most important variables have been included.

As expected, total domestic assets are positively related to the level of economic development as measured by total personal income in each state (Table 3.1, row (a)). However, when the number of foreign corporation employees (FCORP) or the total number of corporations (CORPS) is added, the income variable is no longer significant in the domestic asset equations perhaps because of multicollinearity. The simple correlations between personal income and CORPS and FCORPS are slightly higher than that between personal income and domestic deposits.

However, personal income is positively and significantly related to deposits in domestic banks, independent of the measurement of corporate activity (Table 3.1, rows (f)–(j)). The income variable appears to explain

Table 3.1 The determinants of the level of domestic financial activity in thirty-one states in 1984[a]

	Constant	Income	Imports	Exports	USMN	FCORPS	CORPS	R^2
Assets in domestic banks								
(a)	-101.97 (-0.026)[b]	0.271* (4.09)	2.63* (2.83)	-1.25 (-1.04)	316.97* (3.31)			0.96
(b)	-3333.9 (-0.744)	0.152 (1.44)	5.41* (7.43)	-4.65* (-4.20)		40.56** (2.63)		0.96
(c)	-1794.0 (-0.403)	0.134 (1.09)	4.51* (6.23)	-3.66** (-3.48)			0.328** (2.31)	0.96
(d)	-4443.0 (-1.24)	0.016 (0.157)	3.14* (3.90)	-2.30** (-2.19)	327.50* (3.84)	41.79*** (2.83)	0.026 (0.189)	0.98
(e)	-4414.5 (-1.25)	0.025 (0.282)	3.15* (4.01)	-2.30** (-2.23)	332.53* (4.19)	43.32* (3.59)		0.97
Deposits in domestic banks								
(f)	-865.7 (-0.332)	0.359* (3.27)	0.885 (1.45)	-0.169 (-0.216)	123.10*** (1.97)			0.97
(g)	3147.41 (-1.23)	0.254* (4.22)	2.12 (5.08)	-1.77* (-2.80)		26.16* (2.97)		0.97
(h)	-1552.50 (-0.563)	0.303* (3.98)	1.61* (3.59)	-1.11 (-1.70)			0.131 (1.49)	0.96
(i)	-3510.36 (-1.48)	0.224* (3.35)	1.24** (2.33)	-0.820 (-1.18)	144.69*** (2.57)	30.80* (3.17)	-0.060 (-0.670)	0.98
(j)	-3579.25 (-1.53)	0.204* (3.47)	1.21** (2.31)	-0.829 (-1.21)	132.89** (2.51)	27.26* (3.39)		0.98

* significant at the 1% level
** significant at the 5% level
*** significant at the 10% level

Notes: [a]Dependent variables: Total assets or deposits in domestic banks
[b]t-ratios in parentheses

Table 3.2 The determinants of the level of foreign financial activity in twenty-six states in 1984[a]

	Constant	Income	Imports	Exports	USMN	FCORPS	CORPS	R^2
Assets in foreign banks								
(a)	−1863.4 (−0.450)[b]	−0.245* (−4.17)	3.97* (4.72)	−2.68* (−2.53)	206.22** (2.43)			0.92
(b)	−1011.2 (−0.212)	−0.240** (−2.42)	5.61 (8.28)	−4.50 (−4.42)		9.62 (0.684)		0.90
(c)	−2996.7 (−0.716)	−0.394* (−4.07)	5.15* (8.86)	−4.31* (−5.17)			0.288** (2.57)	0.92
(d)	−3699.7 (−0.910)	−0.396* (−4.06)	3.94 (4.88)	−2.95* (−2.85)	162.16*** (1.95)	−4.60 (−0.325)	0.263*** (2.01)	0.94
(e)	−2963.3 (−0.681)	−0.308* (−3.30)	4.11* (4.77)	−2.96** (−2.66)	209.96** (2.46)	11.17 (0.882)		0.93
Deposits in foreign banks								
(f)	−1377.9 (−0.399)	−0.220* (−4.39)	3.46* (4.80)	−2.29** (−2.52)	191.26** (2.63)			0.93
(g)	−827.2 (−2.06)	−0.222** (−2.60)	5.00* (8.54)	−4.02* (−4.57)		10.14 (0.834)		0.90
(h)	−2376.0 (−0.673)	−0.351* (−4.19)	4.57* (9.06)	−3.80* (−5.27)			0.257** (2.65)	0.93
(i)	−3078.4 (−0.908)	−0.359* (−4.32)	3.46 (5.02)	−2.57* (−2.90)	154.92* (2.18)	−1.63 (−0.134)	0.220*** (1.96)	0.94
(j)	−2497.54 (−0.694)	−0.286* (−3.62)	3.60* (4.93)	−2.57** (−2.73)	195.03** (2.69)	11.54 (1.07)		0.93

* significant at the 1% level
** significant at the 5% level
*** significant at the 10% level

Notes: [a]Dependent variables: Total assets or deposits in foreign banks
[b]t-ratios in parentheses. Maine, Montana, North Dakota, and Vermont have no foreign assets or deposits. South Carolina has no foreign assets

Table 3.3 The determinants of the level of financial employment in thirty-one states in 1984[a]

	Constant	Income	Imports	Exports	USMN	FCORPS	CORPS	R^2
(a)	-89.02 (-0.017)	1.32* (14.79)[b]	3.25** (2.60)	-1.94 (-1.21)	462.79* (3.61)			0.99
(b)	-5640.8 (-0.976)	1.09* (8.04)	7.43* (7.91)	-7.13* (-5.00)		67.66* (3.40)		0.99
(c)	-2331.40 (-0.380)	1.14* (6.72)	6.02* (6.02)	-5.45* (-3.76)			0.448** (2.29)	0.98
(d)	-7147.7 (-1.75)	0.931* (8.06)	4.13* (4.51)	-3.66* (-3.05)	502.38* (5.16)	78.93* (4.51)		0.99
(e)	-7228.0 (-1.79)	0.907* (9.00)	4.10* (4.56)	-3.68* (-3.11)	488.55* (5.37)	71.72* (5.19)	-0.070 (-0.454)	0.99

* significant at the 1% level
** significant at the 5% level
Notes: [a]Dependent variable: Employment in finance, insurance and real estate
[b]t-ratios in parentheses.

the largest component of the variation in domestic deposits. The fact that the relationship between income and deposits is stronger than that between income and assets is consistent with the financial intermediation function of banking.

When employment in finance, insurance, and real estate is used to measure financial activity, personal income is positive and significant in all cases, and it seems to explain the largest component of the variation in the dependent variable. The strength of the income variable's performance in both the deposit and employment equations lends credence to the suspicion that multicollinearity is partly responsible for the income variable's poorer performance in the asset equations.

The most puzzling results occur when foreign assets or deposits are used to measure financial activity. In these cases personal income is significantly negatively related to the dependent variable (Table 3.2). Income appears to explain somewhat less of the total variation in foreign banking than imports, but the negative coefficient for the income variable is unexpected. In terms of total explained variation the 'foreign' equations perform less well than domestic models.

Imports have a positive impact on financial activity; the estimated coefficient is significant in all but one of the twenty-five equations. While imports account for somewhat less of the variation in domestic financial activity, they appear to be the most important explanation for foreign assets and deposits. Generally the estimated coefficient for imports is greater in the asset equations (foreign and domestic) than the deposit equations, while the opposite was true for the income variable. We infer that imports are more vital to the distribution of assets, while personal income is more important for deposits.

What might at first glance seem surprising is that exports were found to have a negative impact on the level of financial activity in all cases (Tables 3.1, 3.2 and 3.3).[5] The estimated coefficients are generally significant except when financial development is measured by domestic deposits. Like imports, exports seem to be more fundamental to the explanation of foreign than domestic financial activity, judging from the size of the respective t-statistics. In all of the estimated equations the absolute value of the coefficient for imports is greater than that for exports, indicating that the balanced trade multiplier for financial services is positive, as expected.

There are sound economic reasons for expecting this result. First, it may be more important for banks to have access to importers since the importer is the one who frequently finances the trade and applies for letters of credit. If information on the importer (e.g. his credit-worthiness) is best revealed by local contacts, then banks may focus in areas where imports dominate. Also it may be useful for banks to be near importers to obtain access to their demands for credit. Second, exporters may have an incentive to conduct financial transactions in their major export markets. By denominating their debt in the importer's currency, exporters can match the currency of payables and receivables, thereby reducing their foreign exchange exposure.[6] This can often be done most expediently by using banks or financial markets in the importer's country. This may lead to fewer banks to concentrate in areas where exports are a larger fraction of international trade.

Turning to the number of US multinational companies (USMN) headquarters in a state, it is positively and significantly related to state financial activity in all the estimated models. Replacing USMN with FCORP increases the proportion of variation explained by the trade variables (rows (a) and (b), Tables 3.1–3.3). While FCORP is significant on its own in the domestic equations, this is not the case in the foreign assets or deposits equations. When all three corporate measures are included (row (d), all Tables 3.1–3.3), FCORP performs satisfactorily except in the foreign equations where the sign is opposite to that expected, but not significant. This is consistent with the argument outlined above: just as US exporters may use foreign banks (i.e. banks in the importer's country) to hedge foreign exchange risk, foreign corporations in the United States may favour American banks over 'their own', for similar reasons.

To summarize to this point, we have found very strong statistical support for the hypothesis that IFC development is closely linked to national financial centre development and therefore to the level of sophistication of the domestic economy, as shown by the great importance of the national income variable. Over and above this, we found that imports are strongly and positively associated with IFC development and that exports are strongly, but negatively, tied to IFC growth. These apparently counter-intuitive results are, however, consistent with the behaviour of importers and exporters as shown, and with the results that follow for nations.

Evidence from national cross-sectional and time-series data

This section presents similar evidence on factors influencing the development of national financial sectors. The dependent variable in this part of the study is the percentage of employment nationally in finance, insurance, real estate, and business services. The use of this broad measure of financial development was necessitated by a lack of more disaggregated data. The choice of explanatory variables was determined as we did earlier by considering first the influences on financial development suggested in the theoretical literature on the growth of nations (versus regions) as financial centres, and then determining the subset for which comparable data are available.

The level of overall economic development, which would be expected to be positively related to the importance of financial employment in a country, can be measured from real GNP per capita. To the extent that the financing and settling of international transactions is more complex than purely domestic transactions, international trade can be expected to have incremental effects on financial sector development over and above those conferred by real per capita GNP. International trade can be measured, separately again, by merchandise imports and merchandise exports, each as a fraction of GNP.[7] The rate of savings should influence the extent of financial intermediation for given levels of real GNP, and of imports and exports. Savings rates are measured by comparing the flow of savings to GNP. The extent of government financing can also be expected to influence the level of financial employment. Thus, countries whose governments borrow relatively heavily should have higher percentages of the labour force in their financial sectors. In the time-series regression for the United States, the United Kingdom and Canada, presented below, government borrowings can be found from the year-to-year changes in national debts. For the cross-sectional study the national debts themselves should be related to financial employment given the work involved in handling the debts.

A factor which could have a temporary effect on financial employment in time-series regressions is the rate of unemployment. This is because there may be a tendency for the financial sector to expand or contract differently from other sectors. This follows because the financial sector lags general economic downturns because of its need to continue to service problem loans and declining business with continuing staff levels until bank profits begin to reflect actual loan losses and actual staff reductions take place. Our time-series regressions therefore include the rate of unemployment. The addition of the variable has little effect on other coefficients, but substantially reduces serial correlation.

In order to capture the above ideas, the following regression equation was estimated using available time-series data for the United States, the United Kingdom and Canada:

$$\frac{FE}{TE} = \beta_0 + \beta_1 \frac{RGNP}{POP} + \beta_2 \frac{IMP}{GNP} + \beta_3 \frac{EXP}{GNP} + \beta_4 \frac{SAV}{GNP}$$

$$+ \beta_5 \frac{DEBT}{GNP} + \beta_6 \, UN + \mu_t$$

In this regression equation:

FE   = Employment in banking, insurance, real estate and business services
TE   = Total employment
RGNP = Real GNP
GNP = Nominal GNP
IMP   = Nominal imports (f.o.b.)
EXP   = Nominal exports (f.o.b.)
SAV   = Nominal savings
DEBT = Change in value of national debt
UN   = Unemployment rate
POP   = Population

United States time-series results

The time-series results for the United States when the previous year's unemployment rate is included are shown in Table 3.4, spanning the period 1907–84. Table 3.4 shows that financial employment as a fraction of total employment grows with real GNP per capita, supporting the view that financial services are a luxury (in economic terms). The size of the standardized coefficient shows that this variable contributes substantially to the overall explanation of the fraction of employment occurring in the financial services sector.

We can see from Table 3.4 that imports have a very significant and positive impact on the size of the financial sector in the United States over time. The standardized coefficient shows imports to be second in importance only to real GNP. Exports have a significantly negative effect on financial service employment in the United States over time. These findings reinforce the previous ones based on US state data and similar reasoning explains these results.[8] Once again, the balanced-trade multiplier is also positive.

Table 3.4 shows that the savings rate does not appear to affect financial employment, nor does annual government borrowing relative to GNP. The amount of borrowing does not appear to influence significantly the demand for financial services (e.g. brokers, agents, jobbers, etc.).

Increases in unemployment however, do have a significantly positive impact on financial employment, as would be expected if the financial sector delays laying off personnel longer than other sectors during contractions, and is also slower in recruiting labour during expansions. Looking specifically at the effect of unemployment on financial employment, only temporary effects should be found because in the long-run the percentage of financial employment in the economy can be expected to return to normal. In order to allow for this possibility a distributed lag on unemployment is included in the regression.[9]

Table 3.4 US time-series regression results, 1907–84

Variable name	Estimated coefficient	t-ratio 70 df	Standardized coefficient
IMP/GNP	0.3629E−03	5.5422	0.4111
EXP/GNP	−0.1892E−03	−3.4324	−0.2693
GNPCAP	0.9139E−05	11.323	0.8649
GOVBORR/GNP	0.7553E−02	0.5922	0.3603E−01
SAV/GNP	0.1126E−02	0.5135E−01	0.3547E−02
UNEMP	0.9208E−03	4.9689	0.4034
CONSTANT	0.1286E−01	2.2892	0.0

Durbin-Watson – 0.2591   Adjusted R^2 – 0.8077

Table 3.5 confirms the results in Table 3.4. Again a significantly positive effect of per capita real GNP is found, and an incremental positive effect of imports as a fraction of GNP. As in Table 3.4, exports as a fraction of GNP are found to exert a negative effect on the share of employment in financial services. Unemployment is found to raise financial employment only for two years, with a correction occurring in subsequent years.

A problem with the results in Tables 3.4 and 3.5 is that there are high degrees of first order serial correlation. This could cause upward bias in *t*-statistics and be responsible for the significance of the results, although serial correlation would not bias the regression coefficients themselves. An analysis of the regression errors shows that the serial correlation occurs principally in the 1920s and 1930s. Therefore, rather than seeking out omitted variables, the same equation was run excluding these years.

Tables 3.6 and 3.7 show the time-series regression results for the United States for the period 1947–84. (Table 3.6 differs from Table 3.7 only by the inclusion of lagged unemployment in Table 3.7.) The Durbin-Watson statistics are improved in the shorter period, while there is very little change in the significance of imports, exports and real GNP. The results therefore still strongly support the view that financial employment moves to the importing country with an overall positive effect of a balanced trade expansion on financial employment. As with the 1907–84 period, unemployment causes a short-run increase in the fraction employed in financial services. The adjusted R^2 continues to show that the bulk of variations in financial employment are being explained.

Table 3.5 US time-series regression results, 1907–84, lagged unemployment

Variable name	Estimated coefficient	t-ratio 68 df	Standardized coefficient
IMP/GNP	0.3881E−03	5.6028	0.4495
EXP/GNP	−0.2033E−03	−3.6477	−0.2952
GNPCAP	0.8600E−05	10.277	0.8268
GOVBORR/GNP	0.1006E−01	0.7496	0.4914E−01
SAV/GNP	0.1363E−01	0.5699	0.4387E−01
UNEMP	0.9559E−03	3.8031	0.4291
UNEMP(−1)	0.1399E−03	1.1259	0.6281E−01
UNEMP(−2)	−0.1746E−03	−1.2207	−0.7838E−01
UNEMP(−3)	−0.1658E−03	−1.8702	−0.7443E−01
UNEMP(−4)	−0.1173E−04	−0.8479E−01	−0.5263E−02
UNEMP(−5)	0.1096E−03	0.9019	0.4918E−01
UNEMP(−6)	0.1998E−04	0.9392E−01	0.8970E−02
CONSTANT	0.1261E−01	2.2727	0.0

Durbin-Watson – 0.3151   Adjusted R^2 – 0.8033

Table 3.6 US time-series regression results, 1947–84

Variable name	Estimated coefficient	t-ratio 32 df	Standardized coefficient
IMP/GNP	0.3639E−03	5.4040	0.6345
EXP/GNP	−0.1359E−04	−2.1961	−0.1811
GNPCAP	0.7910E−05	6.1726	0.6364
GOVBORR/GNP	−0.2263E−01	−0.9424E−01	−0.5661E−02
SAV/GNP	−0.1197	−2.6575	−0.1149
UNEMP	0.5355E−03	1.7449	0.1237
CONSTANT	0.3438E−01	4.6365	0.0

Durbin-Watson − 1.2426   Adjusted R^2 − 0.9491

Table 3.7 US time-series regression results, 1947–84, lagged unemployment

Variable name	Estimated coefficient	t-ratio 32 df	Standardized coefficient
IMP/GNP	0.3954E−03	5.7599	0.6913
EXP/GNP	−0.2023E−03	−2.9310	−0.2698
GNPCAP	0.8137E−05	6.4477	0.6547
GOVBORR/GNP	0.1348E−02	0.0402	0.3373E−02
SAV/GNP	−0.1158	−2.3157	−0.1109
UNEMP	0.3852E−03	1.1346	0.0891
UNEMP(−1)	0.4374E−03	2.2466	0.1011
UNEMP(−2)	0.1693E−03	0.8040	0.0391
UNEMP(−3)	−0.1932E−03	−0.3454	−0.0446
UNEMP(−4)	−0.4241E−03	−2.1573	−0.0980
UNEMP(−5)	0.2975E−03	−1.6147	−0.0687
UNEMP(−6)	0.4128E−03	1.4660	−0.0954
CONSTANT	0.3482E−01	4.1802	0.0

Durbin-Watson − 1.4514   Adjusted R^2 − 0.9517

Results for Canada and the United Kingdom

Estimating the equation over the period 1952–84 for Canada (Tables 3.8 and 3.9) and the United Kingdom (Tables 3.10 and 3.11) produces findings very similar to those for the United States. As in the US results, financial employment is seen to increase as a fraction of total employment with increases in real GNP per capita. Moreover, financial employment is strongly and positively related to the share of imports in GNP and strongly and negatively related to the share of exports in GNP. The size of co-efficients on imports and exports again suggest an overall positive incremental influence of international versus domestic trade. Unemployment has the same effect as in the American case, causing relatively more

Table 3.8 Canadian time-series regression results, 1952–84

Variable name	Estimated coefficient	t-ratio 26 df	Standardized coefficient
IMP/GNP	0.7301E−01	3.5345	0.1729
EXP/GNP	−0.3106E−01	−1.7583	0.1030
GNPCAP	0.2979E−05	13.132	0.7687
GOVBORR/GNP	−0.2554E−05	−1.1526	−0.0522
SAV/GNP	−0.7169E−04	−0.5240	−0.0143
UNEMP	0.1265E−02	5.3440	0.2823
CONSTANT	0.2598E−02	0.6643	0.0

Durbin-Watson − 1.5555 Adjusted R^2 − 0.9865

Table 3.9 Canadian time-series regression results, 1952–84, lagged unemployment

Variable name	Estimated coefficient	t-ratio 26 df	Standardized coefficient
IMP/GNP	0.8232E−01	3.9041	0.3622
EXP/GNP	−0.3428E−01	−1.9418	−0.1597
GNPCAP1	0.2929E−05	9.6005	0.6164
GOVBORR/GNP	−0.3242E−04	−1.4591	−0.0143E−03
SAV/GNP	0.2161E−06	0.0133	0.9615
UNEMP	0.1280E−02	5.2116	0.1779
UNEMP(−1)	0.6665E−04	0.5692	0.9262E−02
UNEMP(−2)	−0.2318E−03	−1.8130	−0.0322
UNEMP(−3)	−0.1962E−04	−0.2210	−0.2726E−02
UNEMP(−4)	0.2989E−03	2.2213	−0.0415
UNEMP(−5)	0.3196E−03	2.6242	−0.0444
UNEMP(−6)	−0.3618E−03	−1.4763	−0.0503
CONSTANT	0.2731E−03	0.0661	0.0

Durbin-Watson − 1.8403 Adjusted R^2 − 0.9871

Table 3.10 United Kingdom time-series regression results, 1952–84

Variable name	Estimated coefficient	t-ratio 26 df	Standardized coefficient
IMP/GNP	0.1240	2.2053	0.1841
EXP/GNP	−0.1431	−1.7210	−0.1731
GNPCAP	0.2605E−04	9.3916	0.8395
GOVBORR/GNP	−0.9384E−05	−0.1449	0.9810E−02
SAV/GNP	0.5527E−01	1.1256	0.5736E−01
UNEMP	0.1478E−02	3.3085	0.2007
CONSTANT	−0.4072E−01	−3.7469	0.0

Durbin-Watson − 0.9401 Adjusted R^2 − 0.9728

Table 3.11 United Kingdom time-series regression results, 1952–84, lagged unemployment

Variable name	Estimated coefficient	t-ratio 26 df	Standardized coefficient
IMP/GNP	0.8476E–01	1.7198	0.1259
EXP/GNP	−0.1708	−2.2393	−0.2066
GNPCAP1	0.2817E–04	9.2372	0.9076
GOVBORR/GNP	0.6501E–01	1.5166	0.0675
SAV/GNP	−0.4296E–04	−0.6337	−0.0449
UNEMP	−0.2678E–03	−0.3295	−0.0364
UNEMP(−1)	−0.7555E–03	−1.7235	−0.1026
UNEMP(−2)	−0.4302E–03	−0.8085	−0.5841
UNEMP(−3)	0.4032E–03	0.7863	−0.5474
UNEMP(−4)	0.1440E–02	2.3566	0.1955
UNEMP(−5)	0.2374E–02	3.5162	0.3224
UNEMP(−6)	0.2920E–02	2.5067	0.3940
CONSTANT	−0.3767E–01	−3.5445	0.0

Durbin-Watson – 1.3207   Adjusted R^2 – 0.9797

employment in the financial sector in the short-run. Government borrowing and the savings rate also do not appear to exert a significant effect as before. The adjusted R^2 shows that a very substantial part of the variation in financial employment is explained.

The real GNP per capita once more appears to exert a significant influence on financial employment's share of total employment in Britain (Tables 3.10 and 3.11). Imports are again positively related and exports negatively related to the size of the financial sector. Unemployment is insignificant, while government borrowing and the savings rate do not have a significant effect on financial employment. The adjusted R^2s are consistently high.

Cross-sectional results

The consistent finding that international trade shifts financial employment from the exporting to the importing country can be checked by relating these variables to cross-sectional data. In order to do this, data were collected on thirty-eight countries.[10]

Cultural and other factors can cause countries to differ in the relative importance of financial services. Therefore, in lieu of employment data for banks, the cross-sectional approach uses data on bank assets and liabilities. These allow an independent examination of the previous finding that the reason for the positive coefficients on imports and negative coefficients on exports is the tendency for exporters to bank in the importer's countries for ease of credit or to reduce foreign exchange exposure. If this is so, relatively higher ratios of foreign assets and liabilities to total assets and liabilities should be found in countries with the highest imports, and lower ratios in countries with the highest exports.

Using either the ratio of foreign assets to total assets, or foreign liabilities to total liabilities it is banks in importing countries which have relatively more foreign business (Tables 3.12 and 3.13). Importing countries seem to attract foreign deposits as well as borrowings.[11]

Tables 3.12 and 3.13 include other variables from the time-series regressions, and a dummy variable for tax havens.[12] As expected, there is a relatively larger fraction of foreign business in tax havens.

CONCLUSIONS, POLICY IMPLICATIONS, AND USEFUL EXTENSIONS

Conclusions

A surprisingly small number of factors have been found to explain the bulk of variation in the importance of financial employment in the United

Table 3.12 Cross-sectional results on the fraction of foreign assets, 1984

Variable name	Estimated coefficient	t-ratio 32 df	Standardized coefficient
IMP/GNP	1.4514	4.0422	10.742
EXP/GNP	−1.4299	−3.9533	−10.562
GNPCAP	0.1904E−04	2.4999	0.3058
GOVBORR/GNP	0.1448	1.8319	0.2333
SAV/GNP	0.1854E−01	0.1332	0.1632E−01
UNEMP	0.2689	2.4984	0.3331
CONSTANT	0.3632E−01	0.4141	0.0

Adjusted R^2 – 0.6244

Table 3.13 Cross-sectional results on the fraction of foreign liabilities, 1984

Variable name	Estimated coefficient	t-ratio 32 df	Standardized coefficient
IMP/GNP	1.5056	3.4953	11.759
EXP/GNP	−1.4889	−3.4313	−11.606
GNPCAP	0.3332E−05	0.3646	0.56461E−01
GOVBORR/GNP	0.9057E−01	0.9549	0.1540
SAV/GNP	−0.9292E−01	−0.5564	−0.8633E

States, Canada, and Britain over extended periods of time. A similar and small number of factors also explain variations in the proportion of international financial assets and liabilities across thirty-eight countries in 1984. Finally, a cross-sectional analysis of thirty US states and the District of Columbia revealed that virtually the same limited set of factors explained variations in financial activity in the states for three different measures of state financial activity. Specifically, foreign trade has been shown to shift financial employment to the importing country or region, with an overall effect of international trade that is mildly positive.[13] Such a shift in financial employment is to be expected as a result of the importance of information on the credit-worthiness of exporters versus importers; the need to be near the buyers of letters of credit; the common practice of exporters to borrow in the currency of their receivables; the need to insure each international shipment, and the resulting considerably greater complexity of an import transaction compared with domestic trade. Lastly, unemployment has a temporary effect on financial employment, causing a short-run increase in the fraction of total employment in the financial sector.

These findings are not only consistent with each other, but are reinforced by findings in the literature on the location of international financial activity. First, the availability of a whole range of producer services (legal, accounting, insurance, real estate, and management consulting) and the associated skilled and unskilled labour market is closely tied to the location of international banks (Reed 1981; Gerakis and Roncesvalles 1983; Daniels 1986; Lee 1986). Second, another important international bank location factor, the presence of advanced transportation and telecommunication systems (Reed 1981; Gerakis and Roncesvalles 1983; Lee 1986), is much more likely to be found in highly developed nations. Third, the locational literature stresses access to foreign bank customers who have located abroad (Ball and Tschoegl 1982; Daniels 1986; Fairlamb 1986). This result strongly supports the foregoing stress on imports as being generators of international financial services.

Fourth, foreign banks are increasingly combing the world for locations where they can find new international customers. Thus, locations with buoyant economies, particularly those with strong international trade ties, are especially sought out by foreign banks (Gerakis and Roncesvalles 1983; Choi *et al.* 1986), reinforcing the primacy of international trade in financial sector development.

Finally, stability is seen as being increasingly important in the location of international financial services. Political stability in the normal sense is important (Gerakis and Roncesvalles 1983). So is the assurance that government actions will be predictable and consistent from the regulatory side (Gerard 1985; Lee 1986), and that regulations will be flexible and competitive in similar jurisdictions around the world (*The Banker* 1983; Fairlamb 1986; Lee 1986; Daly and Logan 1989). Stability of the sort talked about here is closely tied to the degree of economic development as given by real GNP per capita in the findings of the present paper.

Policy implications

A number of policy implications follow from the foregoing findings. First, local and regional policy makers interested in promoting the development of their areas as IFCs must work closely with their national counterparts since such key elements as monetary policy and financial institution regulation are national responsibilities and have enormous impacts on IFC development through impacts on savings behaviour, imports and exports through the exchange rate, and costs of doing business for international firms through the national regulatory and tax environment.

Second, national governments also need to develop both short- and long-term strategies to foster economic growth and stability. Clearly, in the longer term there is no substitute for a buoyant and advanced economic system as a means for promoting IFC development. National political stability, as noted, is also a key element in the IFC literature and clearly closely tied to ultimate economic growth and stability.

Third, the emphasis on export promotion that many nations and regions pursue, is not likely to be helpful for IFC development. Rather, a balanced international trade strategy needs to be followed where both imports and exports might be encouraged. The massive Hong Kong and Singapore re-export entrepot function comes immediately to mind as an international trade strategy likely to produce international financial activity.

Fourth, local governments must ensure that there is adequate room to grow in the urban region and national governments need to provide for state-of-the-art telecommunications and transportation infrastructure, once again a concomitant of high levels of per capita income.

Useful extensions

As noted at the outset, our focus is on IFCs, which are large urban centres which provide international financial services. It would have been desirable to augment the state and national analyses presented here with urban based analyses. Regrettably no consistent equivalent urban database exists that would permit such analyses. Thus, the first extension would be to develop an urban database for the world's forty to fifty IFCs so that the necessary analyses could be done.

Beyond this, it would also be desirable to strengthen our present work with the direct observation of various international assets and liabilities held by both foreign and domestic financial institutions. We had to rely on proxy measures such as financial sector employment, and total assets or liabilities of foreign and/or domestic financial institutions.

ACKNOWLEDGEMENT

Financial support from the British Columbia Ministry of Finance and Corporate Relations is gratefully acknowledged.

NOTES

1. Reed (1981: 4) discusses these 'eminence *theories*'.
2. These models include: Fieleke (1977); *Grubel* (1977); Dean and Grubel (1979); Aliber (1980, 1984); Gray and Gray (1981); Ball and Tschoegl (1982); and Monti (1982).
3. The international finance literature also suggests that variations in savings rates and financial regulations (for example, reserve requirements) should affect the level of financial activity in a country. Since regional variations in the former are unmeasurable, and regional variations in the latter do not exist, these factors are excluded from our analysis.
4. The Commerce Department's *Guide to Foreign Trade Statistics* states (p. 11), 'the Customs district shown is the district of unlading, i.e., the district at which the merchandise is unloaded from the importing vessel or aircraft. The district of unlading may not be the same as the district of entry.'
5. This result also occurs in our time-series and cross-section studies of the determinants of the level of financial activity across nations. See Goldberg *et al.* (1988).
6. Levi (1987) establishes the tendency of exporting firms to hedge foreign exchange risk by borrowing in the currency of their receivables.
7. As noted earlier, there are grounds for expecting exports and imports to affect the financial sector in opposing ways. The evidence bears this out.
8. The history of merchant banking in fact bears out this supposition. Chapman (1984) traces the evolution of merchant banking in England from the importing function and the ability to ascertain the credit-worthiness of importers.
9. See Trivedi (1970) and Schmidt and Wold (1973) for how the absence of end-point constraints avoids bias.
10. The thirty-eight included countries are: Argentina, Australia, Bahamas, Bahrain, Belgium, Brazil, Canada, Chile, Columbia, Denmark, Finland, France, Germany, Indonesia, Italy, Japan, Korea, Malaysia, Mexico, Netherlands Antilles, New Zealand, Norway, Panama, Peru, Philippines, Portugal, Singapore, South Africa, Spain, Sweden, Switzerland, Thailand, United Kingdom, United States, Uruguay, Venezuela and Yugoslavia.
11. The data are not available to check whether this is due to currency of denomination of deposits or loans. 'Foreign' refers here to the residency of the borrower or depositor, not the currency of loans or deposits.
12. The included tax havens are the Bahamas, Bahrain, Panama, the Netherlands Antilles and Singapore. It should be noted that explanatory variables are not lagged one period as they are in the time-series regression because unlike the case of employment, effects might be expected to show up immediately on bank balance sheets.
13. Only for the United Kingdom do we find a slightly larger effect of exports than of imports.

REFERENCES

Aliber, R.Z. (1980) 'The integration of offshore and domestic banking systems', *Journal of Monetary Economics*, 6 October, 509–26.

Aliber, R.Z. (1984) 'International banking: a survey', *Journal of Money, Credit and Banking*, 16 (4), 661–78.

Ball, C.A. and Tschoegl, A.E. (1982) 'The decision to establish a foreign bank branch or subsidiary: an application of binary classification procedures', *Journal of Financial and Quantitative Research*, XVII(3), September.

The Banker (1983) 'Banking Exodus', October 8 and 13.

Blomquist, G.C., Berger, M.C. and Hoehn, J.P. (1988) 'New estimates of quality of life in urban areas', *American Economic Review*, 78(1), 89–107.

Bryant, R.C. (1987) *International Financial Intermediation*, Washington, DC: The Brookings Institution.

Chapman, S. (1984) *The Rise of Merchant Banking*, London, England: Heinemann.

Choi, Sang-Rim, Tschoegl, A.E. and Yu, Chow-Ming (1986) 'Banks and the world's major financial centers, 1970–1980', *Weltwirtschaftliches Archiv*, 122 (1), 48–64.

Daly, M.T. and Logan, M.I. (1989) *The Brittle Rim: Asian Nations of the Pacific Rim. The Revolution in International Finance, and the Global Restructuring of Manufacturing*, Sydney, Australia: Penguin Books.

Daniels, P.W. (ed) (1979) *Spatial Patterns of Office Growth*, Chichester, England: John Wiley and Sons.

Daniels, P.W. (1986) 'Foreign banks and metropolitan development: a comparison of London and New York', *Journal of Economic and Social Geography*, 77 (4), 269–87.

Dean, J. and Grubel, H. (1979) 'Regulatory issues and the theory of multinational banking' in F.R. Edwards (ed) *Issues in Financial Regulation*, New York: McGraw-Hill.

Dunning, J.H. and Norman, G. (1983) 'The theory of the multinational enterprise: an application to multinational office location', *Environment and Planning A*, 15 (4), 675–92.

Economic Development Office (1985) 'Vancouver: an analysis of economic structure, growth and change', Vancouver, BC: Economic Development Office, City of Vancouver, June.

Enderwick, P. (1987) 'The strategy and structure of service-sector multinationals: implications for potential host regions', *Regional Studies*, 21 (3), 215–23.

Fairlamb, D. (1986) 'Foreign banks take an increasing share of the cake', *The Banker*. March, 87–134.

Ferris, P. (1984) *The Master Bankers*, New York, NY: New American Library.

Fieleke, N.S. (1977) 'The growth of US banking abroad: an analytical survey', in *Key Issues in International Banking*, Conference series 18, Boston, Mass.: Federal Reserve Bank of Boston.

Gerakis, A.S. and Roncesvalles, O. (1983) 'Bahrain's offshore banking center', *Economic Development and Cultural Change*, 31 (2), 271–93.

Gerard, K. (1985) 'Why they fled the Big Apple (and do they regret it?)', *Across the Board*, 22, 56–63, May.

Goldberg, M.A. (1985) *The Chinese Connection: Getting Plugged in to Pacific Rim Real Estate, Trade and Capital Markets*, Vancouver, BC: The University of British Columbia Press.

Goldberg, M.A., Helsley, R. and Levi, M. (1988) *Factors Influencing the Development of Financial Centers*, unpublished, Faculty of Commerce and Business Administration, The University of British Columbia.

Goldsmith, R. (1969) *Financial Structure and Development*, New Haven, Conn.: Yale University Press.

Gray, J.M. and Gray, H.P. (1981) 'The multinational bank: a financial MNC?', *Journal of Banking and Finance*, 5, 33–63.

Grubel, H. (1977) 'A theory of multinational banking', *Banca Nazionale del Lavoro*

Quarterly Review, 1–3, 349–63.

Hakim, J. (1986) *The International Investment Banking Revolution: Strategies for Global Securities Trading*, Special Report No. 1065, London, England: The Economist Intelligence Unit, December.

Hamilton, A. (1986) *The Financial Revolution*, New York, NY: The Free Press.

Heenan, D.A. and Perlmutter, H. (1978) *Multinational Organization Development: A Social Architecture Perspective*, Reading, Mass.: Addison-Wesley.

Hutton, T. and Ley, D. (1987) 'Location, linkages and labor: the downtown complex of corporate activities in a medium size city, Vancouver British Columbia', Vancouver, BC: Department of Geography, The University of British Columbia, mimeo, February.

Kaufman, H. (1986) *Interest Rates, the Markets, and the New Financial World*, New York, NY: Times Books.

Kindleberger, C.P. (1974) *The Formation of Financial Centers: A Study in Comparative Economic History*, Princeton Studies in International Finance, 36, Princeton, NJ: Princeton University Press.

Kindleberger, C.P. (1985) 'The functioning of financial centers: Britain in the 19th century, the United States since 1945' in W. Feeathier and R. Marston (eds) *International Financial Markets and Capital Movements*, Princeton, NJ: Princeton University Press.

Lee, Kam-Hon (1986) 'Competition among commercial banks in Hong Kong – a strategic marketing review for local Chinese banks', a paper presented at the International Conference on 'Chinese Banking and Nation Building in Southeast Asia', 10–11 October, Shatin, Hong Kong: Chinese University of Hong Kong.

Lee, Kam-Hon and Vertinsky, I. (1988) 'Strategic adjustment of international financial centres (IFCs) in small economies: a comparative study of Hong Kong and Singapore', Vancouver, BC: Centre for International Business Studies, Faculty of Commerce and Business Administration, The University of British Columbia, mimeo, January.

Levi, M.D. (1987) *Exchange Rates and the Value of the Firm: The Theory and Evidence*, Working Paper, Vancouver, BC: Faculty of Commerce, University of British Columbia, mimeo.

Ley, D. and Hutton, T. (1986) 'Vancouver's corporate complex and producer services sector: linkages and divergence within a provincial staple economy', Vancouver, BC: Department of Geography, The University of British Columbia, mimeo, September.

Leyshon, A., Thrift, N.J. and Daniels, P.W. (1987) 'The urban and regional consequences of the restructuring of world financial markets: the case of the City of London', Working Papers on Producer Services No. 4, Liverpool: Department of Geography, Universities of Bristol and Liverpool, July.

Lund, L. (1979) 'Factors in corporate locational decisions', *Information Bulletin*, 66, Toronto, Ontario: The Conference Board of Canada.

Markusen, J.R. (1986) 'Trade in producer services: issues involving agglomeration economies, human capital, and public inputs', Series on Trade in Services, Victoria, BC: The Institute for Research on Public Policy, December.

Mayer, M. *The Money Bazaars: Understanding the Banking Revolution Around Us*, New York, NY: New American Library.

Mendelsohn, M.S. (1980) *Money on the Move: The Modern International Capital Market*, New York, NY: McGraw-Hill Book Co.

Monti, A. (1982) 'Recent trends in international banking', *Journal of Banking and Finance*, 6, 389–99.

O'Connor, K. and Edgington, D. (1987) 'Producer services and metropolitan development in Australia', Clayton, Victoria: Department of Geography, Monash University.

Reed, H.C. (1981) *The Preeminence of International Financial Centers*, New York, NY: Praeger Publishers.

Schmidt, P. and Wand, R. (1973) 'The Almon Lag technique and the monetary versus fiscal policy debate', *Journal of the American Statistical Association*, 68, 11–19.

Shelp, R.K. (1981) *Beyond Industrialization: Ascendancy of the Global Service Economy*, New York, NY: Praeger.

Tan, A.H.H., and Basant Kapur (1986) *Pacific Growth and Financial Interdependence*, Sydney, Australia: Allen and Unwin.

Thrift, N.J., Leyshon, A. and Daniels, P.W. (1987) "Sexy Greedy": the international financial system, the City of London and the south east of England', Working Papers on Producer Services No. 2, Liverpool, England: Department of Geography, Universities of Bristol and Liverpool, October.

Trivedi, P.K. (1970) 'A note on the application of Almon's method of calculating distributed Lag coefficients', *Metroeconomica*, 22, 281–6.

Wachtel, H. (1986) *The Money Mandarins: The Making of a Supranational Economic Order*, New York, NY: Pantheon Books.

Wood, P.A. (1986) 'The anatomy of job loss and job creation: some speculations on the role of the "producer service" sector', *Regional Studies*, 20 (1), 37–46.

4 The global intelligence corps and world cities: engineering consultancies on the move

Peter J. Rimmer

> The new global 'empire' has a quite different definition from the old. International armies are seen to be fighting for power. Rather than searching for new lands to exploit they are perceived to be locked in economic warfare.... In this new *imperium* financiers are considered to be the 'top dogs', consultants 'the intelligence corps' and contractors 'the imperial army'.
>
> (Rimmer and Black 1982: 135)

Since the economic crisis of the early 1970s, a transition to a new phase has been recognized in the development of the world economy. This economic restructuring is being orchestrated by transnational corporations through an emerging set of world cities – nodal points in transport and communications networks controlling and co-ordinating global economic activity and sites for the production of associated inputs (Friedmann and Wolff 1982; Friedmann 1986; Feagin and Smith 1987). A key role in this new metropolitan hierarchy has been afforded by transnational capital to both construction companies and property developers. Based in top-ranking world cities, the international financial institutions are engaged in re-shuffling ownership and control of productive assets on a global scale. At the behest of their financial overlords, transnational construction companies and property developers, widely-dispersed throughout the metropolitan hierarchy, are employed in transforming the physical fabric of cities and regions with the aid of an international army of peripheralized labour. Before investing in particular construction projects, however, the leading financial agencies draw upon the advice of transnational engineering consultants to assess their location, specifications and likely rate of return.

Unlike the activities of financial corporations and property developers, the spatial organization of transnational engineering consultancies has received little attention in the producer services literature (Daniels 1987). Hence, three basic questions must be raised: what are engineering consultancies; what services do they perform; and what are their national affinities? Then a more specific question on locational behaviour can be put: how are the corporate headquarters of these firms related to the newly-emerging metropolitan hierarchy? A complete analysis of spatial organization, however, requires an answer to: how are the overseas branch offices of the transnational engineering consultancies related to the emergent metropolitan hierarchy?

Before addressing these issues a very brief examination is made of the distinctive characteristics of engineering consultancies compared with other service activities – the prelude to analysing the new urban hierarchy. With this background, a more detailed study is carried out of headquarters locations in relation to the metropolitan hierarchy pivoted on world cities. Then, attention is directed to the distribution of offices established by transnational engineering consultancies in different parts of the world – an examination leading to a comparison of specific country and inter-firm patterns. The results of these exercises allow us to discuss the interdependence between metropolitan restructuring and the headquarters and branch office locations of transnational engineering consultancies – the veritable global 'stalking horses'.

ENGINEERING CONSULTANCIES AND THE NEW URBAN HIERARCHY

At the turn of the century the term 'consulting engineer' was undefined in the United Kingdom. Both the client and the professional engineer were uncertain as to the duties and responsibilities because contracting and consulting work were mixed to the detriment of both. This problem was resolved with the establishment of the Association of Engineers in 1913. Its Articles defined a 'consulting engineer' as:

A person possessing the necessary qualifications to practise in one or more of the various branches of engineering who devotes himself to advising the public on engineering matters or to the designing and supervising the construction of engineering works, and for such purposes occupies and employs his own office and staff, and is not directly or indirectly concerned or interested in commercial or manufacturing interests such as would tend to influence his exercise of independent professional judgement in the matters upon which he advises.

(ACE 1988: 10)

These Articles have been used as a model for a number of similar organizations throughout the world.

Basically, a consulting engineer offers five distinct types of services: pre-investment studies; design and supervision services for the construction of works; specialized design and development; project management; and advisory services. Hence, the stipulation that the consulting engineer may have no direct or indirect interest in any commercial, manufacturing or contracting organization that may influence professional judgement. Remuneration of consulting engineers is based solely on the fees paid by the client. The employment of the consulting engineer is not restricted to a particular country and increasingly their services are being offered world-wide. Consequently, metropolitan areas are continually being re-evaluated as likely sites for the extension of a firm's international network. The latter has been prompted not only by the globalization of manufacturing but through the parallel internationalization of producer services, notably finance, insurance and real estate.

The effect of this booming international trade in engineering con-sultancy services on metropolitan areas, however, is elusive because transactions are information-based and reliant upon an advisory relation-ship being established between consultant and client. Although the delivery of personally-provided and underwritten engineering and architectural services uses tangible inputs, such as computer time, transport and office furniture, a tangible commodity is not exported. Consequently, these services cannot be sampled or pre-tested for performance; they are time and place specific; they cannot be stockpiled for future use; they are custom-built requiring promotion to each client for a specific project; and, invariably involve a time-lag between application for and commissioning a project. Rather than the physical movement of a commodity, the key attributes are data exchange, documentation and personal mobility.

Exchange of project data (i.e. statistics and measurements) has been facilitated by improvements in communications, notably microcomputers and telecommunications networks, which have speeded the processing and analysis of information, increased productivity, facilitated diversity and opened up new and global markets. These information network enter-prises, according to Enis and Roering (1981: 1), purchase neither tangible objects nor intangible features but 'a bundle of benefits – a product' (including ideas, behaviours and concepts). Engineering consulting, however, is different from financial services as it cannot be transmitted down the wire but requires on-site input at the behest of the client. As these client-instigated services have not been well-conceptualized there have been few empirical studies of engineering consultancies and their role in transferring technology from core to peripheral locations. A major problem, however, in relating these elusive transactions to the changing status of metropolitan areas is the absence of an agreed urban hierarchy.

Nomura's world cities

Considerable attention has been given to 'world cities' that have risen to prominence since the late 1960s as part of networks designed by trans-national corporations to articulate the internationalization of production and services on a global scale. Restructuring of the world urban hierarchy has brought in its wake a number of studies aimed at providing taxonomies of the new phenomena (Cohen 1981; Freidmann and Wolff 1982; Friedmann 1986; Yamaguchi 1988). For our purposes, however, the most useful survey of the form and strength of the integration of urban centres into the world capitalist system has been that undertaken by the Nomura Research Institute (1982).

The Nomura Survey reviews some 345 cities in terms of twenty attributes reflecting personal services, commodity and commercial transactions, information flows, and international finance. Initially, 178 cities, including Shanghai, Fukuoka, Dacca, Venice and Bordeaux, were eliminated from the survey because they failed to reach the minimum threshold for recognition as 'international cities' (i.e. five attributes). Thus, 167 cities remained.

Eighty cities, including Akron, Baghdad, Birmingham (UK), Nagoya and Stuttgart, were classified as third-ranking international cities on account of their particular importance in commodity and commercial transactions. Reflecting their additional pull in personal services, Bombay, Osaka, Rotterdam and Taipei were classified among the fifty-seven second-ranking international cities. New York, was in the first ranking group with London, Paris, Singapore, Sydney, Melbourne and Tokyo and twenty-five others – a reflection of their greater strength in information flows and financial transactions. Clearly, there could be challenges to the results of the survey. The classification of Melbourne but not Osaka, for example, as a first-ranked city would be difficult to substantiate; both should be second-ranking. The inclusion of contiguous built-up areas, such as Tokyo and Yokohama (*Keihin*), and Kyoto, Osaka and Kobe (*Keihanshin*), as separate entities is also difficult to justify. Nevertheless, the Nomura survey still stands as the most comprehensive of its type and the best available for our purposes.

When the three levels of international cities were located they were concentrated most heavily in North America, western Europe and, to a lesser extent, in east Asia. Their presence was sparse in Africa, and Central and South America, and in other parts of Asia. Although both China and India had well-developed urban networks they were, for the most part, self-contained serving an internal rather than a global market. Presumably, representation in eastern Europe was weak because the Soviet Bloc had, until recently, its own 'world system' under the hegemony of the Union of Soviet Socialist Republics.

As a means of overcoming the cartographic problems of this imbalance a new method of representation has been developed in Figure 4.1 which shows the distribution of the three ranks of international cities in western Europe, the Middle East and Africa, eastern Europe, Asia, Oceania, North America, and Central and South America. Using these base maps, we can explore the relationship between engineering consultancy services and metropolitan areas.

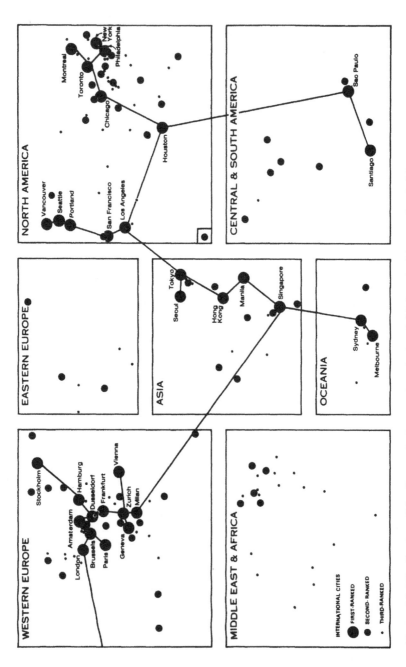

Figure 4.1 Distribution of the first-ranked, second-ranked and third-ranked international cities
Source: Based on Nomura Research Institute (1982)

LOCATION OF NATIONAL HEADQUARTERS

Attention is focused on the headquarters locations of transnational engineering consultancies because the geographical dispersion of plants brings about the need for top-level centralized management to run and control the global network. Originally, engineering consultancies were located to be near the majority of their clients as personal contact is essential; the initial site was invariably retained even after a network of new domestic branch offices was established. The growing trend to operate more international offices, however, has contributed to the importance of central headquarter activities – planning, internal administration, and distribution and marketing. Assuming that the number of transnational engineering consultancies is a key indicator of the form and strength of a city's integration into the world system, interest is centred on their location relative to the Nomura urban hierarchy rather than on the intra-metropolitan sites of firms (the corporate expansion into the suburbs being worthy of study in its own right).

The Fédération's survey, 1983

Before examining the headquarters locations of firms it is pertinent to study the spatial organization of the Fédération Internationale des Ingénieurs-Conseils (FIDIC). Basically, the Fédération is a group of national associations of independent consulting engineers who comply with an agreed code of professional conduct. Although construction contractors are anxious to gain access to the extensive intelligence on markets and product performance, members of the Fédération have argued that their continued success depends on maintaining an impartial stance with their overseas clients. The Fédération's objectives include promoting its organization in countries where it does not exist by encouraging clients to select engineering consultants on technical ability rather than on price competition (i.e. a supply-driven service). Thus, the ultimate aim is to further the economic viability of the profession.

Having shifted its own headquarters from the Netherlands to Lausanne the *International Directory of Consulting Engineers 1983* (FIDIC, 1982) recorded that the Fédération was represented in thirty-eight countries. As shown in Figure 4.2, there is no strong correlation between the location of the headquarters of national associations and the hierarchy developed by the Nomura Research Institute. Generally, the associations have sought representation in national capitals rather than in the highest ranking cities presumably to tap work generated by international aid agencies. Exceptions are in Australia, Brazil and Israel where the economic clout of the coastal cities overrides the political pull of inland capitals.

The spatial organization of the Fédération stands out in strong contrast to the pattern of headquarters locations exhibited by the 215 'major

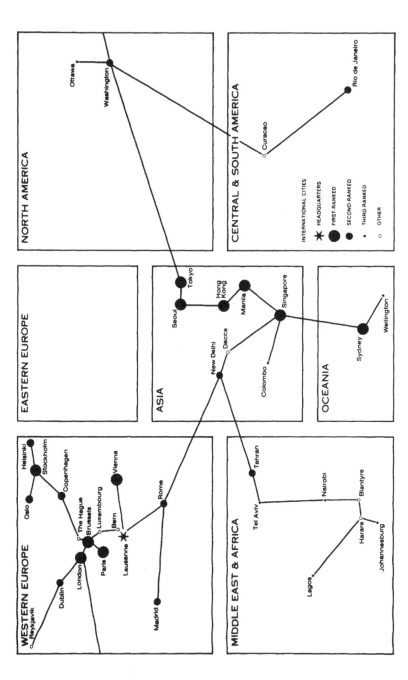

Figure 4.2 The spatial organization of the Fédération Internationale des Ingénieurs-Conseils
Source: Based on data from FIDIC (1982)

consulting firms' listed as 'working internationally' in the *International Directory of Consulting Engineers 1983* 'to assist clients, owners, bankers and international agencies to select consulting engineers' (FIDIC 1982: 3). An examination of the nationality of these firms in Table 4.1 suggests that engineering consulting is very much a West European phenomenon as it accounted for almost two-thirds of all firms. In large part, it was a British fiefdom with over one-quarter of the selected firms being domiciled there. The main challengers were the Scandinavian countries and the Nether-lands. Middle and southern European countries were not well-represented. Although the United States ranked third in number of firms, North America as a whole trailed western Europe. Surprisingly, Oceania outranked Asia, the Middle East and Africa, and Central and South America. There were no firms in eastern Europe.

When the individual firms are matched against the Nomura Research Institute's urban hierarchy in Figure 4.3 two-fifths are found in first-tier cities, one-fifth in second-tier and less than one-tenth in third-tier; the remaining one-third are in unranked cities. There is also marked variation in their preference for first-tier cities. The dominance of London is immediately apparent. Yet, its strength is underplayed because its total does not include the large number of firms swarming around it. If the firms in London's penumbral zone were included, it would outstrip the strong presence of engineering consultancies in the Scandinavian capitals of Copenhagen, Helsinki, Oslo and Stockholm and those focused on

Table 4.1 Nationality of engineering consultants 'working internationally' according to the Fédération Internationale des Ingénieurs-Conseils, 1983

Country	No.	Per cent	Country	No.	Per cent
Canada	10	4.6	France	–	–
USA	24	11.2	West Germany	10	4.6
			Italy	1	0.5
North America	34	15.8	Netherlands	21	9.8
			Scandinavia	35	16.3
Central and South			Switzerland	2	0.9
America	9	4.2	United Kingdom	59	27.4
			Other	9	4.2
Middle East and					
Africa	4	1.9	Western Europe	137	63.7
Japan	8	3.7	Eastern Europe	–	–
Other	4	1.9			
			Oceania	19	8.8
Asia	12	5.6			
			Total	215	100.0

Source: FIDIC (1982)

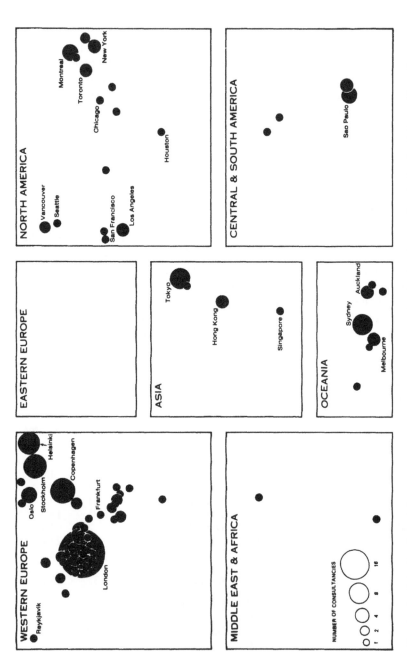

Figure 4.3 Distribution of the headquarters of firms listed by the Fédération Internationale des Ingénieurs-Conseils
Source: Based on data from FIDIC (1982)

Rotterdam and the Hague. Paradoxically, only Stockholm is among the top-ranked cities. Of the other first-ranking cities, Frankfurt has only one firm with headquarters there and Zurich, Geneva, and Vienna, like Paris, have none.

The absence of firms with head offices in Paris raises doubts about the representativeness of the survey. Although it provides a fair reflection of first-ranking Hong Kong, Sao Paulo (Kowarick and Campanario 1986), Singapore and Tokyo (Rimmer 1986), the status of Sydney and Melbourne seems to be overrated. Conversely, North American cities would appear to be under-represented. Although the concentration of firms in New York, Los Angeles, Vancouver, Montreal and Toronto stands out, the significance of other first-tier cities – Chicago, Houston and Philadelphia – seems to have been downplayed. Indeed, the relative strength of metropolitan areas in the United States appears to have been underestimated compared with cities in Oceania and western Europe, and London in particular. The distorted picture provided by the Fédération's survey suggests that *size* of firm has not been taken into account, and that the nexus between headquarters locations, research and development and high-value production in first-ranked cities remains unexplored.

Size of firms, 1980

An opportunity for studying the largest engineering consultancies is offered by the *Engineering News-Record* (ENR 1979–87). Since 1978, it has been ranking transnational engineering consultancies based on their foreign billings derived from pre-feasibility studies, design and planning, construction management and design-construct (i.e. a turnkey arrangement); they include fees from reimbursables, temporary staff transfers and construction inspections. Foreign billings are defined as those received from countries other than where a firm's head office is located. Although the specific rankings of firms on this score have been challenged by individual companies they admit that there is no alternative to the *Engineering News-Record* survey which in 1980 provided information on the head office locations of the 'top 150' firms.

An examination of the nationality of the 'top 150' firms in 1980, as shown in Table 4.2, contradicts the findings of the Fédération Internationale des Ingénieurs-Conseils (FIDIC 1982). It shows that there are two major concentrations of firms in North America and western Europe, with the former rather than the latter having the largest number. The American edge is attributed to its then technical lead in high technology areas. Of the other regions only Asia stands out as there was only token representation in the Middle East and Africa, and Central and South America. Like eastern Europe, Oceania, on this occasion, is unrepresented.

These differences are also apparent when the 'top 150' firms are related

Table 4.2 Nationality of the 'top 150' transnational engineering consultancies, 1980 and 1986

Country	1980 No.	Per cent	1986 No.	Per cent	Country	1980 No.	Per cent	1986 No.	Per cent
Canada	9	6.0	9	6.0	France	18	12.0	13	8.7
USA	62	41.3	34	22.7	West Germany	8	5.3	16	10.7
					Italy	3	2.0	5	3.3
North America	71	47.3	43	28.7	Netherlands	5	3.3	8	5.3
					Scandinavia	7	4.7	8	5.3
					Switzerland	3	2.0	8	5.3
Central and South America	3	2.0	2	1.3	United Kingdom	20	13.3	20	13.3
					Other	2	1.3	4	2.6
Middle East and Africa	2	1.3	3	2.0					
					Western Europe	66	44.0	82	54.7
Japan	4	2.7	10	6.7	Eastern Europe	–	–	1	0.7
South Korea	1	0.7	4	2.7					
Other	3	2.0	2	1.3	Oceania	–	–	3	2.0
Asia	8	5.3	16	10.7					
					Total	150	100.0	150	100.0

Source: ENR (1981–7)

to the urban hierarchy in Figure 4.4. Clearly, the location of the firms recognized by the Fédération Internationale des Ingénieurs-Conseils as 'working internationally' gives a misleading picture of the metropolitan location of transnational engineering consultants. Rather than two-thirds of the engineering consultancies being headquartered in ranked cities the new survey boosts their number to four-fifths – 57 per cent being in first-tier cities, 17 per cent in second-tier, 9 per cent in third-tier and 19 per cent in unranked cities. The marked variation among first-tier cities, however, still persists.

Paris, not London, has the largest concentration of the 'top 150' transnational engineering consultants – though they would be on a par if those located in the Home Counties were included. There is also a truer reflection of the relative strength of Frankfurt and Stockholm as bases for engineering consultancies. Although Brussels, Milan and Zurich have token representation, the other first-tier cities – Dusseldorf, Geneva, Hamburg – have no major international firms. Conversely, Copenhagen belies its second-tier status. The strength of second-tier cities is a feature of the tightly structured European pattern of headquarters locations. It is not apparent in the United States which boasted a larger number of firms in the 'top 150' in 1980.

The number of headquarters of transnational engineering consultancies in first-ranking cities in the United States reinforced the argument that 'Snowbelt' cities were prominent in producer services (Noyelle and Stanback 1984). This phenomenon is reflected in the traditional concentration of activity in the east centred on New York and, to a lesser extent, on Chicago. Still, there was an emerging presence in the south with the banking centre of Houston as the major pivot. The largest number of transnational engineering consultancies in a single centre, however, was in Los Angeles which dominated the western coastal strip. Although Vancouver was a logical extension of this strip, most activity in Canada was centred on Montreal and Toronto respectively. In the east, Philadelphia was the sole first-ranking city without a major engineering consultancy. Washington and Atlanta, however, were the only second-ranking cities with more than two or more consultancies.

The only counterpole to Europe and North America in 1980 was in eastern Asia where Tokyo figured prominently. Of the other first-ranked cities only Seoul had engineering consultancies in the 'top 150'; there were no firms located in Hong Kong, Manila and Singapore. Their presence in Seoul and second-ranked Taipei, however, suggests that the rise of indigenous engineering consultancies based in the newly industrializing countries (NICs) was about to commence – a phenomenon that was not marked outside the European–North American–Asian axis. The only first-tier city with a major engineering consultancy was Sao Paulo; Melbourne, Sydney and Santiago had no representatives. Conversely, there were consultancies located in Beirut and Buenos Aires.

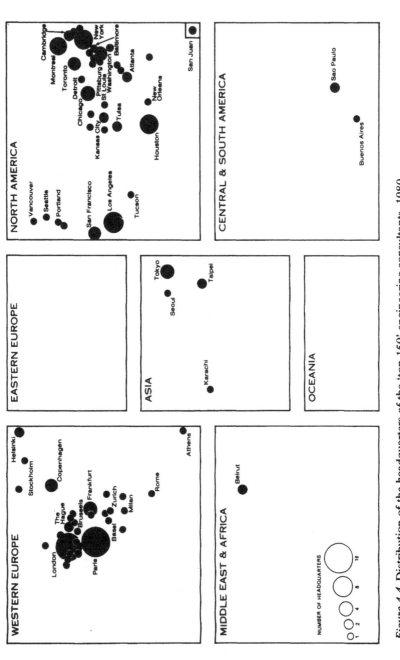

Figure 4.4 Distribution of the headquarters of the 'top 150' engineering consultants, 1980
Source: Based on data from ENR (1981)

Clearly, the distribution of the 'top 150' firms in 1980 offers a different perspective on transnational engineering consultancies from that inherent in the survey conducted by the Fédération Internationale des Ingénieurs-Conseils in 1982. A major problem with both surveys, however, is that they refer to a single year and cannot reflect *changes* in the importance of headquarters locations over time. Fortunately, this difficulty can be resolved for the leading engineering consultants by recourse to subsequent surveys by the *Engineering News-Record.*

Changes 1980–6

The *Engineering News-Record* has charted the key trends in international consulting (or design) markets (ENR 1981–7); though a major drawback has been a shift in the database from the 'top 150' firms in 1980 and 1981 to the 'top 200' firms between 1982 and 1986 (Table 4.2). Therefore, in order to examine changes in headquarter locations there is little alternative but to concentrate on the 'top 150' engineering consultancies to maintain comparability over a longer period. In interpreting these variations it is a moot point as to whether the change is the result of variations in the nature of the firm arising from its performance, takeover or merger, or deep-seated shift in the standing of a particular metropolitan area stemming from its growth or decline. As only a six-year time span is involved the emphasis must be on organizational performance rather than on long-term urban change though the survey does not necessarily specify where a consultancy is part of a conglomerate (e.g. De Leuw, Cather and Company based in Washington is a wholly-owned subsidiary of the Parsons Corporation with headquarters in Pasadena, California).

An examination of the nationality of the 'top 150' consultancies engaged in providing engineering and architectural services between 1980 and 1986 in Table 4.2 shows that the dominant position exercised by North American firms has been eroded (see Figure 4.5); a small loss was also experienced by Central and South America. In contrast, marked gains were recorded in western Europe as a host of firms had sharpened their design skills in key energy and manufacturing fields and moved offshore. Relatively slow-growth countries, such as France and the United Kingdom, had to tap export markets because, unlike the United States, they did not have strong domestic markets. There was also a marked positive showing in Asia with smaller net gains occurring in Oceania, eastern Europe, and the Middle East and Africa.

Within these regions there were differences in the importance of individual countries. The full brunt of the North American loss was experienced by the United States as Canadian consultants maintained their rankings among the 'top 150' firms. Within western Europe, West Germany, Switzerland and, to a lesser extent, Italy, Netherlands and Scandinanvian countries increased their number of representatives; the

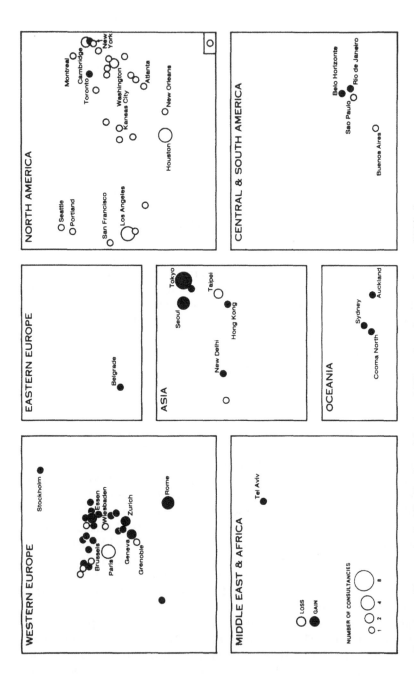

Figure 4.5 Changes in the distribution of the 'top 150' engineering consultants, 1980–6
Source: Based on data from ENR (1981–7)

United Kingdom maintained its position; and France experienced an overall loss. Asia's positive showing was almost entirely due to Japan. Generally, interpretations of these major changes in the distribution of engineering consultancies between countries are attributed to distinctive management cultures (Rimmer 1988). Rather than repeat these arguments, it is preferable to pursue a more searching analysis by relating engineering consultancies to an emerging global network of metropolitan areas.

When the changes are related to the Nomura urban hierarchy no remarkable shifts had occurred between tiers. The number of the 'top 150' firms in first-tier cities had declined slightly from eighty-five to eighty-two; they had increased from twenty-five to twenty-seven in second-tier cities; and they had dropped from fourteen to ten in third-tier cities. Thus, there was a net gain of five firms in unranked cities. These small shifts between tiers, however, masked marked changes in the status of first-ranked cities.

The most striking feature is the net loss of major consultancies in North America which affected all cities. First-tier Toronto, Canada's economic hub, and second-tier Boston, re-focused on high-tech developments, were the only exceptions. Otherwise, this 'hollowing out' of engineering consultancy services was irrespective of either tier or region. Both ranked and unranked cities were singled out; and both 'Snowbelt' and 'Sunbelt' cities were hit. The effects were most pronounced in Los Angeles and Houston – first-tier cities that, hitherto, had epitomized corporate growth in the United States. Much of the loss can be attributed to the propensity of major United States consultancies to engage in mega-projects, particularly in the Middle East. The decline of the Middle East design market due to a decline in oil revenues and political unrest had an adverse affect on the billings of firms from the United States. There were also allegations that the United States firms were saddled with a number of tax and regulatory policies that hindered their competitiveness overseas.

Paris was the only other first-tier city to experience a loss comparable to Los Angeles and Houston – a change supporting arguments for removing strict planning controls over development in the French capital. The few minor losses in London and unranked western European cities were outweighed by widespread gains. Paradoxically, they were not marked in first-ranking cities, such as Amsterdam, Frankfurt, Hamburg, Milan and Vienna but in a host of lowly-ranked or unranked West German cities that had shifted from a traditional heavy industrial base to lighter activities. The desire of engineering consultancies to be near financial services countered this trend as they were attracted to Switzerland's first-tier cities, Geneva and Zurich. The largest growth, however, was in Rome, a second-tier city, which paralleled the rise of the Italian construction contracting industry to world prominence.

A similar happening occurred in the first-tier Asian cities of Tokyo and Seoul where the marked increase in the number of major engineering consultancies coincided with the move of Japanese and Korean construction contractors offshore. Elsewhere in Asia, Hong Kong was the only other first-ranking centre to record an increase. As the gain in second-tier Delhi was more than offset by losses in Taipei there is no evidence of a strong presence of major engineering consultancies in the newly industrial-izing countries (NICs). Other developments included: Sydney's positive showing as part of Oceania's net gain; the loss in Sao Paulo being cancelled out by developments in lower and unranked South American cities; and gains in the third-tier city of Tel Aviv and unranked Belgrade.

Merely examining headquarters locations, however, only provides a partial insight into the way transnational engineering consultancies have complemented the internationalization of production by providing inter-mediate engineering and architectural services to domestic clients. Rather than following in the wake of corporate producers, transnational engineer-ing consultancies have sought out their own opportunities for construction contractors and property developers. Indeed, to fully comprehend their information-gathering activities attention has to be shifted from head-quarters to branch offices located in overseas countries. In addition to the home-based clients, the foreign beach-heads also serve local producers and, therefore, penetrate the local engineering consultancy market and, coincidentally, fuel the demand for local office space.

LOCATION OF BRANCH OFFICES

Originally, overseas work was conducted from head office with the principals of the consultancy being responsible for maintaining personal contact with the client. The growth of the internationalization of produc-tion and services, however, has transformed engineering consultancies into transnational corporations with a network of overseas branch offices. As international work has expanded, the network has been elaborated by the establishment of different categories of branch offices. In particular, a distinction is made between:

(a) project offices which last for the duration of the contract (e.g. Halcrow Fox have a project office in Taiwan);
(b) permanent country-level offices (e.g. W.S. Atkins and Associates have an office in Riyadh to oversee consultancies in Saudi Arabia); and
(c) regional-level offices located in key cities that have developed as pivots of international finance and control (e.g. Sir Alexander Gibb and Partners' base in unranked Canberra, Australia, serves Hong Kong and the Pacific).

As project offices are ephemeral and not necessarily located in major metropolitan areas attention here is focused upon the global network of country-level and regional-level branch offices by contrasting the office patterns of two sets of national consultants, and exploring the spatial distribution of offices within individual firms.

International patterns

As already stated above, attention has to be restricted to an examination of the pattern in 1983. A review of changes in the status of cities following the displacement of the Middle East by Asia as the leading regional market and the replacement of Saudi Arabia by Indonesia as the first-ranking market between 1980 and 1986 is precluded because the *Engineering News-Record* lists only national markets (Table 4.3). Consequently, recourse has to be made to the *International Directory of Consulting Engineers 1983* (FIDIC 1982) to plot the distribution of branch offices of the major firms 'working internationally'.

The Fédération survey 1983 revisited

The offices of the 215 firms in the survey conducted by the Fédération Internationale des Ingénieurs-Conseils (FIDIC 1983) are spread unevenly throughout the world. They are heavily concentrated in the Middle East, Africa and Asia and, to a lesser extent, in Oceania and Central and South America. There is a dispersed pattern in western Europe; a sparse pattern in North America (though later the market became a target for European firms); and no firms in eastern Europe.

An examination of branch offices in relation to the urban hierarchy mapped out by the Nomura Research Institute in Figure 4.6 shows a different pattern from the command locations. Apart from Paris and London none of the other first-ranked centres in western Europe and North America have attracted branch offices. Paradoxically, the Oceanian first-ranked cities of Sydney and Melbourne have been chosen as pivotal

Table 4.3 Regional breakdown of foreign billings, 1980–6

Region	1980	1981	1982	1983	1984	1985	1986
Middle East	0.9	1.1	1.3	1.3	1.1	1.0	0.9
Africa	0.6	0.7	0.9	0.8	0.8	0.8	0.9
Asia	0.4	0.5	0.7	0.8	0.8	0.9	1.0
Central and South America	0.3	0.3	0.4	0.4	0.3	0.5	0.3
Europe	na	na	0.4	0.4	0.4	0.4	0.3
North America	na	0.0	0.1	0.1	0.1	0.1	0.2
Other	0.4	0.5	0.0	0.0	0.0	0.0	0.0
Total	2.6	3.1	3.7	3.8	3.5	3.6	3.5
No. consultants	150	150	201	200	204	200	200

Note: Rounding errors
Source: Based on ENR (1981–7)

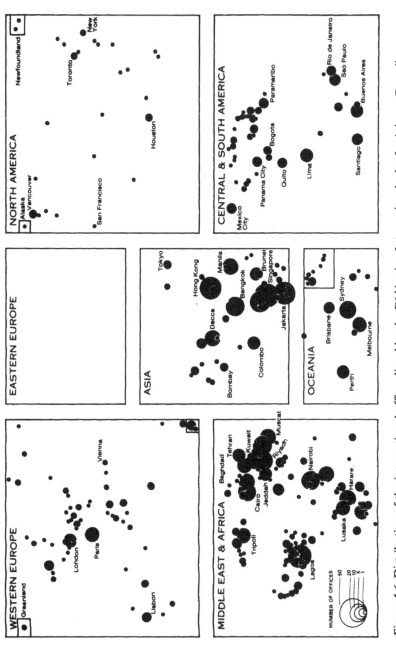

Figure 4.6 Distribution of the international offices listed by the Fédération Internationale des Ingénieurs-Conseils
Source: FIDIC (1982)

locations. Although their selection suggests that Australia may be dependent on overseas command centres it is also an important staging point for work in the Pacific Islands. In the Middle East and Africa, however, it has been the second- and third-ranked cities, and even the unranked that have figured prominently as 'basing points' for transnational engineering consultancies.

In 1983, Riyadh was the prime focus of attention in the Middle East though other second- and third-ranked centres, notably Abu Dhabi, Baghdad, Cairo, Kuwait and Tehran, and the unranked Muscat were important – a reflection of the availability of projects fuelled by oil dollars in what was then the world's largest market for design work. Elsewhere in Africa, petrodollars and large populations had attracted consultants to both Lagos in West Africa and Tripoli in North Africa. Surprisingly, considering the amount of aid that Egypt received from the United States and neighbouring Arab states, Cairo did not figure as prominently. Reflecting that countries with fast-growing populations must double the capacity of their infrastructure every twenty-five years, the other main African concentrations were focused on Nairobi and the unranked centres of Dar es Salaam, Gaborone, Harare and Lusaka. Unranked centres, however, were less prominent in Asia.

As Tokyo, like its first-ranked counterparts in western Europe and North America, did not attract transnational engineering consultancies much of the branch office activity was focused on Hong Kong and Singapore. Articulating the economies of east and South-east Asia respectively, these regional metropolises attracted the largest number of firms. The other main centres for branch offices were Bangkok, Brunei, Kuala Lumpur, Jakarta and Manila which were the capital cities of countries with emerging markets for consulting work. Although the less vibrant south Asian economies needed a vast amount of design help to maintain and improve housing, schools, transport and water supply systems, only Colombo and Dacca were prominent.

On some occasions, market penetration is less than anticipated because of the requirement to work in joint ventures with local engineering companies. Further, expanding in-house engineering services within governments has limited the opportunities available to private sector firms. When public funds are reduced, government work is maintained but that for independent consulting engineers is decreased. Indeed, the relative importance of Asian cities for transnational engineering consultancies depends very much on the strength and vitality of the national economy – a conclusion that has equal applicability in Central and South America.

Representation in Central and South America, however, was more subdued as entry into markets with sizeable development projects, such as Mexico, was difficult without a joint venture partner. Indeed, only where specific know-how is unavailable at the local level can joint ventures be avoided. Also, the first-ranked centres of Santiago and Sao Paulo (Kowarick and Campanario 1986) did not appear to function as regional metropolises. Nevertheless, the distribution of branch offices favoured capital cities, notably Bogota, Buenos Aires, Caracas, Lima, Mexico City, Panama City and Paramaribo, as joint ventures with firms domiciled there made economic sense; this predilection for capital cities was also apparent in the small Caribbean nations.

There is no support for the suggestion, however, that consultants prefer to locate in cities with bases for export production, such as Mexico City, rather than in those which have been by-passed by multinational corporations as production and assembly sites, such as Lima (Wilson 1987). This may reflect the attention paid by consultants to informal sector activities. Although plans have been made to dissolve these activities others recognize their role in supporting the formal sector and recommend their conservation. The overall importance of Central and South America, however, was underplayed in the Fédération survey as it was biased towards western European engineering consultancies, especially those from the United Kingdom, rather than North America – the traditional source of branch offices in the region. Thus, it is important to examine nationality.

National patterns

Nationality has been deemed an important determinant of the spatial organization and locational behaviour of transnational engineering consultancies providing engineering and architectural services. Individual firms have sprung from different national economic systems which are supported internationally by varying degrees of political power. Also, the way that they are operated and controlled has been attributed to cultural preferences.

An examination of consultant nationality in terms of foreign billings recorded by the *Engineering News-Record* between 1980 and 1986 shows that the United States had the highest share throughout the period (Table 4.4). It was followed by the United Kingdom while a number of countries contested third position. The other striking feature was the emergence of the Japanese and Korean consultants. As there is not scope to examine all national groupings in relation to the urban hierarchy, attention is concentrated on the United Kingdom and Japan – the significance of trade in engineering consulting services being greater for the former than the latter. Projects accounted for 7 per cent of the United Kingdom's exports in 1981–2 and 14 per cent of manufacturing exports (NEDC 1983).

United Kingdom

A prime reason for choosing the United Kingdom over the United States is that considerable ancillary data is available on an activity still dominated by private companies. For instance, the *Association of Consulting Engineers*

Table 4.4 Consultant nationality by percentage share of foreign billings, 1980–6

Nationality	1980	1981	1982	1983	1984	1985	1986
American	34.0	37.0	33.8	31.3	29.9	32.0	25.9
Canadian	na	na	7.8	7.0	8.3	7.3	5.8
British	15.7	12.6	13.8	15.4	13.1	12.7	13.6
French	9.7	7.9	8.7	9.4	6.7	6.6	8.6
German	na	na	6.1	6.6	7.2	6.3	8.0
Other	na	na	19.5	20.1	21.8	21.4	25.1
European	na	na	48.1	51.5	48.8	47.0	55.3
Japanese	na	3.1	3.0	3.3	4.8	6.2	6.2
Korean	na	na	na	na	1.5	1.3	1.5
Others	na	na	7.3	6.9	6.9	6.2	5.3
No. consultants	150	150	201	200	204	200	200

Note: Rounding errors
Source: Based on ENR (1981–7)

Who's Who and Yearbook 1988 (ACE 1988) provides information for an examination of the branch offices of United Kingdom firms (see also BCB 1983). When this material is mapped in Figure 4.7 it emphasizes that the British consultants have not sought strong representation in major world cities. Rather they have capitalized on their economic affiliations with former colonies and singled out cities in Australia, Nigeria, East Africa, South-east Asia, and a host of Middle Eastern countries (Guttman 1976). This penetration has been assisted by the British government providing United Kingdom-based firms with a number of benefits for doing work abroad, including credit guarantees and reimbursement as part of the cost of preparing unsuccessful bids. Also, the staffs of British consultants do not pay income taxes in the United Kingdom if they are working under long-term contracts overseas.

The stability of the pattern of overseas branches is in some doubt with the emergence of mega-firms among the 220 British consultants analysed by the *NCE Consultants' File* (1988). As shown by the employment figures for the 'top ten' in Table 4.5 some large firms, such as the Ove Arup Partnership, have emerged. Their strength is also underlined in terms of the number of staff employed overseas. It reveals the remarkable strength of the British consulting industry in overseas markets and the emergence of 'super firms'. With strong financial backing and economies of scale these firms will be able to supply a wide range of services in overseas markets. While there still may be scope for small specialist firms, slower growing medium-sized firms are under threat of merger or dissolution. For instance,

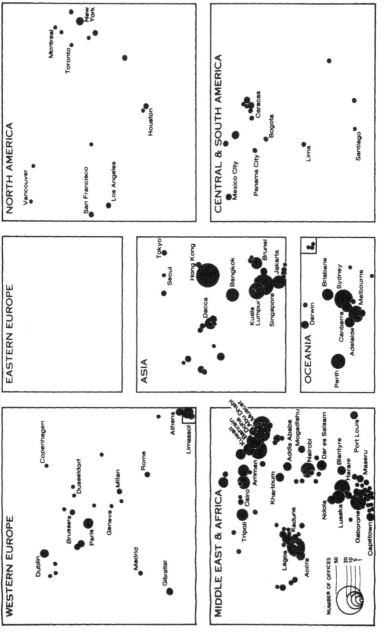

Figure 4.7 Distribution of country-level offices of engineering consultancies with headquarters in the United Kingdom
Source: ACE (1988)

Table 4.5 The 'top 10' firms in the United Kingdom ranked by number of staff

Company	Total employment no.	Company	Working overseas no.
1. Ove Arup Partnership	3,738	1. Ove Arup Partnership	961
2. W. S. Atkins & Partners	1,900	2. Maunsell Group	920
3. Maunsell Group	1,352	3. Sir Alexander Gibb Partnership	781
4. Sir Wm Halcrow & Partners	1,346	4. Scott Wilson Kirkpatrick	600
5. Building Design Partnership	1,310	5. Binnie & Partners	580
6. Mott Hay & Anderson	1,274	6. Ewbank Preece Group Ltd	535
7. Scott Wilson Kirkpatrick	1,250	7. Reinforced Earth Company Ltd	525
8. Ewbank Preece Group Ltd	1,062	8. Mott Hay & Anderson	478
9. Acer Consultants Ltd	1,050	9. Sir Wm Halcrow & Partners	415
10. Sir Alexander Gibb Partners	977	10. Acer Consultants Ltd	410
50. ...	130	50. ...	30

Source: NCE Consultants File (1988)

John Taylor and Freeman Fox merged in 1986 to form Acer Consultants. Although this merger was to build up their regional network in the United Kingdom similar organizational changes also led to a re-evaluation of the global urban hierarchy by transnational engineering consultants.

This reassessment of overseas branch office locations by British firms is also being prompted by the decline in the world construction market, particularly in the Middle East. While Asia and Africa still remain buoyant there is a trend among British consultants to transfer their attention from traditional infrastructure projects in the Third World towards the growth areas in North American and continental cities in readiness for European harmonization in 1992. An unknown factor, however, is whether the emerging British super firms will be able to counter the anticipated growth of the Japanese engineering consultancies, particularly with the growth of competitive bidding as part of a general challenge against the recommended fees and advertising methods used by members of groups within the Fédération Internationale des Ingénieurs-Conseils.

Japan

The international branch office network of Japanese consultants is less extensive than that of their British counterparts. When the information from the yearbook of Japan's Engineering Consulting Firms Association

(ECFA 1984) is mapped in Figure 4.8, it reveals that their subsidiary offices are concentrated on a small number of cities in their 'backyard market' of South-east Asia. For the most part, these offices are engaged in directing water resource projects – hydro-power generation, transmission facilities, irrigation and flood control – and road projects. Low representation in Middle Eastern cities stems from the fading of the market from its high point in 1977. Although the African and Latin American markets showed fitful promise in the 1980s, particularly when the Japanese government funded Nigeria's 'Operation Feed the Nation', a strong consulting presence was not warranted.

The explanation of this fledgling network has been attributed to the low level of internationalization among Japanese engineering consultancies formed since the Pacific war. The pool of international engineering consultancies has been expanded through the efforts of the Engineering Consulting Firms Association (ECFA), sponsored by the Ministry of International Trade and Industry (MITI), based on the premise that it is important to secure the contract for project leadership. Nevertheless, the consultancies are still heavily reliant on the 'public sector' for foreign commissions. Overseas work is still dominated by two firms – Nippon Koei and Pacific Consultants – which account for 80 per cent of Japanese foreign billings (Rimmer 1988). With the surge in the value of the yen the number of Japanese engineering consultancies working overseas is expected to grow. Following the success of the country's producers, construction companies and developers in the United States it is not surprising that their share of the North American market is targeted to increase. Indeed, the growing importance of this market is a feature of the analysis of the spatial organization of individual companies.

Company patterns

A further dimension of the spatial consequences and responses to the restructuring of the world economy exhibited by transnational engineering consultancies involves an examination of the locational behaviour of individual firms. As a means of bringing out the varying spatial outcomes of the restructuring of enterprises attention is focused initially on a range of firms drawn from different countries to illustrate how engineering consultancies have internationalized since the Second World War. Selected consultancies are then examined to highlight the variety of spatial strategies that have been incorporated within organizational networks.

Internationalization

Engineering consultants have internationalized their corporate structures, particularly since the late 1960s when fact-finding surveys gave way to the planning and implementation of projects. Most of the corporate activity,

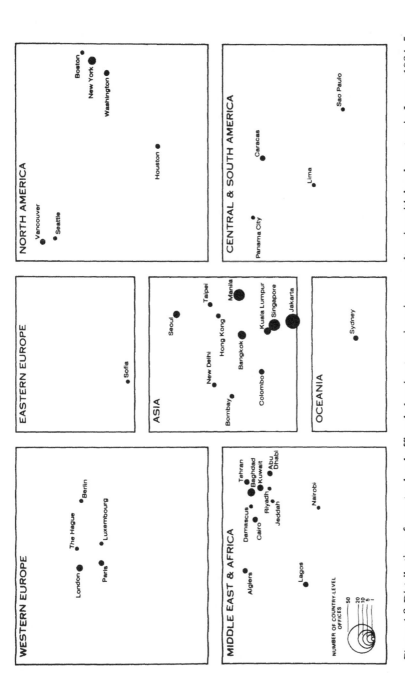

Figure 4.8 Distribution of country-level offices belonging to engineering consultancies with headquarters in Japan, 1984–5
Source: Based on data in ECFA (1984)

however, has not been centred on the first-ranked centres but on second- and third-tier cities, especially with the shift to a greater consideration of organizational and management problems, the monitoring and evaluation of completed projects and the environmental impacts of development from the early 1980s. The firms located there have replicated the services offered by the first-ranked cities, though on a smaller scale, to provide facilities for institution building, and manpower development and training. This pattern of internationalization is illustrated by reference to six engineering consultancies – four based in the United Kingdom, one in the United States and one in the Netherlands.

The pace and direction of internationalization, however, as shown in Table 4.6, has varied markedly between firms. Among the four firms based in Britain, Scott, Wilson, Kirkpatrick and Partners commenced their overseas operations in 1956; had five overseas offices by 1968 including Hong Kong; nine more were added during the 1970s; and the total augmented by a further eight between 1980 and 1987. After initial development in Africa during 1955, the Ove Arup Partnership expanded overseas in the mid-1960s, followed by infilling and extensions during the 1970s which included offices in the first-ranking cities of Paris, Singapore and Sydney. The overseas push by the Binnie Group was delayed until the 1960s with Hong Kong and Singapore – first-tier cities – being targeted in the 1970s and unranked Fresno being added in 1987 to signal the increased interest of British transnationals in the United States (The latter reflected the opportunities offered by large-scale urban redevelopment and renewal projects and new forms of public-private partnerships). Discussion of the remaining firms turns to the location of affiliates and subsidiaries.

The fourth British firm, Sir M. McDonald and Partners, has over twenty overseas offices but affiliates and subsidiaries have been restricted to a mix of all three tiers, Hong Kong, Brisbane, Lagos and Karachi – a distribution highlighting that irrigation and drainage is one of the group's prime specialities. Similarly, Metcalf & Eddy International Inc.'s interests in waste-water, water, hazardous waste, solid waste, environmental, and civil engineering projects are reflected in the choice of its bases for affiliates and subsidiaries offshore, but also they have been dictated, at least in part, by contract work awarded by the defence needs of the United States with Seoul and Manila figuring prominently. Finally, Euroconsult, founded in 1974 from the amalgamation of two well-established Dutch firms, Grontmij International Inc. (founded 1951) and ILACO (1954), has maintained its pre-existing bases in first-ranking Sao Paulo, third-ranking Nairobi and unranked Dar es Salaam and Paramaribo. Subsequently, it has extended its rural development work in Africa and South America by establishing branch offices in Asia and Oceania to tackle problems arising from rapid urban growth – it now has a total of twenty branch offices. The geographical consequences of internationalization are explored further by examining other characteristics of individual firms to demonstrate how cities are linked into the organizational web of transnational engineering consultancies and their subsidiaries and affiliates.

Spatial variants

The attributes of a further seven firms – Louis Berger International, Inc., East Orange, USA; Dar Al-Handasah Consultants, Cairo, Egypt; Sir Alexander Gibb & Partners, Reading, United Kingdom; Maunsell Consultants, Melbourne, Australia; DHV Consulting Engineers, Amersfoort, Netherlands; Electrowatt Engineering Services Ltd, Zurich, Switzerland; and Daniel, Mann, Johnson & Mendenhall (DMJM), Los Angeles, USA – are discussed to focus on their organizational structure and managerial functions in relation to the urban hierarchy outlined by the Nomura Research Institute. As it is not possible to canvass all facets of these firms, interest is centred on comparison between Louis Berger International and Dar Al-Handasah to bring out contrasting regional distributions, between Sir Alexander Gibb & Partners and Maunsell Consultants to illustrate differing organizational structures, and DMJM, DHV and Electrowatt to highlight how consultants are involved in changing the relative status of cities within the urban hierarchy.

Louis Berger International, Inc., founded by Dr Louis Berger in 1940, has the widest spread of branch offices of any engineering consultancy with offices in first-tier Stockholm, Paris, Frankfurt, Zurich, Seoul, Hong Kong, Singapore and Sydney and a strong presence in Africa, South Asia and Central America (Figure 4.9a). Conversely, Dar Al-Handasah Consultants, established in Beirut during 1956 and transferred to Cairo in 1987, is an 'indigenous multinational' with both its head office and branch offices in the Third World with the exception of London – the only first-tier city in which it has a base (Figure 4.9b). Whereas Louis Berger operates seventy-five offices that are strategically located to minimize the time and cost of mobilization on a global scale Dar Al-Handasah's thirty-eight offices have a similar objective but on a macro-regional scale. Both Louis Berger's and Dar Al-Handasah's rationale is that the principals of these offices will be familiar with local standards and practices and fully conversant with national planning programmes and needs. Unlike Louis Berger, however, Dar Al-Handasah has elevated one of its offices – London – to satellite status. A better example of differences in managerial organization, however, is provided by a comparison of a British and a British-Australian consultant.

Sir Alexander Gibb & Partners is distinguished in Figure 4.10a by being an amalgam of the headquarters firm in Reading and associated firms in second-tier Athens, third-tier Nairobi and Riyadh, and unranked Canberra with a permanent professional and technical staff of 880 (personal communication). Maunsell Consultants, with an overseas staff of 920, is unique in that it is an international partnership comprising the partners of

Table 4.6 Pattern of internationalization within selected firms, 1955–87

Date	Scott Wilson Kirkpatrick	Ove Arup	Binnie Group	Sir M. McDonald	Metcalf & Eddy	Euro-Consult
1955		Lagos(3)				
1956	Lilongwe (–)					
1957						
1958						
1959	Hong Kong(1)					
1960	Nairobi(3)					Dar es Salaam(–)
1961						Paramaibo(–)
1962						
1963		Freetown(–)				
1964		Sydney(1)				
1965	Lagos(3)	Accra(–)	Kuala Lumpur(2)			
1966		Kuala Lumpur(2)				Nairobi(3)
1967						
1968	Accra(–)		Lima(2)			
			Brunei(–)			
1969						
1970	Abu Dhabi(3)				Bombay(2)	
	Dubai(–)					
	Athens(2)					
	Harare(–)					
	Ruwi(–)					
1971			Hong Kong(1)			
			Singapore(1)			

Year						
1972		Dublin(2) Monrovia(−)	Bombay(2)	Lagos(3)	Al-Khobar(−)	Sao Paulo(1)
1973			Maseru(−)			
1974			Melbourne(1)			
1975	Tripoli(−) Gaboroner(−) Mt Hagen(−)	Paris(1) Port Louis(−)				
1976			Baghdad(3)		Bangkok(2) Cairo(2)	
1977	Baghdad(3)	Kuching(−) Brunei(−)				
1978			Cairo(2)			
1979						
1980	Bahrain(3) Riyadh(3) Doha(−)		Jakarta(2)	Hong Kong(1)	Alexandria(−) Riyadh(3)	
1981						
1982	Maseru(−)			Brisbane(2)	Saipan(−)	
1983	Colombo(−)					
1984						
1985	Kathmandu(−) Antananarivo(−)					
1986	Sharjah(−)		Fresno(−)	Karachi(3)	Manila(1)	
1987						

Note: Figure in parentheses refers to the city's Nomura ranking; a dash denotes an unranked city

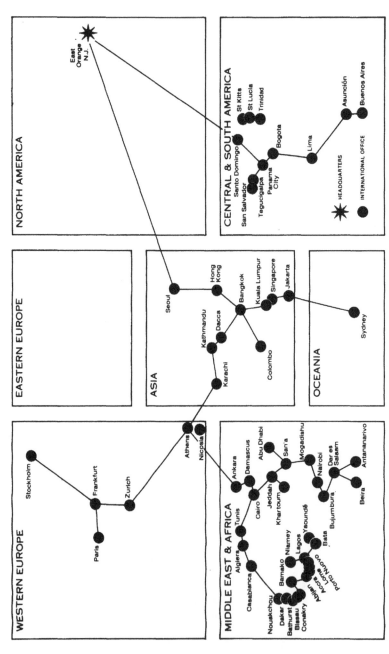

Figure 4.9 (a) Distribution of international offices of Louis Berger International, 1983; (b) Distribution of international offices of Dar Al-Handasah Consultants, 1987
Source: FIDIC (1982)

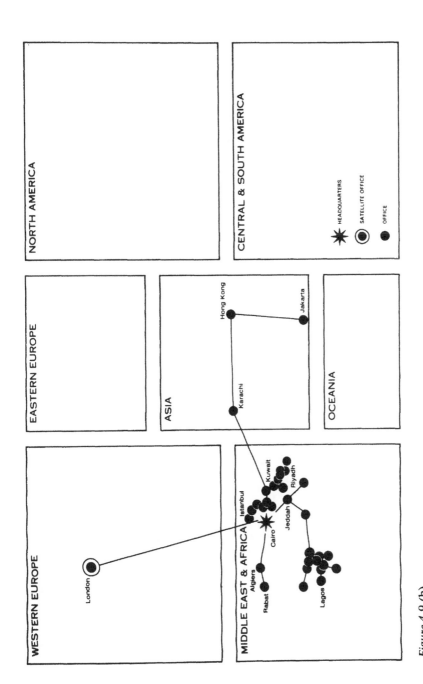

Figure 4.9 (b)
Source: personal communication

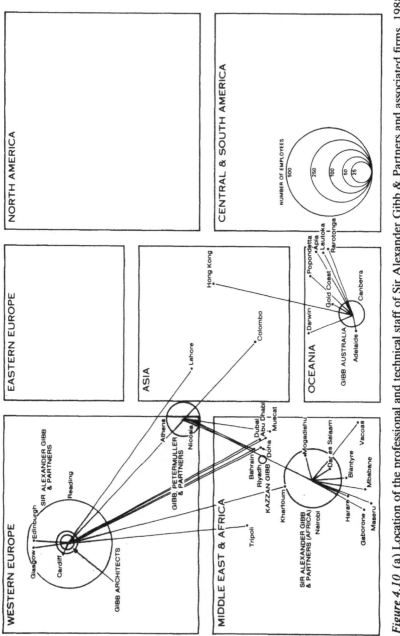

Figure 4.10 (a) Location of the professional and technical staff of Sir Alexander Gibb & Partners and associated firms, 1988;
(b) Distribution of the international offices of Maunsell Consultants, 1983
Source: personal communication

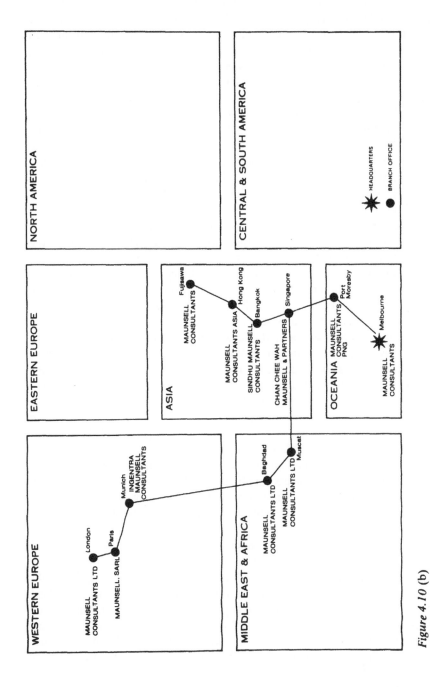

Figure 4.10 (b)
Source: FIDIC (1982)

G. Maunsell & Partners, London, and the directors of Maunsell & Partners Pty Ltd, Melbourne (Figure 4.10b). A key feature of the Gibb Group's distribution is the overlap of responsibilities between the parent and associated offices in controlling offices in a further twenty countries. Both components of Maunsell's, on the other hand, undertake assignments in all parts of the world other than in the United Kingdom and Australia which are served by appropriate constituent firms. A marked feature of the Gibb distribution is that it has eschewed first-ranked cities whereas Maunsell has offices in Hong Kong, Singapore and Paris. Rather than concentrate on the choice of overseas office locations, however, it is important to recognize that engineering consultants are contributing to changes in the urban fabric and, ultimately, the ranking of cities as illustrated by reference to three firms.

Daniel, Mann, Johnson & Mendenhall (DMJM), Los Angeles, USA, DHV Consulting Engineers, Amersfoort, the Netherlands, and Electrowatt Engineering Services Ltd, Zurich, Switzerland, have been chosen to demonstrate the type of activities undertaken by transnational engineering consultancies over the full range of ranked and unranked cities. The Tokyo base of DMJM, for instance, is engaged in military work in eastern Asia; its Manila base on architectural work in Asia and the Middle East, including the King Faisal University in Saudi Arabia; its Taipei base on the city's rapid transit system; its Cairo base on medical clinics; and its Riyadh and Al-Khobar bases on medical and educational facilities and housing (Figure 4.11a). Covering ranked and unranked cities, DHV Consulting Engineers is engaged on its own account and through the Dutch consultancy foundation, NEDECO Engineering Consultants, in developing master-plans for new towns in Cameroon, Indonesia, Sudan and Suraname and a mass of port studies from Busan through Chittagong to Paramaribo (Figure 4.11b). Electrowatt Engineering Services Ltd is engaged in rail-based mass transit systems including Manila's Light Rail Transit, Singapore's MRT (Mass Rapid Transit) and Metro Medellin, Colombia (Figure 4.11c). Investigation of these projects highlights the fact that decisions to locate these fixed capital activities are moulded by an array of economic and policy factors that, in turn, are sensitive to trends in the world economy but not city status.

Thus, the studies of overseas branch offices, selected national groups of consultants and individual firms have emphasized the importance of going beyond the scrutiny of headquarters locations. In particular, they highlight the critical role of first-tier cities as regional metropolises in the Third World, notably Hong Kong and Singapore, and the changing importance of second- and third-ranked cities and the growth of unranked cities in an emergent world economy; this is underlined by the activities of trans-national engineering consultancies. The integration of these lower-order centres within the world economy and the functions assigned to them in the new spatial division of labour is being re-assessed continually by trans-national engineering consultants in terms of the global system of markets

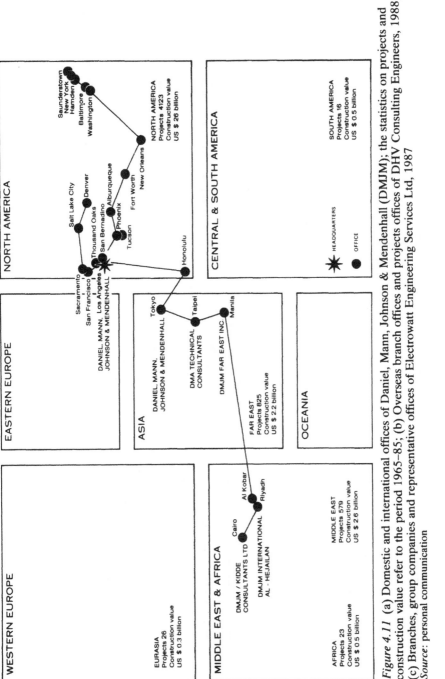

Figure 4.11 (a) Domestic and international offices of Daniel, Mann, Johnson & Mendenhall (DMJM); the statistics on projects and construction value refer to the period 1965–85; (b) Overseas branch offices and projects offices of DHV Consulting Engineers, 1988; (c) Branches, group companies and representative offices of Electrowatt Engineering Services Ltd, 1987
Source: personal communication

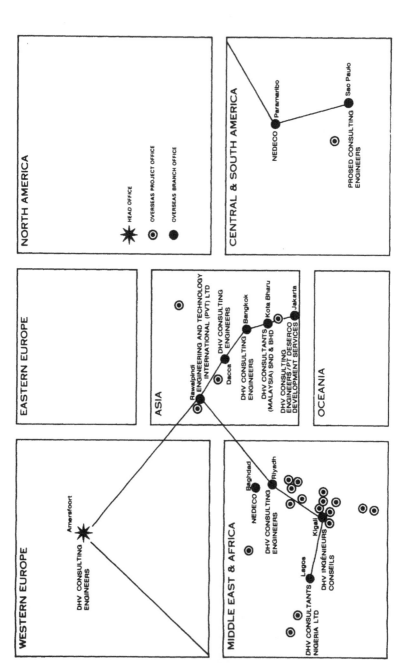

Figure 4.11 (b)
Source: personal communication

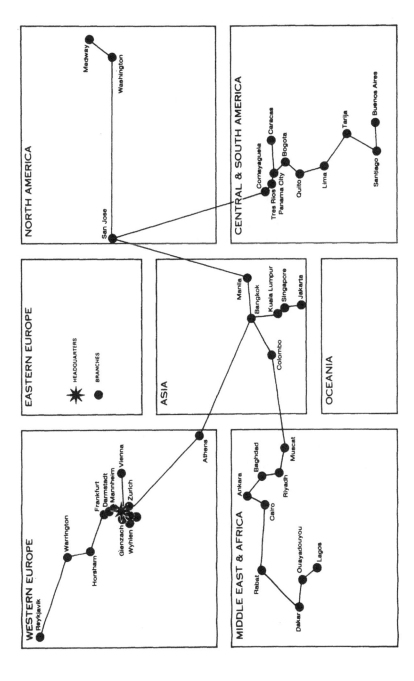

Figure 4.11 (c)
Source: personal communication

for capital, labour and commodities as impending changes in urban status generate a heightened need for their services.

CONCLUSION

Engineering consultancies alone, however, do not make or break metropolitan areas. They are merely part of a set of general business services that have underpinned the internationalization of production. Although engineering consultancies have not attracted the attention given to finance, insurance and real estate, they are, together with advertising, legal offices and management consulting, an essential part of the suite of activities that facilitate productive and other service operations. Consequently, movements in the location and importance of transnational engineering consultancies are faithful barometers of shifts in a city's international status and the processes of economic restructuring.

The examination of headquarters and branch locations highlights that there are marked divisions among the thirty-two first-ranked cities. This evidence suggests that a 'superclass' of cities is emerging facilitated by global electronic networks which are allowing information to be centralized in London, Paris, Tokyo and New York and transmitted to branch offices around the world for further action (i.e. a one-to-many network). A second group of world-scale regional pivots can also be recognized – Frankfurt, Seoul, Montreal, Toronto, Los Angeles and Chicago. In terms of engineering consultancies, Rome, and possibly Copenhagen, deserve to be included among the first tier and Manila and Santiago, and possible other cities, be reduced in rank. Boston has good grounds for being promoted from the third to the second tier.

Explorations of the relationship between engineering consultancies and world cities, however, have been hampered by a fundamental lack of knowledge – a reflection of inadequacies in both theoretical and empirical studies. Consequently, the next step is to provide a conceptual framework that links together such disparate influences as financial institutions, engineering consultancies and construction contractors, the role of the state, and the social consequences of their activities. Particular attention needs to be paid to linking metropolitan-forming processes and the agents (engineering consultancies) to the larger historical movements in the world economy. Empirical inadequacies, however, can only be resolved through the increased availability of data on the issues raised in this chapter. We cannot, however, be too sanguine on this point because, after all, we are dealing with the global intelligence corps.

ACKNOWLEDGEMENTS

The assistance afforded by individual engineering firms to requests for information is very much appreciated. Dr R. Grotz, Geographisches

Institut der Universitat Bonn also provided specific information on Germany. Barbara Banks has commented on the paper and Ian Heyward, Cartographic Unit, Research School of Pacific Studies, the Australian National University, Canberra, drew the diagrams.

REFERENCES

ACE (1988) *Association of Consulting Engineers Who's Who and Year Book, 1988*, London: Municipal Journal Limited.

BCB (1983) *Directory 1983–84*, London: British Consultants Bureau.

Cohen, R.B. (1981) 'The new international division of labour, multinational corporations and urban hierarchy', in M. Dear and A.J. Scott (eds) *Urbanisation and Urban Planning in Capitalist Society*: London: Methuen.

Daniels, P.W. (1987) 'The geography of services', *Progress in Human Geography*, 11 (3), 433–47.

ECFA (1984) *Engineering Consulting Firms Association, Japan*, Tokyo: Engineering Consulting Firms Association.

Enis, B.M. and Roering, K.J. (1981) 'Services marketing: different products, different strategy', in J. Donelly and W. George (eds) *Marketing of Services*, Chicago: American Marketing Association.

ENR (1979–87) *Engineering News-Record* (The McGraw-Hill Construction Weekly), New York: McGraw-Hill.

Feagin, J.R. and Smith, M.P. (1987) 'Cities and the new international division of labor: an overview', in M.P. Smith and J.R. Feagin (eds) The Capitalist City: *Global Restructuring and Community Politics*, Oxford: Basil Blackwell.

FIDIC (1982) *International Directory of Consulting Engineers 1983*, Lausanne: Fédération Internationale des Ingénieurs-Conseils.

Friedmann, J. (1986) 'The world city hypothesis', *Development and Change*, 17 (1): 69–83.

Friedmann, J. and Wolff, G. (1982) 'World city formation: an agenda for research and action', *International Journal for Urban and Regional Research*, 6 (3): 309–44.

Guttman, H.P. (1976) *The International Consultant*, New York: McGraw-Hill.

Kowarick, L. and Campanario, M. (1986) 'Sao Paulo: the price of world city status', *Development and Change*, 17 (1): 159–74.

NCE (New Civil Engineer) Consultants' File (1988) 'Record year in the private sector', April: 4–9.

NEDC (1983) 'Overseas capital projects. Memorandum by Director General of the National Economic Development Council, London, August 1983', NEDC (83) 29.

Nomura Research Institute (1982) *Bunsankei keizai shakai o mezashita kotsu, tsushintai ni kansuru chosa (II) – kokusai kotsu yuso kiban no seiritsu joken kiso chosa* (Investigation of the transport and communications structure with the emphasis on the dispersion model of society – fundamental research into the conditions for establishing an international base in transport and communications), Tokyo: Nomura Sogo Kenkyujo.

Noyelle, T.J. and Stanback T.M. Jr (1984) *The Economic Transformation of Cities*, Totowa, NJ: Rowman and Allenheld.

Rimmer, P.J. and Black, J.A. (1982) 'Global financiers, consultants and contractors in the Southwest Pacific since 1970', *Pacific Viewpoint*, 24 (2): 112–39.

Rimmer, P.J. (1986) 'Japan's world cities: Tokyo, Osaka, Nagoya or Tokaido Megalopolis?', *Development and Change*, 17 (1): 121–58.

Rimmer, P.J. (1988) 'The internationalisation of engineering consultancies: problems of breaking into the club', *Environment and Planning A*, 20: 761–88.

Wilson, P.A. (1987) 'Lima and the new international division of labor', in M.P. Smith and J.R. Feagin (eds) *The Capitalist City: Global Restructuring and Community Politics*, Oxford: Basil Blackwell: 3–34.

Yamaguchi Takashi (1988) 'Factorial ecology of world cities'. Unpublished paper prepared for the Commission on Urban System in Transition, International Geographical Union, Sydney, August, 1988.

5 Services and counterurbanization: the case of central Europe

Carlo Jaeger and Gregor Dürrenberger[1]

INTRODUCTION

There is considerable evidence suggesting that the spatial dynamics of service industries is presently shaping a new system of central places along the lines indicated by Christaller (1937). But whereas Christaller's original idea was derived from a regional context, that is historical market places in southern Germany, the updated version reflects processes and patterns on a global scale. Thereby, processes like the concentration of producer services in large agglomerations are regarded as contributing gradually to a new and world-wide system of central places. At the top of this system we would find a small number of world cities (Friedmann 1986; Moss 1987). Of these, three stand out, playing a relay race around the clock and around the globe. In alphabetical order they are London, New York, and Tokyo. In addition to this triad there are some other places which aspire to the status of a global metropolis; Los Angeles, Hong Kong, Paris, Singapore, and a few other cities would belong to this category. Together with the first three they would form a rather exclusive city system linked to larger networks of continental, national and regional centres.

With the magnification of Christaller's idea from regional dimensions to the size of global electronic market places the somewhat idyllic character of the original model is lost. Indeed, fantastic wealth and cruel poverty coexist not only at a global scale but also within the world cities themselves. Also buoyant economic growth of a world city may parallel depressing perspectives for peripheral regions in the same nation. Other contributions in this volume stress such striking spatial disparities associated with present trends towards a world-city system. To design policies aimed at the correction of such forms of unequal development seems an arduous task with limited chances of success.

In this chapter, we try to highlight a path of development that is not so directly associated with Christaller's notion of central places. We suggest that the urban size hierarchy does not necessarily coincide with a coherent cluster of socially relevant hierarchies like those of political power, of economic leverage, of the quality of education, and the like. To give a simple example: Zurich ranks first in the Swiss urban size hierarchy, followed by several relatively large cities like Basle, Berne or Geneva. Looking at the economic and political strength of the Swiss cities Bailly (1985) has shown that as soon as some more variables than just population size are considered Zurich is no longer the uncontested number one. It shares the first place with Basle and Berne, whereas Geneva ranks on the next lower level together with four medium-sized cities. The crucial point of this proposition is that the distribution of political and economic strength is not at all congruent: Zurich ranks on top mainly due to the financial services industry, Basle mainly because it has a strong and internationally orientated manufacturing industry (chemicals), Berne, finally, because it is the capital and concentrates most of the national bureaucracy. We may therefore distinguish several dimensions in such a way that a given place may be central on one dimension but peripheral on another. Under such conditions the rank of a city in a size hierarchy would be only very loosely associated with its position in different socially defined hierarchies and each one of these hierarchies would involve specific centre-periphery relations.

THE CENTRAL EUROPEAN SETTING

A pattern of diverging hierarchies both in social and in spatial terms is by no means a Swiss peculiarity. As we will see, it could be relevant for large parts of Europe: the Benelux countries, western Germany (prior to unification in 1990), Switzerland, northern and central Italy and to some extent France seem to be similar in this respect. The corresponding territory may be visualized as a core area of the European space economy (see Figure 5.1).

This area is surrounded by a maritime periphery ranging from Greece to Spain and Portugal, by the UK and Ireland, by Scandinavia and by eastern Europe. Parts of the latter, especially eastern Germany, Poland, Czechoslovakia and Hungary are usually included in discussions about the cultural heritage of central Europe – and it would be dangerously naive to ignore the special link between the two Germanies in discussions about political geography. In the present context, however, we define central Europe in a narrow sense because we want to focus on trends and possibilities in the spatial dynamics of service industries.

The rationale for the boundary which we have drawn in Figure 5.1 is given by historical developments. With the territory so defined, primarily we want to stress the existence of something like a European core area that has dominated so much of the history of this continent and that will also play a major role in the integration of the European Community and the future of Europe at large. Clearly the integration of this core area presents a major challenge for the United Kingdom. Historically, the United Kingdom has tried to develop its global relations while keeping the nations of this area at distance. Meanwhile a rather different process has begun

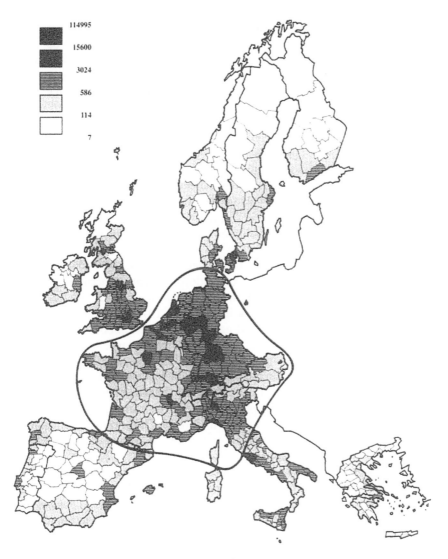

Figure 5.1 Economic weight of narrowly defined central European area
Source: After Sortia, Vandermotten and Vanlaer 1986: 164

which in the long run may blur the demarcation between central Europe
and the UK. The exclusion of southern Italy from the area considered has
to do with cultural differences that go far back into history. In fact, the
northern and central parts of Italy have developed a vocational culture and
a socio-economic structure which they share with countries on the north of
the Alps. These cultural and structural traits on the other hand are largely
absent in the Italian Mezzogiorno as well as in Spain and Portugal. The
allocation of Denmark to northern Europe is somewhat equivocal. But

with regard to the subsequent discussion, an important difference between the northern countries and the European core area is to be noted: the sparsely populated Scandinavian regions have developed a less complex settlement structure than the central European regions. Finally, the reason why we have excluded eastern Europe is her different economic and political structure. It should be noted from the outset that our main argument is perfectly consistent with the possibility that central Europe can be delimited only in a fuzzy way.

The third section of the present contribution uses a comparison between Italy and the United Kingdom to emphasize the varieties of spatial patterns which may emerge in connection with service dynamics (see Lash and Bagguely (1988) for a related argument based on a comparison of five countries). In this context we want to explore the possibility that the service industries in central Europe could contribute to complex spatial patterns eluding a single hierarchical order.

The fourth section is devoted to a discussion of social and economic trends relevant for this possibility. A first development concerns migration. It has been abundantly shown that the trend towards urbanization has been reversed in the whole OECD area in the early 1970s. Migration streams running from peripheral to central areas have been temporarily reversed and then stabilized at lower levels. This holds both for large scale areas like the nations of western Europe and for small scale areas like regions and counties. This migration turnaround, often labelled as *counterurbanization*, has clearly modified the growth-structure of the urban system. Although this process is still largely unexplored, both in its causes and effects, it is sensible to state that established urban hierarchies have come into flux and have opened space, quite literally, for settlement structures beyond the original notion of central places.

The impact of counterurbanization on services seems to depend to a considerable extent on the fate of *manufacturing*. Where services, and especially producer services, develop in a situation of industrial decline and shrinking domestic markets, flourishing service-firms are inevitably forced to operate in an international environment. A good example is the financial sector in Britain. The split between the sluggish manufacturing industry and the prospering producer services is additionally supported by the social and historical fabric of Britain's society (Thrift *et al.* 1987b). Probably the most important factor for the concentration of the financial services in London is its traditional international role, a legacy of the British Empire: it is London where the multinational firms' headquarters are located, where people meet, where contacts and contracts around the world are organized. The buoyant growth of the banking business in recent years has entailed a further concentration not just of the financial services, but, due to multiple effects, of producer and consumer services in general. London is indeed a world city whose development is sometimes linked more closely to world markets than to national ones. Under such circumstances the decentralizing impact of counterurbanization on service industries may be of minor importance.

But in central Europe manufacturing tends to be in more synergetic relationships with services and a clear domination of the space-economy of service industries by a single centre is less likely. The relatively decentralized structure of today's manufacturing industry in central Europe coincides with an urban system and an economic history that have developed in a rather diffused shape. The service industries have strong links to an internationally successful domestic production-sector and the prosperity of both depends to a considerable extent on the complementary character of production and services within the same region. Services do not just fit in but even reinforce to some degree dispersed and decentralized economic structures which exist in this area.

A further development which has to be considered in this context refers to technological and organizational changes within firms. There is general agreement that the harsh economic climate since the 1970s has generally hit most the large manufacturing firms. Their highly centralized organizations, their standardized products and their spatially concentrated production systems were not flexible enough to accommodate quickly to an increased competition on the world-markets. Small- and medium-sized firms on the other hand benefited from their generally less rigid and less bureaucratic organization. In addition to this, many small- and medium-sized firms took advantage of the flexibility of the new technologies and produced more specialized products and of higher quality than their large-firm competitors.

In a later section we discuss these developments in the context of a socio-economic paradigm which is sometimes called *flexible specialization*. We argue that flexible specialization is an important option for central European firms, because it contributes to the general process identified above as a trend from a single centre-periphery hierarchy towards complex interferences between several such hierarchies. Although the great post-war economic boom under the American influence has shaped Europe's as well as the world's socio-economic structure for half a century, the specific cultural traits of central Europe have never ceased to be one of the most important assets of the region. Flexible specialization seems tailored to fit in to the cultural heritage of an area that has decisively influenced modern society.

TWO CONTRASTING CASES

To understand the growing importance of services in general and their spatial dynamics in particular it is essential to relate the development of services to manufacturing. Therefore we next discuss processes of de-industrialization and of spatial decentralization in manufacturing. These are then related to service dynamics. The comparison between Britain and Italy should illustrate that no simple uniform pattern captures the variety of real developments.

Britain's industrial decline (Massey and Meegan 1978) fits well into the general framework of de-industrialization (Bluestone and Harrison 1982; Martin and Rowthorn 1986) and complementary tertiarization (Gershuny and Miles 1983; Daniels 1985). The decline of Britain's manufacturing industry was especially marked in the 1960s and 1970s. The big cities were hit first and London lost most of its industrial employment in this period. Italy on the other hand has experienced a quite different development. Except for the north-western parts around the industrial centres of Turin, Genoa and Milan all of the country has experienced ongoing industrialization processes during the 1960s and 1970s.

Figure 5.2 shows the relationship between industrialization and population concentration for the heartlands of England (the Midlands) and Italy (north-western regions) and the emergent areas of both countries (in England: south of the heartland, except London area; in Italy: north-eastern and central parts; for details see Coombes and Dalla Longa (1987)). The horizontal axis measures the change in the scale of industrialization by the percentage point change of the proportion of population employed in manufacturing. Positive figures indicate an increase in the proportion of people employed in manufacturing during a specific decade, negative figures can be interpreted as a de-industrialization tendency

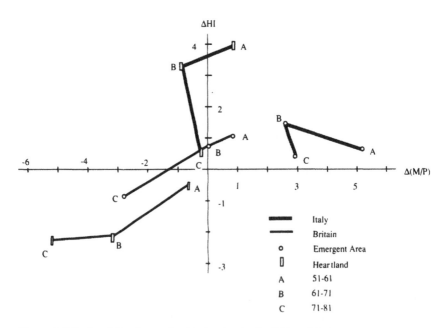

Figure 5.2 Industrialization and urbanization in the UK and Italy
Source: After Coombes and Dalla Longa 1987: 46–8

during a selected ten year period. The vertical axis shows the change of the Hoover index for population in each decade. The Hoover index measures the concentration of the population with respect to a chosen level of regionalization. Positive figures refer to a process of population concentration for the period selected, negative figures show that the population has de-concentrated over the decade.

In the British case, there is a strong relationship between industrialization and urbanization. After the post-war period of simultaneous industrialization and urbanization, the decline in industrial activities is associated with a process of population de-concentration in the regions considered. In Italy, there is no obvious relationship between industrialization and urbanization. In Italy's heartland the process of population concentration slowed down while industrialization came to an end. In Italy's emergent area a relatively slow process of population concentration took place together with sustained industrial development. At a national scale no simple spatial hierarchy was reinforced. This is in strong contrast to the British experience. There, the distinctly uneven space-economy was maintained.

That the hierarchical spatial order in Britain was sustained despite the general decentralization of production is chiefly attributable to a concentration of producer services, more specifically of globally orientated financial services, in the London area (Thrift *et al.* 1987a). Thereby, London's service industries benefited not just from the universal trend towards tertiarization, but also from their traditional function as the mercantile Capital of the Empire and not primarily of the Kingdom. The decline of the domestic manufacturing industry could not devalue substantially the portfolios of London's financial aristocracy. On the contrary, the world recession cut investment opportunities and diverted huge sums of money from the manufacturing sector to the financial services. London, as one of the most important turntables of international service activities, profited extraordinarily from this global development and could attract enormous funds from all over the world that fuelled the growth of the producer services.

The downfall of urban industries has been caused to a substantial degree by the combination of land-shortages and unrest of industrial labour. The success of the latter resulted in wage costs (not necessarily net incomes) that rose faster than productivity growth. This fact has to be considered as one of the major reasons for the subsequent decentralization of production (Keeble 1978; Fothergill and Gudgin 1982). By far the most diffuse industrial pattern has been realized in the emergent southern England around London. Decentralization was achieved especially by subcontracting particular stages of production to small firms without strong unions and in a relatively weak economic position or by establishing branch plants.

The decentralization of manufacturing was not just a British but also an international phenomenon (Keeble *et al.* 1983) and it is also well documented for Italy (Paci 1973; Vinci 1974). Again, decentralization is seen to be in part a response of management to rapidly rising wage costs in the late 1960s (Jaeger and Weber 1988) which have been most marked for large firms in major industrial centres. But while in the UK the decentralization of manufacturing is paralleled by a shrinking workforce, Italy shows a very different pattern. Employment figures have steadily increased in Italy. Especially the north-eastern and central regions have experienced remarkable industrial success since the 1960s. This has led several authors to claim that the traditional opposition between northern and southern Italy is made somewhat obsolete by the emergence of a 'third Italy', a territory previously referred to as an 'emergent area'.

In the UK the decentralization of industrial activities has been characterized to a considerable extent by a spatial separation between head offices and branch plants. This internal reorganization of functions has promoted a spatial division of labour between the southern areas with an over-representation of head-offices (Crum and Gudgin 1977) that attracted further services, and the industrial north that was cut off from this boom. This spatial organization can be related to a rather clear urban hierarchy (Goddard and Smith 1978). Headquarters and services are concentrated in the London area. Intermediate levels of corporate organizations are located in other major towns while material production and to a lesser extent consumer services such as retailers are more dispersed. Rajan (1987) has stressed the impact of external factors like the harsh economic climate of the 1970s recession on decisions by large firms to externalize producer services in order to maintain a tighter grip on cost control. Rajan and Pearson (1986) have estimated that almost half of all jobs created in the service industries in the UK between 1979 and 1985 were due to such contracting out.

Brusco (1982) on the other hand has described the pattern of industrialization in the central parts of Italy as diffuse and strongly related to the social and historical settings of the region. He pointed out that small- and medium-sized firms constitute the basic industrial structure. Brusco and Sabel (1981) distinguish three models of small firms. First, what they call the 'traditional artisan', a skilled craftsman supplying the local market, and second, the 'dependent sub-contractor', a small firm selling mainly to a single larger enterprise. The third case, which needs some further discussion, is 'the small firm in the industrial district'. The market of these firms is national or international. Production is vertically disintegrated in highly competitive units. There is no domination by a single, large firm. Sub-contractors are relatively independent because they supply several firms simultaneously. Machinery is rather sophisticated and in part requires high skilled workers. However, there are also a lot of simple tasks to be performed. The flexibility of the new technologies is used to produce a wide range of very specific, high quality goods in short series. This model of the small firm in the industrial district, Brusco (1982) called it the

Emilian-model, is relevant for Piore and Sabel's model of flexible specialization (Piore and Sabel 1984), for Scott's model of vertical disintegration
(Scott 1986a) and for Cooke and Da Rosa Pires' (1985) framework of
productive decentralization.

It seems that service industries may take a very different course in
situations of general industrial decline as in the UK and in situations of new
industrial successes as in Italy. Although producer services are somewhat
neglected in the literature on the Italian case, their role is of crucial
significance. Small firms hardly have the resources for efficient marketing
and for research and development activities. Such services are often
purchased from external firms that mediate between foreign markets and
domestic producers. These firms tend to be located in the provincial towns,
where the contacts to local firms are guaranteed and the benefits of
agglomeration economies can be skimmed off. American studies have
confirmed a somewhat similar centralization of producer services in areas
with a vertically disintegrated production structure (Storper and Christopherson 1987). But, perhaps due to the relatively diffuse Italian urbanization pattern, the centralization of producer services in Italy is far more
moderate and less clearly tied to the urban hierarchy (Cappelin 1986).

De Bernardy and Boisgontier (1986) have also emphasized that
producer services promote important synergetic effects between small
firms. Their analysis of the Grenoble region has much in common with
what has been reported from central Italy. However, the Grenoble area
seems to have profited much more from high-tech industries than was
the case in the third Italy. Public investments in the CENG (Centre
d'études nucléaires Grenoble) and the local university are considered to be
crucial external factors that have greatly stimulated economic growth in this
area. In this respect there are close parallels to other successful high-tech
regions such as Orange County near Los Angeles (Scott 1986b).

However, public spending is also very important in the case of the third
Italy. The 'economic miracle' of this region is far less spontaneous and
endogenous than is commonly believed. In a comprehensive study about
the Italian development Nanetti (1988) has worked out how strongly
economic decentralization, the social-historical fabric and institutional
decentralization have been tied together. The increasingly chaotic conditions of the national political system during the 1960s and 1970s
threatened not only the functioning of the national government but also
that of the social and economic system as a whole. The transfer of political
power from the national level to regional and local authorities and the
integration of social and economic groups within these areas in the decision
process have modified the whole Italian social structure, a development
that has not occurred at all in Britain where the domestic labour movement
has been politically defeated in a context of globalized financial markets.
The established social system has remained more or less untouched (Thrift
et al. 1987b).

In Italy, the decentralization of political power has allowed for territorially specific and differentiated public policies. They were controlled by municipal authorities responding to regional needs not only for physical infrastructure and social development but for all kinds of services. In a similar vein Bailly *et al.* (1987) recently suggested working out proper territorial policies of endogenous development where improvement of the environment for the service workforce should become a primary target. Similar ideas are put forward by French studies (Perrin 1983; Aydalot 1986) or by Illeris (1986) in the context of central Europe. They all stress the importance of relatively high order producer services for the regional development of areas with small towns.

THE IMPACT OF COUNTERURBANIZATION

It seems that service dynamics in central Europe could contribute to more complex spatial patterns than the monolithic hierarchy suggested by, for example, recent British developments. In this section, we will show how this alternative is linked to different outcomes of the international phenomenon of counterurbanization. In the literature on the geography of services this phenomenon has been somewhat neglected. This is especially unfortunate with regard to central Europe because in this area the spatial dynamics of services is strongly related to counterurbanization.

In the mid-1970s, American authors presented evidence for a fundamental reversal of the secular urbanization trend (Beale 1975; Berry 1976). One of the most interesting findings was the turnaround of the classical rural–urban migration pattern (Brown and Wardwell 1980). Peripheral and small communities with persistent population losses in the decades prior to 1970 began to exhibit increases in net in-migration. These were the first indications that a rural renaissance was taking place and it was soon confirmed for nearly all OECD countries (Vining and Kontuly 1978; Fielding 1982). It became clear that the new pattern was really 'a clean break with the past' (Vining and Pallone 1982) that could not simply be attributed to an increased range of well-known suburbanization processes.

In fact, there is a great difference between counterurbanization and suburbanization. A simple model of agglomeration processes may help to clarify the distinction. In a first phase (urbanization) the agglomeration gains population from the outer (rural) areas. Population growth is particularly high in the city. Suburbanization is the second phase. The agglomeration as a whole is still growing, attracting migrants from the rural periphery. Population gains in the city however have declined, or have even reversed. The suburbs in the fringe of the agglomeration are now the fast-growing zones. In the third phase, the city loses more people than the suburbs gain. The population in the agglomeration is declining whereas outer rural areas may increase in inhabitants. This pattern of disurbanization may be generated by an ongoing tendency of the agglomeration to expand (periurbanization), but it may also be caused by a counterurbanization process, where peripheral, rural places remote from the agglomeration attract urban migrants. A fourth phase concerns reurbanization processes. Such processes have been described in situations where the agglomeration is still characterized by population losses while population in the city is stabilized.

Recent studies have shown a decline in the pace of counterurbanization and seem to suggest that the process has passed its peak in several OECD countries (Engels 1986; Champion 1987). But the changes brought about by counterurbanization will have effects lasting for decades. The built environment has been reshaped to a considerable extent, and the cultural meaning of urban life and metropolitan places has experienced a far reaching transformation which is still largely unexplored. Moreover recent findings indicate that at least in western Germany the counterurbanization process itself is far from over (Kontuly and Vogelsang 1988).

The present understanding of the relationship between counterurbanization and economic changes like the shift from manufacturing to services, the growing importance of small and medium-sized firms or processes of externalization and internalization of services is quite limited. As Keeble and Wever (1986: 30) put it: 'The significance of such activities for so-called counter-urbanization processes and rural region/small town population and employment growth is as yet largely unexplored by urban and regional researchers.'

For obvious reasons the location of consumer services tends to follow to a considerable extent population distribution (Marquand 1983). The role of producer services on the other hand seems to be less clear-cut. Generally, producer services are expected to concentrate in the big agglomerations (Noyelle and Stanback 1984). Some recent studies however indicate that this trend towards concentration has slowed down. Kirn (1987) for example has shown that producer services in the USA became spatially more evenly distributed during the 1960s and 1970s as a substantial proportion of producer services filtered downwards through the urban size hierarchy. Gillespie and Green (1987) have drawn a somewhat similar picture for the UK, where the over-representation of producer services in the highest levels of the UK settlement system has clearly decreased during the 1970s:

> The pattern of change in the spatial distribution of producer services over the decade was characterized by a maintenance of spatial concentration of producer services employment in the South in national core-periphery terms, and a marked relative de-concentration within 'Metropolitan Regions' in both the North and the South.
>
> (Gillespie and Green 1987: 405)

Interestingly, on the intra-regional level investigated by Gillespie and Green, notably the most relevant level of counterurbanization in the UK, the locational trends of producer services employment match very well with the decentralizing effects of counterurbanization. But on a national scale this is not at all the case because of the split between manufacturing and financial services: the latter's buoyant growth is clearly linked to London's role as a world city and largely disconnected from a decentralizing but declining manufacturing sector.

German studies suggest a rather different situation. Rates of firm creation, which are generally considered to be a good indicator for local economic dynamism (Keeble and Wever 1986), seem to be particularly high in regions gaining population in the course of counterurbanization (Bade 1986). More generally metropolitan areas show stronger rates of industrial decline but weaker rates of services growth than non-metro-politan areas (Table 5.1). Clearly in a peripheral area an additional job in producer services will represent a much greater percentage change than in a metropolitan area. But what chiefly matters here is the ranking of such changes within areas of the same type. Producer services show the fastest growth in metropolitan as well as in intermediate and peripheral areas. They are followed by consumer services and financial services while construction and manufacturing are in decline. It should be remembered that financial services include both producer and consumer services; because of the high savings ratio of private households in western Germany the consumer services component is especially important in this country (Oberbeck and Baethge, forthcoming).

There are no signs in western Germany of a split between financial services centred on a single world city and decentralized but declining manufacturing which characterizes uneven regional development in the UK (Marshall *et al.* 1988). One may ask whether such a split is nevertheless appearing with the remarkable reurbanization process taking place in Germany's financial centre in Frankfurt. But the Frankfurt area is also the

Table 5.1 Sectoral changes of employment in different areas: Federal Republic of Germany, 1976–83

	Metropolitan areas	Intermediate areas	Peripheral areas
Manufacturing	−9.2	−4.0	−0.8
Construction	−5.3	−1.1	−0.5
Consumer services	17.4	21.0	25.3
Producer services	25.7	33.5	40.9
Financial services	6.0	15.1	20.0
Total employment	−0.7	3.0	5.7

Note: Percentage change
Source: Bade (1987: 230)

location of highly competitive manufacturing industries (e.g. chemicals). In fact, three metropolitan areas in western Germany are escaping the fate of both population and job losses experienced by the other metropolitan areas. The exceptions are Frankfurt, Stuttgart and Munich (Häussermann and Siebel 1987).

These cities are very dissimilar with regard to the importance of the financial services industry for the local economy, but they are all characterized by producer services which are strongly orientated towards manufacturing, such as research and development or engineering consultancy (Bade 1987). These services are particularly well represented and still rapidly growing. Under such conditions no simple central-place structure captures the complexity of services and metropolitan development which in the German case leads not to a single centre but to three different cities where processes of reurbanization are taking place. Last but not least, it is important to notice that the political capital, Bonn, is by no means at the top of urban hierarchies defined in terms of population size or economic performance.

In France several studies highlight that 'the urban–rural shift is perhaps a measure of moving the development from totally urbanized and industrialized areas towards a mixed environment where the manufacturing firm becomes rural for the most part but continues to get support from urban centres' (Aydalot 1986: 122; see also Planque 1983). The concentration of producer services in relatively small or medium-sized towns and cities in regions mostly favoured by the effects of counterurbanization (such as the départements around the Paris region and in southern France) is identified as a major ingredient of prosperous regional development (for a somewhat different view see van Dinteren 1987 in the case of the Netherlands). This also appears to be confirmed by the results of research in other countries. In Italy for example, there is the diffuse growth of the smaller towns in the central parts of the country. These towns play no major international role, but are very important in a regional context as suppliers of producer services and sustain relatively high levels of international exports both in consumer services (tourism) and in manufacturing. Prosperous areas in Italy (King 1986) as in western Germany (Bähr and Gans 1986) – and interesting parallels can be found in France (Ogden 1985; Aydalot 1986) – have been major regions of in-migration since the 1970s.

THE POSSIBILITY OF FLEXIBLE SPECIALIZATION

We are not describing an established trend. Rather we try to identify a possibility. This possibility may be conceptualized with the help of the notion of flexible specialization introduced by Piore and Sabel (1984). The background for such a possibility is given by the persistent economic crisis that has superseded the long boom of the world economy after the Second World War. During that period management tended to use technical progress to organize the labour process in a strongly hierarchical manner and to substitute human labour as far as possible by machinery. This was done following a socio-economic paradigm of mass production of standardized goods. Following Gramsci (1971) and a series of authors inspired by his work (Aglietta 1979; Blackburn *et al.* 1985; Lipietz 1986; Lüscher 1988) we denote this paradigm as Fordism. Urban life under Fordism was characterized by growing cities and increasing differentiation between areas of work, housing and leisure. Within the labour process Fordism emphasized Taylorist schemes of organization. Among other things this meant that business was organized in a strongly hierarchical way with a mass of unskilled workers at the bottom of the hierarchy. Technical progress was linked to systematic processes of deskilling for large categories of occupations (Braverman 1974).

Meanwhile, new information technologies have opened up important new possibilities. Spatially distributed branch networks can be co-ordinated much more effectively, service activities can be reshaped by the introduction of electronic machinery, systems of flexible automation in manufacturing combine heavy mechanization with the production of small batches. Indeed, basic features of Fordism can be renewed under contemporary conditions. A corresponding scenario could be sketched roughly as follows. Manufacturing further loses importance in terms of occupation. Enduring unemployment as well as organizational and spatial decentralization are used by management to weaken the bargaining power of employees and thereby to control wage costs. Occupational deskilling and steep organizational hierarchies persist. Deskilling and strongly hierarchical organization also characterize service industries but these may experience both economic expansion and spatial concentration. Financial services play a major role as productive investment is likely to remain sluggish while speculative financial operations flourish. It is reasonable to consider the combination of such elements as a possible development which can be labelled as neo-Fordism.

Neo-Fordism would drop several characteristics of Fordism but would retain the tendency to establish a dominant social hierarchy rooted in the labour process and reproduced in a single territorial centre–periphery structure. This means that the hierarchies of education, income, etc. in society at large would persist in the context of a comprehensive spatial hierarchy. What we described in the introduction as a global central-place system may be seen as the world-wide spatial dimension of neo-Fordism. In such a framework unskilled migrants would still move from peripheral regions to industrial or service centres in an effort to raise their standard of living and to start a process of upward mobility for themselves and their children. However, this pattern of spatial and social mobility had partially ceased to be effective in the situations of counterurbanization which were common during the 1970s. A major cause of those situations probably was the fact that Fordism got entangled into a series of difficulties towards the end of the 1960s (Armstrong *et al.* 1984; Jaeger and Weber 1988). This is not the place to discuss these difficulties in detail, not to elaborate how far the contemporary world economy or various national economies have already progressed in a neo-Fordist direction. There can be little doubt that in central Europe too such tendencies can be observed (Boyer 1987). But there are also tendencies which point to another possibility which we denote as flexible specialization.

In the past decades many firms had to realize that productivity was growing considerably slower than wage costs. This concerned not only the big manufacturing firms, although by comparison small firms seem to have some specific advantages. One concerns the hierarchical structure. The small firms in the third Italy for instance display just a fractional number of hierarchical levels in comparison with the structure familiar from Fordist corporations in northern Italy as well as elsewhere. It is essential for many such firms that there are only two hierarchies, namely the workers and the entrepreneur (Lazerson 1988). Instead of growing to a size requiring more hierarchical levels and additional administrative costs, small firms may prefer to engage in satellite firms of similar size. This is relevant for the renewed interest in small firms and especially for decisions by large firms to externalize service activities.

The success of small firms in Italy is not simply based on cheap labour (Lazerson 1988). In fact small firms are not flourishing in the poorest regions of the country while they are very successful in regions with strong trade unions and relatively high wages. Their basic strength is their high productivity under conditions of rapidly changing market conditions. This strength seems to be directly related to a specific way of handling organizational hierarchies. Workers and entrepreneurs live in the same social setting, they may be relatives or friends and take care to avoid low trust relationships. This is favoured by mutual reliance on vocational skills and experience. Instead of the unskilled labour and steep hierarchy of Fordist factories those small firms offer flatter hierarchies in the context of what may be called an informal professionalization of work.

While an emphasis on small firms is common in the literature on flexible specialization the theme of professionalization is often ignored. The importance of the latter theme is especially clear in western Germany. While in Italy processes of professionalization take place mainly in the informal setting of small firms, in Germany similar processes have been documented in major firms both in manufacturing (Kern and Schumann 1984) and in services (Berger 1984; Baethge and Oberbeck 1986; see also Baethge and Oberbeck 1988). There can be little doubt that this yields considerable advances in productivity (Daly *et al.* 1985). An important factor in Germany is a remarkable system of vocational education which is lacking in Italy. This enables large firms to increase systematically the complexity of the tasks performed at the very bottom of the occupational hierarchy. For this purpose specific new vocational careers have been established.

It should be noted that a renewed emphasis on vocational skills is characteristic of recent management approaches both in manufacturing and in services. This holds especially for Germany but also to some extent for Switzerland (Jaeger *et al.* 1987) or the Netherlands (Huppes 1987). A comprehensive view integrating the so-called secondary and tertiary sectors could be a major strength of this approach. This probably means that there is little scope in central Europe for financial services performed so to speak in splendid isolation from a country's manufacturing sector. In a western European context, capitalism divided along such lines could be a specifically British phenomenon (Ingham 1984).

CONCLUSION

Many authors describe the spatial dynamics of contemporary service industries along the lines of a global central-place system. Further growth, internationalization and diversification of advanced services, especially producer services, is considered to promote an uneven spatial division of labour. Highly internationalized service activities may concentrate in a few world cities that dominate international markets, whereas the rest of the cities will be curtailed to domestic or local markets. Moreover, this hierarchical structure of the space-economy is likely to coincide to a considerable extent with similar structures of the polity and of private firms.

We have argued that there are some tendencies in central Europe that may favour a social, economic and territorial situation that differs substantially from the global central-place model. This situation can be characterized as one of multiple hierarchies. The basic idea may be illustrated with the central-place grids of Figure 5.3 (to simplify the presentation we have chosen a quadratic pattern). On one hand we find the elegant spatial pattern which has fascinated generations of geographers. Empirically, this kind of hierarchy means that a whole set of variables is distributed hierarchically in a coherent way. If city A ranks higher than city B with regard to variable y then it will rank higher also with regard to variable z. To say that city A is more central in an absolute sense than city B is only meaningful under such conditions.

On the other hand we find the case of multiple hierarchies: city A may rank higher than city B with regard to variable x while the ranking is reversed with regard to variable z. We therefore have two central-place grids interfering with each other. Whether one place is more central than another may now depend on the variables under consideration. The arrows in Figure 5.3 represent interactions in a network with central places as nodes. The arrows point from a peripheral to a central place. With multiple hierarchies we get a new type of relationship, namely interdependency between places which are central with regard to different dimensions. With multiple hierarchies the number of centres increases. As Figure 5.3 shows,

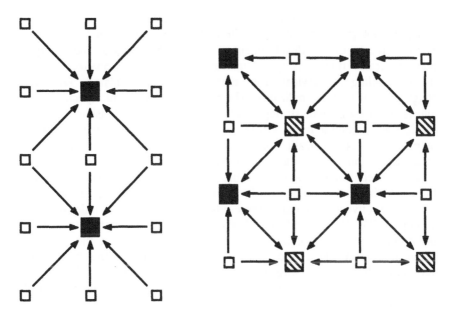

Figure 5.3 Central places with coherent hierarchies (left) and with multiple
hierarchies (right)

this in turn implies an even greater increase in the number of interactions
with such centres.

Now we have argued that in Central Europe counterurbanization seems
to foster situations of multiple hierarchies (see Frey (1987) for a somewhat
similar argument with regard to the United States). We have also argued
that the hierarchical organization of labour plays a pivotal role in the case
of coherent hierarchies. But in recent years a different socio-economic
paradigm with less emphasis on clear-cut hierarchies has been described.
This is the paradigm of flexible specialization. It often involves a shift
from large hierarchical organizations to market relations between smaller
units. Moreover it seems to include processes of professionalization at the
very bottom of organizational hierarchies. Both tendencies often take
advantage of the availability of new information technologies. At least in
central Europe there are good reasons to suggest that tendencies towards
flexible specialization promote situations of multiple hierarchies.

The comparison with the United Kingdom suggests that the development of multiple hierarchies may be strongly hindered if services are so to
speak decoupled from manufacturing. This is the case if financial services
prosper in a context of manufacturing decline. Demand for producer
services will be low on the side of manufacturing while it will be high on the
side of financial services. Consumer services in turn will find limited
demand in regional economies which are focused on manufacturing while
they will contribute to the importance of service industries in metropolitan

areas where economic growth is driven by financial services.

Figure 5.4 gives an outline of these three processes which may have major influence on the development of multiple hierarchies. The question mark indicates that it is quite unclear how flexible specialization and the decoupling of services from manufacturing are related. We have suggested that professionalization which is crucial for flexible specialization allows for management strategies which cut across the divisions between services and manufacturing. On the other hand flexible specialization may involve heavy externalization of services by manufacturing firms and this in turn may deepen the separation between such firms and service industries.

The present analysis points to some possibilities of considerable interest for researchers and maybe also for a larger public. However, it does not offer a definitive assessment of these possibilities. More than a decade ago, Young (1978: 71) claimed that: 'It should then be illuminating to explore the connections, to speculate, and to ask questions, about the relations between central place and general hierarchy theory (the latter also variously called levels-of-organization or levels-of-integration theory).' At least in the case of central Europe such an approach seems indeed appropriate.

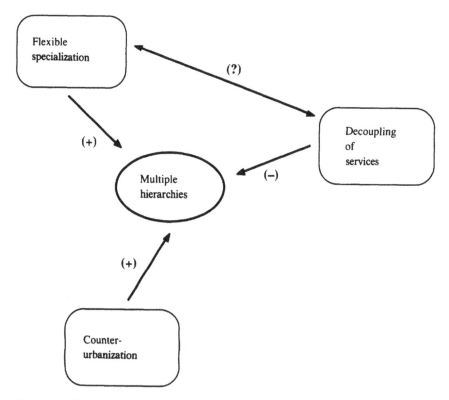

Figure 5.4 Services and multiple hierarchies: major causal relations

The polarity of centre and periphery refers to units at the same level of analysis, usually places. But these places are nodes of larger interaction networks. The relation between different levels of analysis therefore must be integrated in the study of centre–periphery patterns. In the context of European integration the relationship between European cities and Europe at large requires a careful investigation. How is the hierarchy between a system and its subsystems connected with hierarchical relations between these subsystems? We do not want to discuss this issue here. We simply want to point out that with regard to European integration old and unsettled questions of social theory can and must be studied in new ways.

NOTE

1. The present contribution is based on research which began in the framework of the Swiss National Research Programme on working life. This research was financed by grants of the Swiss National Research Council and the British Academy. The hospitality of the Centre for Urban and Regional Development Studies in Newcastle upon Tyne where Gregor Dürrenberger was a visiting academic is gratefully acknowledged.

REFERENCES

Aglietta, M. (1979) *A Theory of Capitalist Regulation – The US Experience*, London: New Left Books.

Armstrong, P., Glyn, A., and Harrison, J. (1984) *Capitalism since World War II: the Making and Breakup of the Great Boom*, London: Fontana.

Aydalot, P. (1984) *Le dynamisme économique a-t-il déserté la ville?*, Dossier de Centre Économique Espace Environment, 39, Paris: Université de Paris I-Panthéon Sorbonne.

Aydalot, P. (1986) 'The location of new firm creation: the French case', in D. Keeble and E. Wever (eds) *New Firms and Regional Development in Europe*, London: Croom Helm.

Bade, F.J. (1986) 'The economic importance of small and medium-sized firms in the Federal Republic of Germany', in D. Keeble and E. Wever (eds) *New Firms and Regional Development in Europe*, London: Croom Helm.

Bade, F.J. (1987) *Regionale Beschäftigungsentwicklung und produktionsorientierte Dienstleistungen*, Berlin: Duncker & Humblot.

Baethge, M. and Oberbeck, H. (1986) *Zukunft der Angestellten: Neue Technologien und berufliche Perspektiven in Büro und Verwatlung*, Frankfurt: Campus.

Baethge, M. and Oberbeck, H. (1988) 'Service society and trade unions', *The Service Industries Journal* 8 (3): 389–400.

Bähr, J. and Gans, P. (1986) 'Federal Republic of Germany', in A. Findlay and P. White (eds) *West European Population Change*, London: Croom Helm.

Bailly A.S. (1985) 'The service sector as a motor for endogenous development', in M. Bassand, E. Brugger, J. Bryden, J. Friedmann and B. Stuckey (eds) *Self-reliant Development in Europe: Theory, Problems, Action*, Aldershot: Gower.

Bailly, A.S., Maillat, D., and Coffey, W.J. (1987) 'Service activities and regional development: some European examples', *Environment and Planning A* 19: 653–68.

Beale, C.L. (1975) *The Revival of Population Growth in Non-metropolitan*

America, Economic Research Service, 605, Washington: US Department of Agriculture.

Berger, U. (1984) *Wachstum und Rationalisierung der industriellen Dienstleistungs-gesellschaft,* Frankfurt: Campus.

Bernardy, de, M. and Boisgontier, P. (1986) 'Les "μ.e" de la région grenobloise et leurs relations au tissu local d'activités économiques', *Revue d'Économie Régionale et Urbaine* 5: 645–62.

Berry, B.J.L. (1976) 'The counterurbanisation process: urban America since 1970', in B.J.L. Berry (ed.) *Urbanisation and Counterurbanisation,* Beverly Hills: Sage.

Blackburn, P., Coombs, R., and Green, K. (1985) *Technology, Economic Growth and the Labour Process,* New York: St. Martin's Press.

Bluestone, B. and Harrison, B.T. (1982) *The Deindustrialization of America,* New York: The Free Press.

Boyer, R. (1987) *La flexibilité du travail en Europe,* Paris: Éditions la Découverte.

Braverman, H. (1974) *Labour and Monopoly Capital,* New York: Monthly Review Press.

Brown, D.R. and Wardwell, J.M. (eds) (1980) *New Directions in Urban–Rural Migration,* New York: Academic Press.

Brusco, S. (1982) 'The Emilian model: productive decentralisation and social integration', *Cambridge Journal of Economics* 6: 167–84.

Brusco, S. and Sabel, C. (1981) 'Artisan production and economic growth', in F. Wilkinson (ed.) *The Dynamics of Labour Market Segmentation,* London: Academic Press.

Cappellin, R. (1986) 'The development of service activities in the Italian urban system', in S. Illeris (ed.) *The Present and Future Role of Services in Regional Development,* FAST Programme, Occasional Paper 74, Brussels: Commission of the European Communities.

Carpenter, E.H. (1977) 'The potential for population dispersal: a closer look at residential preferences', *Rural Sociology* 42: 352–70.

Champion, A.G. (1986) 'Great Britain', in A. Findlay and P. White (eds) *West European Population Change,* London: Croom Helm.

Champion, A.G. (1987) 'Recent changes in the pace of population deconcentration in Britain', *Geoforum* 18: 379–401.

Christaller, W. (1937) *Die Zentralen Orte in Suddeutschland,* Jena: Fischer; trans. Biskin, C.W. (1966) *Central Places in Southern Germany,* Englewood Cliffs, NJ: Prentice-Hall.

Cooke, P. and Da Rosa Pires, A. (1985) 'Productive decentralisation in three European regions', *Environment and Planning A* 17: 527–54.

Coombes, M.G. and Dalla Longa, R. (1987) *Counterurbanisation in Britain and Italy: a comparative critique of the concept, causation and evidence,* CURDS Discussion Paper, 84, Newcastle upon Tyne: University of Newcastle upon Tyne, CURDS.

Crum, R.E. and Gudgin, G. (1977) *Non-production Activities in UK Manu-facturing Industry,* Regional Policy Series 3, Brussels: Commission of the European Communities.

Daly, A., Hitchens, D.M.W.N., and Wagner, K. (1985) 'Productivity, machinery and skills in a sample of British and German manufacturing plants', *National Institute Economic Review,* February: 48–61.

Daniels, P.W. (1985) *Service Industries: A Geographical Appraisal,* London: Methuen.

Engels, R.A. (1986) *The Metropolitan/Non-metropolitan Population at Mid-decade,* paper presented at the Population Association of America Annual Meeting at San Francisco: mimeo.

Ernste, H. and Jaeger, C. (eds) (1990) *Information Society and Spatial Structure,* London: Belhaven Press.

Fielding, A.J. (1982) 'Counterurbanisation in western Europe', *Progress in Planning* 17: 1–52.
Findlay, A. and White P. (eds) (1986) *West European Population Change*, London: Croom Helm.
Fothergill, S. and Gudgin, G. (1982) *Unequal Growth: Urban and Regional Employment Change in the UK*, London: Heinemann Educational Books.
Frey, W.H. (1987) 'Migration and depopulation of the metropolis: regional restructuring or rural renaissance?', *American Sociological Review* 52: 240–57.
Friedmann, J. (1986) 'The world city hypothesis', *Development and Change* 17: 69–83.
Gershuny, J.I. and Miles, I.D. (1983) *The New Service Economy: The Transformation of Employment in Industrial Society*, London: Francis Pinter.
Gillespie, A. and Green, A. (1987) 'The changing geography of producer services in Britain', *Regional Studies* 21: 397–412.
Goddard, J.B. and Smith, I.J. (1978) 'Changes in corporate control in the British urban system', *Environment and Planning A* 10: 1073–84.
Gramsci, A. (1977) 'Americanism and Fordism', in Gramsci, A., *Prison Notebooks*, New York: International Publishers.
Häussermann, H. and Siebel, W. (1987) *Neue Urbanität*, Frankfurt: Suhrkamp.
Huppes, T. (1987) *The Western Edge: Work and Management in the Information Age*, Dordrecht: Kluwer.
Illeris, S. (ed.) (1986) *The Present and Future Role of Services in Regional Development*, FAST Programme, Occasional Paper 74, Brussels: Commission of the European Communities.
Ingham, G. (1984) *Capitalism Divided*, London: Macmillan.
Jaeger, C., Bieri, L., and Dürrenberger, G. (1987) 'Berufsethik und humanisierung der arbeit', *Schweizerische Zeitschrift für Soziologie* 1: 47–62.
Jaeger, C. and Weber, A. (1988) *Lohndynamik und Arbeitslosigkeit*, Kyklos 41 (3): 479–506.
Keeble, D.E. (1978) 'Industrial decline in the inner city and conurbation', *Transactions of the Institute of British Geographers* 3: 101–14.
Keeble, D. and Wever, E. (1986) *New Firms and Regional Development in Europe*, London: Croom Helm.
Keeble, D., Owens, P., and Thompson, C. (1983) 'The urban–rural manufacturing shift in the European Community', *Urban Studies* 20: 401–12.
Kern, H. and Schumann, M. (1984) *Das Ende der Arbeitsteilung?*, München: Beck.
King, R. (1986) 'Italy', in A. Findlay and P. White (eds) *West European Population Change*, London: Croom Helm.
Kirn, T.J. (1987) 'Growth and change in the service sector of the US: a spatial perspective', *Annals of the Association of American Geographers* 77: 353–72.
Kontuly, T. and Vogelsang, R. (1986) 'Explanations for the intensification of counterurbanisation in the Federal Republic of Germany', *Professional Geographer* 40 (1): 42–54.
Lash, S. and Bagguely, P. (1988) 'Labour relations in disorganized capitalism: a five-nation comparison', *Society and Space* 6: 321–38.
Lazerson, M.H. (1988) 'Organizational growth of small firms: an outcome of markets and hierarchies?', *American Sociological Review* 53: 330–42.
Lipietz, A. (1986) 'New tendencies in the international division of labour: regimes of accumulation and modes of regulation', in A. Scott and M. Storper (eds) *Production, Work, Territory*, Boston: Allen & Unwin.
Lüscher, R. (1988) *Henry und die Krümelmonster*, Tübingen: Konkursbuch Verlag.
Marquand, J. (1983) 'The changing distribution of service employment', in J.B. Goddard and A.G. Champion (eds) *The Urban and Regional Transformation of Britain*, London: Methuen.
Marshall, J.N., Damesick, P. and Wood, P. (1987) 'Understanding the location and

role of producer services in the United Kingdom', *Environment and Planning A* 19: 575–95.

Martin, R. and Rowthorn, B. (1986) (eds) *The Geography of De-industrialisation,* Basingstoke, Hants.: Macmillan.

Massey, D. and Meegan, R.A. (1978) 'Industrial restructuring versus the cities', *Urban Studies,* 15: 273–88.

Moss, M. (1987) 'Telecommunications, world cities and urban policy', *Urban Studies* 24: 534–46.

Nanetti, R. (1988) *Growth and Territorial Policies,* London: Pinter.

Noyelle, T.J. and Stanback, T.M. (1984) *The Economic Transformation of American Cities,* Totowa: Rowman and Allanheld.

Oberbeck, H. and Baethge, M. (forthcoming) 'Computer und Nadelstreifen, Die deutschen Finanzinstitute zwischen dominanter Marktsteuerung und konservativer Geschäftspolitik', in P.J. Katzenstein, (ed.) *The Third West German Republic,* Ithaca, NY: Cornell University Press.

Ogden, P.E. (1985) 'Counterurbanisation in France: the results of the 1982 Population Census', *Geography* 70: 24–35.

Paci, M. (1973) *Mercato del lavoro e classi sociali in Italia,* Bologna: Il Mulino.

Perrin, J.C. (1983) *La reconversion du bassin d'Alès,* Aix-en-Provence: Université d'Aix-en-Provence, Centre de'Économie Régionale.

Piore, M. and Sabel, C. (1984) *The Second Industrial Divide,* New York: Basic Books.

Planque, B. (1983) *Le développement décentralisé,* Paris: Litec.

Rajan, A. (1987) *Services: A New Revolution?,* London: Butterworths.

Rajan, A. and Pearson, R. (1986) *UK Occupation and Employment Trends to 1990,* London: Butterworths.

Scott, A.J. (1986a) 'Industrial organization and location: division of labor, the firm, and spatial process, *Economic Geography* 62: 215–31.

Scott, A.J. (1986b) 'High technology industry and territorial development: the rise of the Orange County complex, 1955–1984', *Urban Geography* 7: 3–45.

Sortia, J.R., Vandermotten, C.H., and Vanlaer, J. (1986) *Atlas économique d'Europe,* Bruxelles: Société royale belge de géographie.

Storper, M. and Christopherson, S. (1987) 'Flexible specialization and regional industrial agglomerations: the case of the US motion picture industry', *Annals of the Association of American Geographers* 77: 104–17.

Thrift, N., Leyshon, A. and Daniels, P. (1987a) *The Urban and Regional Consequences of the Restructuring of World Financial Markets: the Case of the City of London,* Working Papers on Producer Services, No. 4, Liverpool and Bristol: University of Bristol and University of Liverpool, Department of Geography.

Thrift, N., Leyshon, A., and Daniels, P. (1987b) *'Sexy Greedy': The New International Financial System, the City of London and the South East of England,* Working Papers on Producer Services, No. 8, Liverpool and Bristol: University of Bristol and University of Liverpool.

Van Dinteren, J.H.J. (1987) 'The role of business-service offices in the economy of medium-sized cities', *Environment and Planning A* 19, 669–86.

Vinci, S. (1974) *Il mercato del lavoro in Italia,* Milano: Angeli.

Vining, D.R. and Kontuly, T. (1978) 'Population dispersal from major metropolitan regions: an international comparison', *International Regional Science Review* 3: 49–73.

Vining, D.R. and Pallone, R. (1982) 'Migration between core and peripheral regions: a description and tentative explanation of the patterns in 22 countries', *Geoforum* 13: 339–410.

Young, G.L. (1978) 'Hierarchy and central place: some questions of more general theory', *Geografiska Annnaler B* 60: 71–8.

6 Service activities and regional metropolitan development: a comparative study

Antoine Bailly and Denis Maillat

During the past decades the study of the service sector has progressed considerably. Two main issues have been developed: the first concerns the growing interdependence between the secondary and tertiary sectors; the second relates to the role of the service sector in promoting economic development, mainly in metropolitan areas. The purpose of this chapter is to examine the increasing overlap between the service and production sectors, to propose a new conceptual framework for analysing the structure of economic activity and to review a series of empirical studies carried out in Canada, Denmark, France and Switzerland (and including their metropolitan regions) using this framework.

A NEW NOMENCLATURE

Changes in economic systems

There exist more and more linkages between the tertiary and manufacturing sectors, as the fabrication of goods incorporates growing quantities of service activities. Research and development, legal and financial services, banking, advertising, and marketing, for example, are now part of the productive systems of modern economies. However, traditional nomenclatures, in which the tertiary sector is regarded as distinct from the secondary sector, are still commonly used for the sake of statistical convenience; analysts rely on census data based on the primary, secondary and tertiary sector classification (Clark 1951). Consequently, no account is taken of the increasing overlap between service and production sectors due to the growing diversification of economic activities and the increasing substitution of goods and services (Bailly and Maillat 1988). The evolution of production systems combining industrial and service activities has been one of the most significant features of modern economic development (Gershuny and Miles 1983).

Barcet *et al.* (1984) refer to a goods–service continuum in order to explain the production of goods in conjunction with service activities or service support, in the case of highly service-dependent goods. All enterprises are situated somewhere along the continuum, according to the characteristics of their production systems.

The interdependence between manufacturing and service activities manifests itself in two ways. Service employment in the manufacturing sector may be located either inside the firm (internal services) or outside the firm in independent service enterprises (external services).

The internalization of services forms the basis for the process of tertiarization of the manufacturing sector, that is, an increase in the proportion of employment in manufacturing not directly concerned with production. In most industrial countries, the share of internal service-related employment is approximately 30 per cent or more. In Switzerland, the relative share of internal service-related employment in manufacturing went from 25.1 per cent to 29.8 per cent between 1970 and 1981 (Jeanneret 1983). The relative share runs as high as 50.4 per cent in chemicals and petroleum industries, and 36.3 per cent in machinery manufacturing. Over time, the proportion of internal service-related employment has increased and has become a strategic factor within the organizational structure of firms. Such employment is allocated to activities such as management, marketing, planning and R&D, that is, the functions upon which a company's dynamism depends.

External services, provided by independent companies, have also experienced sustained growth over the past two decades. The share of external producer services has risen from 4.5 per cent of service sector employment in France in 1954 to 8.1 per cent in the early 1980s (Braibant 1982), and stood at 10 per cent to 12 per cent in 1988 in most developed countries. The purchase of external services is motivated by the desire to acquire new ideas or methods, or for reasons of profitability, flexibility, efficiency and innovation.

The coherence of the production system

We need to avoid two misconceptions which frequently are associated with traditional economic typologies. First, services do not grow at the expense of manufacturing activities; on the contrary, the development of circulation, distribution and regulation activities reflects the need of firms to devote increasing amounts of resources to services in order to increase their productivity and their innovation capacity. Second, this development of service activity should not be viewed as a new stage of economic growth. Rather, it reflects a constant evolution of productive systems. Tertiarization cannot be properly considered as a separate phenomena, even if most economists tend to relate tertiarization to the process of de-industrialization. This perspective arises principally due to the fact that the extent of the tertiary sector is often calculated as the residual of the primary and secondary sectors. In our view, distribution, circulation and regulation activities play an essential role in the dynamism of production systems and must be considered as production-related activities.

There are three sets of classical explanation advanced to explain the growth of service activities (Momigliano and Siniscalco 1982; Cuadrado Roura and Del Rio Gomez 1987). The first set concentrates on analysing the reasons for the changes in the relative and absolute share of tertiary sector employment. In short, it describes the phenomenon of tertiarization as a process leading to the service society. These explanations are based on a stages theory, according to which the demand for services outstrips the growth of disposable household income. Other hypotheses are also put forward, in particular that of productivity differences between services and industry, and that of the tertiary sector as a pool for surplus manpower in the goods production sector.

The second set of explanations analyses de-industrialization, stressing that tertiarization is a result of the absolute and relative decline in employment in the secondary sector, subsequent to the development of new, more productive technologies. The decline in the absolute and relative share of industrial employment is also attributed to the effects of decreased consumption of industrial goods. This type of explanation relies on an interpretation based on the effects of the stages of technological innovation (e.g. microelectronics). According to this point of view, de-industrialization is a consequence of recent technological innovations; employment decreases because of technological and productivity progress, and investment is made more in machines than in blue collar jobs. It is also argued that the resultant reduction in industrial employment corresponds to a growth of employment in the tertiary sector because of the necessity of reabsorbing the manpower laid off and/or because of the reallocation of capital to the service sector with higher returns and profitability.

Finally, a third point of view posits that falling employment in the secondary sector is due to rising employment in the public sector, a consequence of increased demand for collective services. Again, by calculating the secondary sector as the difference in employment with the tertiary sector, its relative share is shrinking. But this does not help us understand the links between regulation activities and manufacturing.

These hypotheses have been elaborated on the basis of employment statistics broken down into the traditional three sectors. Since each sector is considered separately, a gain in one results in a loss in another. This tends to lead to evolutionary explanations showing a shift towards a tertiary society, or even the coining of new terms; e.g. one speaks of a move towards a 'quaternary' society. In reality, however, these traditional typologies only show that the share of the tertiary sector is increasing, but do not provide a satisfactory explanation. Ultimately, this leads to a statistical illusion to the extent that the activities of the tertiary and secondary sectors are arbitrarily separated at the time of classification. Such diverse activities as teaching, retail trade, public sector employment, and also transportation and electric power stations are lumped together under the 'services' heading. Also under this heading come banks, the medical and paramedical sectors, and so forth. The one thing that all these different branches have in common is that they all constitute a residual of primary and secondary activity. Indeed, as we have already seen, services do not develop on their own; rather, they interact to an ever-increasing degree with the productive organization. This point needs to be emphasized and explained through the use of a new nomenclature.

A new classification of economic activities

It is therefore necessary to revise the entire Clark typology, not just part of it. The global production system must be defined in a coherent and interrelated manner so that the role of each activity can be properly determined.

The new classification of economic activities that we are proposing analyses production within the context of an interactive supply and demand system in which the individual establishments of a firm are classified according to their functional role. The overall production system consists of individual establishments whose aim is to guarantee supply through the performance of major functions, namely, manufacturing, circulation, distribution and regulation (Table 6.1). Ideally, one would be able to separate the employment within an individual establishment of a firm into these four functions:

1. Manufacturing – functions involving the processing of raw materials.
2. Circulation – functions performing an intermediary role in the physical flow of persons, as well as in information, communication and financial flows.
3. Distribution – functions providing goods and services directly to end users.
4. Regulation – functions ensuring the overall smooth operation of the production system; in particular, maintenance, modifications, regulation and monitoring.

Due to data limitations, however, the distinction can only be made between a firm's establishments, and not within a given establishment.

It should be pointed out that the 'manufacturing' category does not correspond to the traditional primary and secondary sectors. As the new classification is establishment-based, primary and secondary sector activities not involving manufacturing fall within the circulation and distribution categories.

Each establishment is thus identified according to its major activity and not according to product; this new and more accurate method reflects the role of a firm and of its various establishments within the economic system. Moreover, establishment functions are less likely to change than the manner in which they are performed (variations in products, technology, etc.); this extends the lifespan of the classification.

The typology that we propose is not a mere remodelling of the tertiary sector, but rather a reorganization of the classification of economic activities on the basis of the functions of establishments and their logical ordering in the production system. By avoiding the notion of the heterogeneous composition of the tertiary sector, this typology ensures the coherence of each category of activity when viewed in the context of the system as a whole, where each category has a specific role to play.

THE DEVELOPMENT OF ECONOMIC SYSTEMS IN CANADA, DENMARK, FRANCE AND SWITZERLAND

Four case studies, involving both the national and the metropolitan-regional scales, demonstrate the advantages of this typology. The new formulation provides information on the relative importance of four major classes of activity which constitute the basic elements of modern production systems in Canada, Denmark, France and Switzerland. We can then analyse and identify the roles played by metropolitan regions in these countries, and contrast them with those played by peripheral regions. The goal of our study is to describe, to explain and to understand metropolitan specialization in the national production systems.

Statistical information on the four countries is available for two periods in time: 1971 and 1981 for Canada and Denmark, 1975 and 1982 for France, and 1975 and 1985 for Switzerland. During these periods, manufacturing's share of employment fell by 11.2 per cent in Denmark, 6 per cent in France, 4.8 per cent in Switzerland, and 4.2 per cent in Canada (Table 6.2); even if the figures for Denmark are inflated by the 1981 crisis, the trend is widespread (Table 6.3). However, manufacturing still represents 45.2 per cent of total employment in Switzerland and 41.4 per cent in France.

Circulation activities increased by 3.2 per cent in Canada, 3.1 per cent in Switzerland, 2.5 per cent in Denmark and 2.2 per cent in France (Table 6.2). In Switzerland, the high percentage is mainly linked to the growth of financial and information activities; in Denmark, to the growth of flows of goods and persons; and in Canada and France, to information flows. These different patterns illustrate the changes in the production systems of these four countries and their specializations, with circulation accounting for 22.6 per cent of total employment in Canada, 21.5 per cent in Switzerland, 21.2 per cent in Denmark, and 19.3 per cent in France.

Increased circulation activities, in particular information flows, is widespread and reflects a shift to economic systems designed to respond to increased information and communication requirements. The need for these new roles is leading to far-reaching changes in productive processes. In particular, non-information-trained manpower is being replaced in all

Table 6.1 A modified typology of economic activity

Functional subdiv. of enterprise/ major activity of enterprise	R&D supply stock	Organi- zation	Execut product	Admin monitor	Main- tenance	Market sales	Total
1. Manufacturing							
1.1 Use of natural resources							
1.1.1 Agriculture							
1.1.2 Horticulture							
1.1.3 Silviculture							
1.1.4 Fishing							
1.1.5 Electricity, water, gas							
1.1.6 Mining							
1.2/1.3 Processing of natural resources and manufacturing goods							
1.2.1 Food products							
1.2.2 Drinks							
1.2.3 Tobacco							
1.2.4 Textiles							
1.2.5 Clothing, linen							
1.2.6 Wood, furniture							
1.2.7 Paper							
1.2.8 Graphic arts							
1.2.9 Leather, footwear							
1.3.1 Chemicals							
1.3.2 Plastic, rubber							
1.3.3 Non-metallic mineral products							
1.3.4 Metallurgy							
1.3.5 Machines, vehicles							
1.3.6 Electrical and electronic goods							
1.3.7 Other manufacturing industries							
1.4 Construction and civil engineering							
1.4.1 Construction							
1.4.2 Outfitting, finishing							

2.	Circulation
2.1	Physical flows, flows of persons
2.1.1/2.1.2	Wholesale trade (19 subcategories)
2.1.3	Brokerage (5 subcategories)
2.1.4	Transportation (5 subcategories)
2.2	Information and communication flows
2.2.1	Information transmission (incl. post office)
2.2.2	Information processing
2.3	Financial flows
2.3.1	Banking
2.3.2	Insurance
2.3.3	Financial companies
3.	Distribution
3.1	Health
3.2	Education
3.3	Retail trade (14 subcategories)
3.4	Hotel and restaurant trade
3.5	Repair of consumer items/vehicles
3.6	Personal services
3.7	Culture, sports, leisure
3.8	Domestic services, other
4.	Regulation
4.1	Public: administrations (in the strict sense)
4.1.1	Confederation
4.1.2	Canton
4.1.3	Commune
4.1.4	Other public organizations
4.2	Private: organizations
4.2.1	Social welfare organizations
4.2.2	Religious, social, cultural organizations
4.2.3	Community services, common interest groups
4.2.4	Private road works and sanitation
4.3	International: diplomatic organizations

Source: CEAT (1987)

Table 6.2 Percentage of employment in the major economic activities in four countries

	Switzerland (%)		France (%)		Canada (%)		Denmark (%)	
	1975	1985	1975	1982	1971	1981	1971	1981
Fabrication	50.0	45.2	47.4	41.4	38.3	34.1	45.8	34.6
Circulation	18.4	21.5	17.1	19.3	19.4	22.6	18.7	21.2
Distribution	27.5	28.9	27.1	30.2	32.6	33.5	30.3	34.7
Regulation	4.1	4.4	8.4	9.1	9.5	9.9	5.2	9.4

Table 6.3 Evolution of major activities in four countries

Evolution (%)	Switzerland	France	Canada	Denmark
1. Fabrication	−4.8	−6.0	−4.3	−11.2
1.1 Natural resources	−1.3	−1.8	−2.1	−2.9
1.2 Building ind.	0.7	−0.9	−0.3	−1.6
1.3 Manufacturing	−4.2	−3.3	−1.9	−6.7
2. Circulation	3.1	2.2	3.2	2.5
2.1 Flows of persons	0.7	0.8	0.8	1.6
2.2 Information flows	1.2	1.2	1.8	0.4
2.3 Financial flows	1.2	0.3	0.5	0.5
3. Distribution	1.4	3.1	0.9	4.4
3.1 Private	0.2	−0.2	0.9	−3.2
3.2 Public	1.2	3.3	−0.1	6.7
4. Regulation	0.3	0.7	0.4	4.2

Source: CEAT (1987)

activities by information-trained manpower as a result of a growing demand for activities relating to organization, co-ordination, supervision and access to technology and markets. Today, these activities are becoming driving forces in production systems which are obliged to seek their information both upstream and downstream.

Distribution activities also grew rapidly in Denmark (4.4 per cent) and France (3.1 per cent), due to collective distribution (health, education and leisure). In Canada and Switzerland, growth was slower (0.9 per cent and 1.4 per cent, respectively). The reason for the slow growth of distribution in Switzerland (from 27.5 per cent to 28.9 per cent) involves the fact that individual distribution (retail trade, repairs, personal services) already accounted for nearly seven out of ten jobs in this category in 1975. In Denmark, the share of distribution rose from 30.3 per cent to 34.7 per cent for completely different reasons: collective distribution (health, education, culture-leisure) went from four to six jobs out of ten, while individual distribution gained ground thanks to commercial rationalization. This 'distribution' category differentiates countries where the collectivity has increased its share of the delivery of services. In particular, the distribution role pattern reflects options chosen with regard to health and education, i.e. public services which exhibit a growing need for investment in human capital.

The 'regulation' activities confirm the trend towards increased intervention in the economic system. This role is twice as large in Canada (9.9 per cent) as in Switzerland (4.4 per cent) (Table 6.2). Yet the burden of this regulatory role may well act as a handicap in international competition. In this context, the four-tiered, function-based nomenclature system introduced here is particularly useful, since it makes us ask new questions. For example, what are the advantages or disadvantages of a given function for the productive organization and economic development of a nation, or at a finer spatial scale, of a regional or a metropolitan area? A dual phenomenon emerges when we compare the evolution of the production systems of our four case studies (Figure 6.1): some develop their circulation-regulation activities more highly, while others retain a dominant manufacturing economic base. Does the relative share of these activities lead to an improved level of development and regional or metropolitan organization? Here the issue is no longer tertiarization or de-industrialization, but rather finding the productive organization which is most likely to ensure long-term growth for all regions. This type of analysis would certainly facilitate our understanding of countries such as Switzerland, Great Britain, the Federal Republic of Germany or Japan, which retain large secondary sectors according to traditional typologies, but nevertheless have dynamic economies with rapidly developing metropolitan regions. The development patterns depicted in the triangular diagram (Figure 6.1) show how the different countries reorganized their production systems during the 1970–80 period.

In terms of development, Switzerland and Canada are shifting more and more towards circulation, i.e. towards activities designed to ensure that information circulates and is incorporated in manufacturing. Information activities are the most highly developed in these two countries, reflecting a tendency to reduce transaction costs, i.e. the costs of bringing the supply and demand sides of a market together (Lakshmanan 1988).

France and Denmark are moving towards distribution, more specifically towards collective distribution (education and health). The trend in these two countries is to reorganize production systems through human investment and the collective assumption of adjustment costs, i.e. costs incurred by increasing the supply of resources with long gestation periods. In addition, public regulation has become a major factor in these two countries, which indicates the relative importance of government activity.

The differences in the organization of the four countries' economic

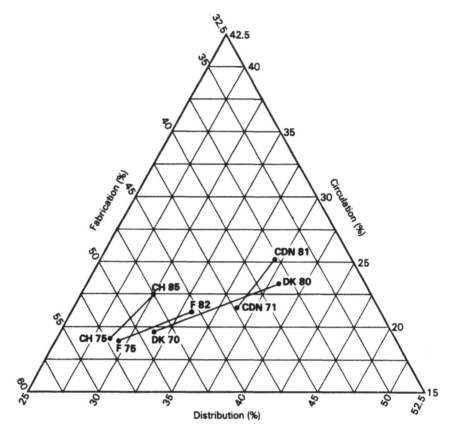

Figure 6.1 Development of production systems in France, Denmark, Canada and Switzerland
Source: CEAT (1987)

systems are clear, and are even more pronounced when it comes to metropolitan and peripheral regions.

REGIONAL ECONOMIC SYSTEMS IN CANADA, FRANCE AND SWITZERLAND

The use of the typology at the metropolitan and regional levels – French regions, Canadian provinces, Swiss cantons – permits the identification of the roles played by the various regions in the national production system.

Are the countries shifting towards circulation more urbanized than those moving towards distribution? In fact the regional-metropolitan triangular diagram (Figure 6.2) reveals three major types of regions. Owing to their reduced geographic dimension, the different Swiss cantons provide a background against which one may position the regions of the other countries. Apportioning the regions into precise, specialized classes highlights regional specificities which help to identify the different types of production systems.

A first group of regions is heavily orientated towards manufacturing (over 50 per cent). These are primarily regions which have traditionally been dominated by the production of manufactured goods, and are now tending to orientate their production systems toward circulation. With the exception of Basel (chemical industry), these regions do not have metropolitan areas. Most of them possess middle-size cities like Neuchatel,

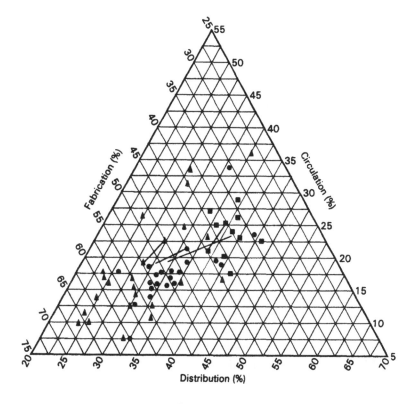

▲   Swiss cantons 1985

●   French regions 1982

■   Canadian provinces 1981

Figure 6.2 Production systems in Swiss, French and Canadian regions

Fribourg or Besançon. The economic problem of these regions is to improve the delivery of services which are necessary for the production system, both upstream and downstream, thereby contributing to their economic renewal. But most of these services come from the two other groups of regions, mainly the metropolitan group. This shows that their production systems do not automatically contain the services necessary for their development.

A second opposed group is easy to identify: it is dominated by circulation (30 per cent). The Swiss city cantons (Geneva, Basel and Zurich) and the Ile-de-France (Paris) are found in this group. It should be noted that this model departs from that of traditional central places which were characterized by the predominance of distribution. Very close to the metropolitan cantons and to metropolitan Paris we find urbanized provinces like Alberta (Edmonton and Calgary) and British Columbia (Vancouver). There is metropolitan specialization in controlling flows of information and financial flows, but these regions have lost most of their fabrication activities.

The third group, which is more heterogeneous, is made up of regions whose production systems are shifting towards distribution and, to a lesser extent, towards circulation. This group corresponds to the French and Danish model rather than to the Swiss and Canadian model, and covers intermediary or transitional development comprising several sub-groups. Among these sub-groups, some are specialized in tourist activities (Grisons, Corsica, Languedoc-Rousillon), and others in service-related activities (Vaud, Rhône-Alpes). This latter group comprises regions whose economic systems are undergoing changes and where manufacturing activities incorporate many service-related activities.

Rather than speaking of regional inequalities in production systems, it is more precise to refer to regional specialization at the base of different national productive systems. In all our case studies, the role of metropolitan areas appears to be strongest in the circulation and regulation functions; these metropolitan areas are 'hubs' necessary for the efficient functioning of the national productive systems. But some regions specialized in manufacturing developed their circulation and distribution activities (in particular information processing) slowly over the last ten years and enter our third group at the stage in which their economy is more diversified. Each regional production system has its own organization, but the presence of three main groups, in slow evolution, indicates national hierarchies of functions, with dynamic metropolitan areas evolving towards a post-manufacturing society and industrial regions integrating more circulation and distribution activities than before. This naturally poses the problem of localizing the different types of activities, and more specifically, the problem of the organization and reorganization of regional productive systems.

TOWARDS A POST-MANUFACTURING SOCIETY?

Can we conclude from this analysis that production systems are evolving towards a post-industrial system? Although circulation–distribution and regulation roles have increased considerably, the manufacturing role remains important in old industrial regions. Consequently, the shift is not towards a service society as traditionally defined, but rather towards a post-manufacturing society in these regions. On the other hand, metropolitan and touristic regions are recently evolving towards what may be called information and consumption society.

In reality, regional differences are much more pronounced than differences between countries. For example, metropolitan areas are characterized by a small share of manufacturing but by a large share of circulation–regulation. In some regions, productive systems are heavily dependent on manufacturing, which constitutes their economic base. In other regions, new employment in circulation and information is directly or indirectly linked to manufacturing. Given the nature of this development, it is clear that the link between industry and circulation–distribution is growing stronger in these areas. This leads us to qualify the characteristics of the post-manufacturing society. On the basis of our analyses, it is possible to distinguish three sub-groups in this post-manufacturing society (Figure 6.3).

Regardless of the sub-group towards which a country or region is shifting, the essential is to identify the organizational structure of its production system in terms of function. In particular, it is necessary to distinguish carefully between the specializations of the various regions and those of the different countries. For example, even though regions characterized by manufacturing need to develop their circulation and distribution activities, manufacturing remains their comparative advantage. The same phenomenon applies to regions characterized by distribution or circulation.

In sum, our alternative typology gives us a better understanding of the composition of the regional and national productive systems. Each region or country can be positioned with reference to three types of models (Figure 6.4):

1. The industrial model, where manufacturing accounts for over 50 per cent of total employment.
2. The consumer society model, where employment in distribution reaches 50 per cent.
3. The information society model, where circulation represents over 40 per cent of total employment.

Most regions are developing somewhere in between these three extreme models, but the industrial model is still widespread and the most different from the information society model which characterizes metropolitan

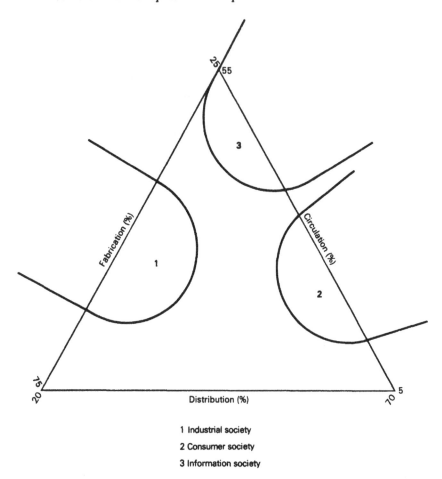

Figure 6.3 Three types of models: 1. Industrial society 2. Consumer society
3. Information society
Source: CEAT (1987)

regions. Even if some regions are moving towards the consumption model,
the industrial and information models still prevail.

Yet, all regions do not possess the same potential, even where information systems are concerned. Some regions, marked by their industrial
past, consolidate their position by increasing productivity in traditional
branches (textiles, iron and steel industry, etc.); however, others in this
same category no longer have the qualified services or manpower needed
for this revitalization.

In yet other regions, a set of complex activities forms the basis of the
production system. Granted, their economies are marked by tourist-related
and administrative specializations, but they are far from the relative
importance of metropoles when it comes to information and circulation

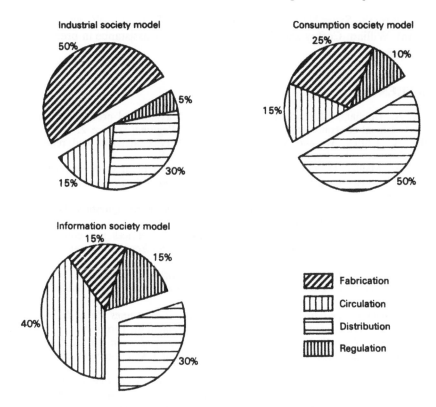

Figure 6.4 Composition of three types of model
Source: CEAT (1987)

networks which are precisely the locus of the relations of power that are essential to the control of economic life.

Thus, according to national and regional positions on a continuum ranging from industrial regions to metropoles, ability to adapt varies from region to region. Regional policies should be formulated on the basis of regions' positions and recent development, rather than on the basis of the secondary-tertiary opposition. Given the nature of the evolution of productive systems, it is clear that the relationship between circulation activities and metropolitan development is an area worthy of intensive analysis. One must not forget, however, the increasing importance of circulation and distribution in other regions, and the role of these activities in maintaining the local industrial base of peripheral regions.

ACKNOWLEDGEMENTS

A major portion of this paper is the result of a group research project carried out with Michel Rey, Louis Boulianne and Nicolas Mettan, and

financed by the Swiss National Fund for Scientific Research. We would like
to thank William Coffey and Peter Daniels for their assistance in preparing
this paper.

REFERENCES

Bailly, A., Boulianne, L. and Maillat, D. (1987) 'Services and production: for a
reassessment of economic sectors', *The Annals of Regional Science*, 2: 45–59.
Bailly, A., Coffey, W. and Maillat, D. (1987) 'Service activities and regional
development', *Environment and Planning*, 19: 653–8.
Bailly, A., Maillat, D. and Rey, M. (1984) 'Tertiaire moteur et developpement
regional: le cas des petites et moyennes villes', *Revue d'Economie Regionale et
Urbaine*, 5: 757–76.
Bailly, A. and Maillat, D. (1988) *Le Secteur Tertiaire En Question*, Geneva:
Editions Regionales Européennes Economica.
Barcet, A., Bonamy, J. and Mayere, A. (1984) 'Les services aux entreprises:
problèmes theoriques et methodologiques', *Recherches Economiques et Sociales*,
9: 119–35.
Bienayme, A. (1980) *Strategie de l'Entreprise Competitive*, Paris: Masson.
Braibant, M. (1982) 'Le tertiaire insaisissable', *Revue d'Economie et de Statistiques*,
1: 3–17.
CEAT (1984) *Le Tertiaire Moteur dans les Petites et Moyennes Villes de Suisse*,
Lausanne: Communauté d'Etudes pour l'Aménagement du Territoire.
CEAT (1985) *Nouvelle Nomenclature du Système de Production et Rôle Activités
de Service*, Lausanne: Communauté d'Etudes pour l'Aménagement du Terri-
toire.
CEAT (1987) *Nouvelle Articulation des Systèmes de Production et Rôle des
Services*, Lausanne: Communauté d'Etudes pour l'Aménagement du Territoire.
Clark, C. (1951) *The Conditions of Economic Progress*, London: Macmillan.
Coffey, W. and Polese, M. (1986) 'The interurban location of office activities: a
framework for analysis', in D.J. Savoie (ed.) *The Canadian Economy*, 85–103,
Toronto: Methuen.
Coffey, W. and Polese, M. (1987) 'Trade and location of producer services',
Environment and Planning A, 19, 5: 597–611.
Cuadrado Roura, J.R. and Del Rio Gomez, C. (1987) 'Structural change and
evolution of the services in the OECD area', *Documentos de Trabajo: Fundacion
Fondo para la Investigacion Economica y Social 3*: 1–36.
Daniels, P.W. (1983) *Service Industries: Supporting Role or Centre Stage?* Annual
Meeting of Institute of British Geographers, Edinburgh.
Fuchs, V. (1968) *The Service Economy*, New York: Columbia University Press.
Gershuny, J. and Miles, L. (1983) *The New Service Economy: The Transformation
of Employment in Industrial Societies*, London: Frances Pinter.
Greenfield, H. (1966) *Manpower and the Growth of Producer Services*, New York:
Columbia University Press.
Jeanneret, P. (1983) *L'emploi Tertiaire dans l'Industrie Suisse*, Neuchatel: Groupe
d'Etudes Economiques Université de Neuchatel.
Kempf, H. (1984) 'Croissance des services et reaménagement du système productif',
La Documentation Française, 1882: 14–19.
Lakshmanan, T.R. (1988) *Technological and Institutional Innovations in the
Service Sector*, Boston: Boston University Press.
Momigliano, F. and Siniscalco, D. (1982) 'The growth of service employment: a
reappraisal', *Banca Nazionale del Lavoro Quarterly Review*, 142: 270–306.

Noyelle, T. and Stanback, T. (1984) *The Economic Transformation of American Cities*, Totawa NJ: Rowman and Allanheld.

Nussbaumer, J. (1984) *Les Services: Nouvelle Donnée de l'Economie*, Paris: Economica.

Singleman, J. (1978) *From Agriculture to Services: The Transformation of Industrial Employment*, London: Sage Publications.

Stanback, T. (1979) *Understanding the Service Economy*, Baltimore: Johns Hopkins University Press.

Stanback, T., Bearse, P. and Noyelle, T. (1984) *Services: The New Economy*, Totawa, NJ: Rowman and Allenheld.

7 Trends in the producer services in the USA: the last decade

William B. Beyers

INTRODUCTION

The American economy has added 20 million jobs over the past decade, an expansion of almost 25 per cent. This is an enormous rate of growth of employment, an expansion accompanied by entry of the post-Second World War 'baby boom' population into the labour force, and by rising labour force participation rates by women (Bureau of Labor Statistics (BLS) 1988). Most of this growth of employment has come in the services, and this has not been without controversy. Concern has been expressed about the part-time and low wage nature of much services employment, and doubt has been expressed about the ability of the economy to have real economic growth without continued expansion of goods producing sectors (Rose 1986; Zysman and Cohen 1987).

The services encompass a number of different economic activities, including retailing, consumer, social and health, governmental, wholesaling, transportation, utilities and the producer services. The producer services are a broad collection of activities, including financial, insurance, real estate, business, legal and professional services, as well as the central administrative units of businesses in all divisions of the economy. They supply service inputs to the production process in all sectors of the economy: the transformative sector, distributive, retail, central administrative, consumer, government, non-profit and producer services. The producer services accounted for about one-third of the job expansion in the United States in the last decade. The dramatic job growth in the producer services has been at a rate double that of the overall rate of job creation, and projections show no abatement of this robust growth (BLS 1988).

The producer services have been regarded as an attractive component of the service economy that now surrounds us, because they pay relatively well, are growing rapidly and have a traded component thereby making them a part of the economic base of communities (Beyers and Alvine 1985). Although the producer services have grown rapidly, and are strongly concentrated in large metropolitan areas, there has been a paucity of research documenting regional trends in their recent development (see Noyelle and Stanback (1983) for an earlier analysis of metropolitan areas).

To help remedy this situation, this paper focuses upon the growth of the producer services in the United States over the past decade, and emphasizes trends in metropolitan areas.

BACKGROUND

Service industries have generally been regarded as the handmaidens of 'basic' industry. Regionally, their employment level has traditionally been viewed as functionally related to employment levels in goods producing sectors (Reifler 1976). However, with the rise of information-orientated producer services sectors, whose client base is more removed from the production, distribution, and consumption of goods, the role of these information-orientated services in the regional development process has demanded reassessment. The recent record of job creation has also demanded a re-examination of traditional views, for producer services employment has grown rapidly in the face of stagnation, decline, and restructuring of manufacturing employment (see Piore and Sabel 1984; Zysman and Cohen 1987). This is not to say that there are no vibrant components within manufacturing or agriculture, or to argue that these sectors are unimportant in the contemporary economy. It is only to make the point that we have yet to explain how and why the producer services have performed so spectacularly.

The producer services must be viewed as a set of highly differentiated economic activities. Some sectors have outputs resembling physical goods (software is a good example), while others are almost entirely involved with information exchange (like the telephone company). However, almost all of these service sectors share a common dependence upon technically sophisticated manufactured goods. Baumol has made a cogent argument about how creatively the 'services' have made use of 'goods', emphasizing the co-mingling of production activities in different sectors of the economy with the work functions that take place within each of these sectors (Baumol 1985).

Within the producer services, the same smorgasbord of functions also exists, with different degrees of routinization, niche articulation, multi-establishment organizational structures, capital intensity or labour organizational arrangements. Occupational structures are highly differentiated, ranging from professionally dominated sectors such as law and engineering, to those with a relatively large proportion of low-skill labour (such as banking tellers). Within all of these sectors, there are highly skilled people, as well as some employees doing low-skill work. In short, simplistic categorizations of work activities in the producer services need to be replaced with a more sophisticated and better researched understanding of how these vital sectors of our economy are structured, and what factors are leading to their expansion and future developmental role in the economy.

While this development has been taking place in our national economic environment, interestingly there has been little in the way of regional analysis of these trends in recent years. This paper specifically addresses regional patterns of growth. It provides information on patterns of employment change within the producer services among the 183 BEA economic regions for the 1974–85 time period. The BEA economic areas are defined on the basis of commuting data. Standard Metropolitan Statistical Area (SMSA) counties form the 'core' of these BEA economic areas, while non-SMSA counties constitute the balance of the territory in each of these regions. This paper provides information for the SMSA and non-SMSA components of these economic regions, so that urban/metropolitan and rural/nonmetropolitan differences in trends can be observed.

A BRIEF BACKGROUND ON RELATED RESEARCH

Prior research on the spatial pattern of growth of the services in the United States has used other time periods and different kinds of regionalization for its analyses. For example, the excellent work of Noyelle and Stanback focused on larger SMSAs for the 1959–76 time period (Noyelle and Stanback 1983), while Kirn looked at employment trends in a sample of 96 metropolitan and nonmetropolitan areas distributed all over the United States for the 1958–77 time period (Kirn 1987). O'hUallacháin recently reported on an analysis of 264 SMSAs for a selection of twenty-eight service sectors over the 1977–84 time period (O'hUallacháin 1987). In contrast, this paper uses a regionalization of the entire United States. Given the differences in time periods, sectoring, and regionalization among these various studies, there is little reason to expect identical findings.

Noyelle and Stanback found that large SMSAs in the United States had quite different industrial structures. Using various statistical techniques, they grouped SMSAs into four broad categories: (a) diversified service centres; (b) specialized service centres; (c) production centres; and (d) consumer orientated centres. They analysed trends in employment in these groups, and developed a classification showing differences in changes of structure (transformation in their parlance), most notably towards the increased presence of the producer services and 'the complex of corporate activities'. They found tendencies for strong transformation (most dramatically a decline in manufacturing with rapid growth in services, especially producer services) to be occurring in many of the slower growth diversified or specialized service centres, and in some slow growth production (manufacturing) centres. There was little evidence showing decentralization of the producer services within the set of large SMSAs that they examined in the United States (Noyelle and Stanback, 1983: 51–92, 222–33).

Kirn looked at four sets of places in his analysis of service sector development: (a) large SMSAs; (b) small SMSAs; (c) large nonmetropolitan areas (e.g. counties not designated as an SMSA with at least 50,000 population); and (d) small nonmetropolitan areas (counties with less than 50,000 population). Over the 1958–67 and 1967–77 time periods he found growth in employment in one key producer service (finance, insurance, and real estate) was greatest in the smallest nonmetropolitan areas, and slowest in the larger places. Business services (SIC 73) were found to be more rapidly growing in the small SMSAs and large nonmetropolitan areas than in the large SMSA areas, while legal services showed relatively strong growth in larger SMSAs. Professional services were observed to have strong relative growth in small SMSAs and large nonmetropolitan areas between 1958 and 1967, but showed stagnation in SMSAs between 1967 and 1977, while large nonmetropolitan areas showed strong growth, particularly in accounting (Kirn 1987).

O'hUallacháin analysed changes in employment in a set of twenty-eight rapidly growing services, including a number of producer services sectors. His methodology differed from that of Kirn and Noyelle and Stanback. Kirn's measures of change were presented in percentage terms, while O'hUallacháin used absolute employment data to model differences in the importance of localization and urbanization economies. In explaining the growth of the producer service sectors, O'hUallacháin determined that localization economies were overwhelmingly more important than urbanization economies, or alternatively expressed, places that had large employment in a given producer service sector in 1977 had large absolute employment growth in the 1977–84 time period. When viewed in relative terms, his analysis concluded that 'size counts', that agglomeration tendencies were strong, and have not diminished in their strength. Ó hUallacháin concluded that little decentralization of producer services had occurred within the nation's metropolitan areas (O'hUallacháin 1987).

These analyses have produced contrasting views of trends in the growth of producer service employment. Kirn's work suggests that nonmetropolitan growth is strong, and if this is the case, given the difficulties of many transformative sectors in rural America in recent years, it would seem important to learn more about trends in the producer services in nonmetropolitan territory. Noyelle and Stanback find important changes in economic structure, implying relatively rapid growth of producer service employment in those places experiencing structural transformation, but they find that size of place is not related to the strength of this transformation. It is difficult to say whether this conclusion is at odds with O'hUallacháin's findings, because of differences in methodology and time periods being analysed. These analyses do provide some guidance on the spatial dimensions as well as some of the causes of the recent expansion of producer services, but much research remains to be done.

RESEARCH FOCUS

Some key questions that remain to be addressed in analysing the contemporary burst of employment in the producer services include the following:

1. What has happened to the geography of employment in the United States since the mid-1970s, a period with several deep recessions and increasing international competition that has severely impacted some sectors of American manufacturing and some regions strongly dependent upon manufacturing? How have the producer services grown regionally, given differences in the performance of other sectors in regional economies?
2. Accepting the fact that most new jobs have been in the services, how have different service sectors performed? How do their markets differ structurally and spatially, and how are these markets changing over time?
3. Accepting for the moment the popular notion that producer services have strong manufacturing markets (see Scott 1988), what has happened to producer service employment in regions where manufacturing employment has declined, stagnated, or expanded? Expressed alternatively, can the performance of local manufacturing be related at all to the performance of local producer services?
4. Assuming that producer services have some 'exportable' component, and that their markets to other service sectors are also strong and growing, how have developments in export markets influenced the regional distribution of producer service activities?
5. The producer services are a complex set of differentiated business activities; they are not a homogeneous group of sectors. How have trends in various sectors differed, between regions, and between metropolitan and nonmetropolitan territory?
6. Given that there is no broadly accepted theory of producer service location, can we begin to identify key variables that ought to be encompassed in such a theory? (Daniels 1985; Marshall *et al.* 1988).

This chapter addresses a number of these questions briefly; the author has recently completed a much more detailed analysis of the growth of the producer services in the United States (Beyers 1989).

APPROACH

The present analysis has been conducted using recent data for employment by county in the United States: the US Department of Commerce County Business Patterns (CBP) data tapes for 1985, together with CBP data tapes for 1974. These tapes contain considerable industrial detail for each county in the United States, but exclude government and agricultural employment. A database comprising ninety-eight sectors was developed for each region, including all two-digit sectors, administrative and auxiliary employment for all major industrial divisions, and detailed estimates of employment within the producer services (especially in SIC 73 and 89). The Census Bureau applies a disclosure rule to these data, but provides suppression codes indicating an employment range for suppressed employment. Therefore, it is possible to know of the existence of employment in a sector in a particular county, even if exact employment data are not available.

Figure 7.1 shows the procedure used to estimate the employment series used in this research. For each state the following procedure was used for the years 1974 and 1985. From the data tapes, known values (i.e. disclosed employment estimates) for each sector by county were placed in a data file, and the suppression codes were placed in another file. The magnitude of the missing values could then be calculated by summing the total known values, by sector and by county, and subtracting these estimates from the county total employment and the statewide sectoral employment values.

Then the differences between the disclosed and the total values were calculated, and used as constraints for a biproportional matrix adjustment procedure (see Bacharach 1970). Initial values for the suppression codes were given within the employment range of the suppression codes. Through iterative procedures, estimates for each sector in each county were derived that summed up to the total value of the missing county and sectoral employment levels. The resulting matrix of estimated employment was then merged with the matrix of known values, to obtain the state by state estimates of employment by county and sector. Finally, the county files were merged into the BEA regional files, either as SMSA or non-SMSA territory. SMSA territory will be considered metropolitan in this paper, while non-SMSA territory will be considered nonmetropolitan.

RESULTS

National overview

Total US employment in the CBP data series totalled 63 million in 1974, rising to 81 million by 1985, an increase of 28.5 per cent, as shown in Table 7.1. SMSA territory increased slightly more rapidly than non-SMSA territory during this time period (29.6 per cent versus 23.7 per cent respectively), and accounted for about 83 per cent of total national employment in both years. Expressed alternatively, about four out of five Americans work in metropolitan areas. These areas, as well as nonmetropolitan territory, had an expansion of about one job between 1974 and 1985 for every 3.5 jobs that existed in 1974.

Producer service employment grew more than twice as fast as aggregate national employment, expanding by 60 per cent in both SMSA and nonmetropolitan territory. In contrast, manufacturing posted a small decline in both SMSA and nonmetropolitan territory, although the percentage rate of decline was somewhat higher in SMSA areas than in nonmetropolitan areas. Other employment grew more slowly than the

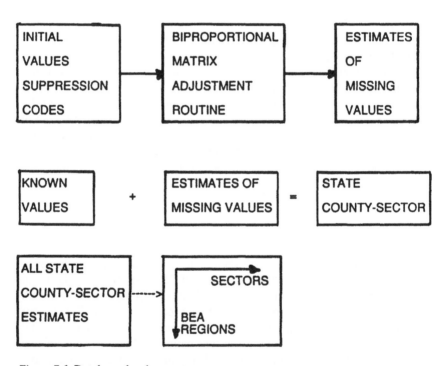

Figure 7.1 Database development

producer services, but again it expanded slightly more rapidly in SMSAs than in nonmetropolitan areas. The highly centralized employment pattern of the producer services is also evident in Table 7.1. In 1974 the producer services comprised 18.2 per cent of total SMSA employment, manufacturing accounted for 28.8 per cent of total SMSA employment, and other sectors accounted for the remaining 53 per cent. In contrast, in

Table 7.1 Summary of employment change, United States, 1974 and 1985

	Total employment (millions)		
	1974	*1985*	*Change (%)*
Total employment			
SMSA	52.2	67.7	29.6
Non-SMSA	10.8	13.3	23.7
Manufacturing			
SMSA	15.0	14.2	−5.2
Non-SMSA	4.1	4.0	−1.7
Producer services			
SMSA	9.5	15.2	60.0
Non-SMSA	0.9	1.4	59.0
Other sectors			
SMSA	27.7	38.3	38.0
Non-SMSA	5.8	7.9	36.0

Source: US Census Bureau, *County Business Patterns*

nonmetropolitan areas the producer services accounted for only 8.3 per cent of total employment, manufacturing for 38 per cent of the total, and other employment 54 per cent (about the same as in SMSA). (It should be re-emphasized that these data exclude agriculture and government; if they were included somewhat different proportions would be evident, but their inclusion would not significantly change this image of dramatic differences in industrial structure.)

The fact that producer services growth rates in SMSA and nonmetropolitan areas were identical over the study period means that in the aggregate there was no significant SMSA to nonmetropolitan decentralization of the producer services. Because producer services grew more rapidly than the whole economy, their share of total employment in both SMSA and nonmetropolitan territory increased. Their share of nonmetropolitan employment had risen to 10.8 per cent of total employment in 1985, up from 8.3 per cent in 1974. In contrast, within SMSAs their share rose from 18.2 per cent to 22.5 per cent of total employment. Ninety-one per cent of producer services employment was located in SMSAs in 1974 and 1985, compared to about 83 per cent of total national employment found in SMSAs.

Sectoral trends

Producer service sectors had widely varying rates of growth over the study time period, as shown in Table 7.2. A number of sectors experienced quite rapid growth, and had very large absolute gains in employment, such as

Table 7.2 Producer services employment growth in the United States, 1974–85

SIC	Sector	Employment (1974) ('000s)	Change in employment ('000s)	Change (%) (1974–85)
Rapidly growing sectors				
4700	Services to transport	133	145	109
6200	Security brokers	184	151	83
7330	Mail/steno/reprogr.	92	98	106
7360	Labour supply services	355	507	143
7370	Computing services	148	365	247
7391	R&D laboratories	69	73	106
7392	Consulting	177	334	189
7394	Equipment rental	76	147	187
8100	Legal	318	366	115
8910	Arch/engineering	343	330	96
8930	Accounting	191	213	111
8990	Other prof. services	14	130	928
149/	Mining/admin/auxiliary	69	73	105
497	Trans/comm/util auxiliary	101	94	93
679/	FIRE admin/auxiliary	59	108	184
899/	Services admin/auxiliary	126	116	92
Moderate growth sectors				
6100	Other credit agents	443	300	68
6400	Insurance agents	345	200	60
6700	Holding and investment	115	78	68
7310	Advertising	113	70	62
7340	Building services	378	227	60
7393	Detective services	248	164	66
7397	Commercial test labs	25	18	71
7399	Other business services	211	158	75
8600	Membership organizations	997	550	55
8920	Noncomm research organizations	47	23	50
098/	Resources admin/auxiliary	2	1	46
599/	Retail admin/auxiliary	509	392	52
Slow growth sectors				
6000	Banking	1190	391	33
6300	Insurance carriers	1111	135	12
6500	Real estate	843	300	36
6600	Combi. RE and insur.	30	−6	−20
7320	Credit reporting	64	9	14
7350	News syndicates	6	1	22
7395	Photofinishing labs	54	21	40
7396	Trading stamp services	1	−0.2	−17
179/	Construction admin/auxiliary	15	5	36
399/	Mfg admin and auxiliary	1142	135	12
519	Wholesale admin and auxiliary	224	68	30
	Total	10568	6371	60

Source: US Census Bureau, *County Business Patterns*

computing and data processing services or accounting. These rapidly growing sectors generally had at least a doubling of the number of employees over the study period. Other sectors grew at a rate similar to producer services as a whole, such as advertising or insurance agents. Still other sectors had growth rates below the overall national rate of employment, such as insurance carriers or the administrative offices of manufacturing firms.

Spatial patterns

This section reports on a spatially disaggregate analysis of trends in the producer services. The preceding picture of a highly centralized producer services sector should not lead us to conclude that all sectors within the producer services are highly centralized or that some sectors are not decentralizing. To understand trends at the sectoral level, various approaches to description have been undertaken, including shift-share analysis, estimation of location quotients and changing location quotients and estimates of coefficients of industrial concentration.

Location quotients were calculated for producer services by using total employment in all sectors as the base and are shown in Table 7.3. For almost every sector the SMSA location quotients were above 1.0 in 1974, but almost every one of these sectors experienced modest decentralization between 1974 and 1985, as evidenced by generally negative changes in location quotients in SMSAs and positive changes in location quotients for nonmetropolitan territory. While aggregate location quotients are above 1.0 for most sectors in all SMSA territory, when location quotients are computed for aggregate producer services employment in each BEA economic area another picture emerges. Only about one-sixth of the nation's BEA economic areas had aggregate producer services location quotients above 1.0 in 1974, and these places tended to be the nation's largest cities. Thus, producer services were not only concentrated in SMSA territory, but they were also differentially concentrated in the largest SMSAs. Over the study period, the concentration in the largest metropolitan areas weakened slightly, while the level of concentration tended to increase in the bulk of the smaller BEA economic areas. However, in 1985 aggregate producer services location quotients were still less than 1.0 for five out of six of the 183 BEA economic areas.

Several sectors emerge as being relatively widely distributed between SMSA and nonmetropolitan territory, given their relatively high location quotients in nonmetropolitan territory. These include banking, credit agencies other than banks, insurance agents, legal services, and accounting services. These activities tend to have a mix of business and consumer clients, and might be regarded as the 'retail' end of the producer services, found not only in nonmetropolitan territory, but also within the typical large SMSA serving the same types of mixed consumer and business

Table 7.3 United States producer services location quotients 1974, and change 1974–85

SIC	Sector name	Location quotient (1974)		Change in location quotient (1974–85)	
		SMSA	Non-SMSA	SMSA	Non-SMSA
4700	Transport services	1.13	0.37	−0.02	0.08
6000	Banking	1.0	0.99	−0.02	0.10
6100	Credit agents	1.06	0.73	0.01	−0.04
6200	Sec/com brokers	1.19	0.09	−0.02	0.07
6300	Insurance carrier	1.15	0.25	−0.02	0.06
6400	Insurance agents	1.05	0.77	−0.01	0.03
6500	Real estate	1.10	0.49	−0.01	0.04
7310	Advertising	1.17	0.19	−0.01	0.01
7320	Con cred agents	1.05	0.74	−0.02	0.08
7330	Mail/sten/reprogr.	1.15	0.27	−0.01	0.02
7340	Building services	1.14	0.34	−0.02	0.07
7350	News syndicates	1.12	0.03	−0.05	0.21
7360	Labour supply services	1.17	0.19	−0.01	0.00
7370	Computing	1.17	0.16	−0.01	0.03
7391	R&D services	1.13	0.35	−0.02	0.09
7392	Mgt., cons, PR	1.16	0.23	−0.02	0.08
7393	Detection/protection	1.18	0.13	−0.04	0.13
7394	Eq. rent/lease	1.10	0.50	−0.04	0.18
7395	Photo labs	1.12	0.41	−0.02	0.08
7396	Trade stamps	1.08	0.59	−0.12	0.60
7397	Comm test labs	1.13	0.39	−0.06	0.28
7399	Misc. business services	1.12	0.41	0.00	−0.03
8100	Legal services	1.06	0.73	0.02	−0.11
8910	Arch. and eng.	1.12	0.42	0.00	−0.03
8920	NC ed com RD	1.13	0.35	−0.03	0.00
8930	Accounting	1.06	0.71	−0.02	0.07
8990	Misc prof. services	1.10	0.51	−0.04	0.20
098/	Primary admin	0.80	1.95	0.32	−1.58
149/	Mining admin	1.01	0.94	−0.01	0.03
179/	Constr. admin	1.18	0.15	−0.01	0.02
399/	Mfg. admin	1.13	0.38	0.01	−0.07
497/	TCU admin	1.16	0.20	−0.02	0.06
519/	Wholesale admin	1.15	0.26	−0.04	0.17
599/	Retail admin	1.15	0.27	−0.01	0.00
679/	FIRE admin	1.15	0.26	0.00	0.00
899/	Services admin	1.14	0.32	0.01	−0.09

Source: US Census Bureau, *County Business Patterns*

markets. However, within SMSAs we could expect to find much more complex/sophisticated segments of these industries than in nonmetropolitan areas, serving primarily business or government clients, performing intrasectoral business activity, and often engaged in interregional trade of their services.

Spatial patterns in several individual producer services will now be examined, to illustrate the diversity of locational patterns in the producer services. First, some cartograms are presented which illustrate variations in the evenness of the distribution of producer services employment among the BEA economic areas. Then, selected results of shift-share analyses will be presented, illustrating trends for rapidly growing sectors, as well as for moderate and slow growth sectors.

The data in Table 7.2 showed aggregate differences in the concentration of producer services in SMSA and non-SMSA territory. However, individual producer services are also distributed in a varying manner among the BEA economic areas. Coefficients of industrial concentration are one measure that can be used to convey the evenness of the distribution of sectors among a set of regions. However, another alternative means is visual, through the use of cartograms. To illustrate this approach for several producer services, Figures 7.2 to 7.5 show cartograms for employment in each BEA economic area. The area of each polygon in these figures is proportional to the level of employment in the region. Figure 7.2 provides a reference point, by showing total employment in 1985. Figure 7.3 shows the pattern for legal services, which is distributed among the BEA economic areas in a manner similar to total employment. In contrast, Figures 7.4 and 7.5 show two sectors with more uneven patterns of employment, security brokers and research and development laboratories. The coefficients of industrial concentration for these three producer

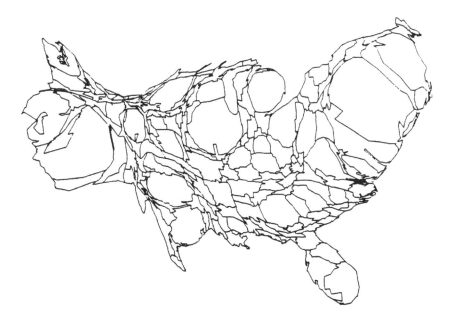

Figure 7.2 Total employment, by BEA, 1985

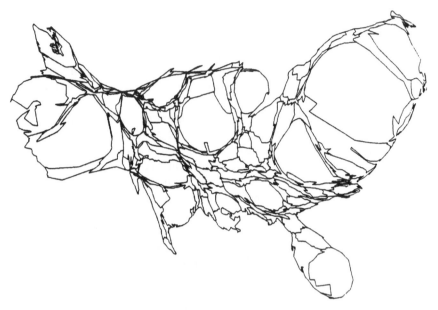

Figure 7.3 Legal services employment, by BEA, 1985

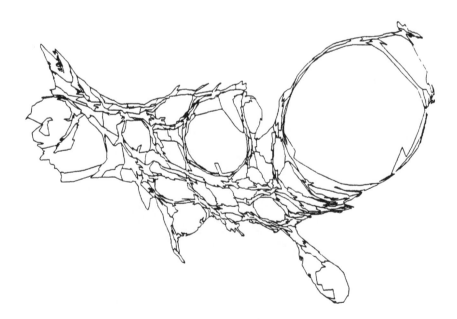

Figure 7.4 Securities employment, by BEA, 1985

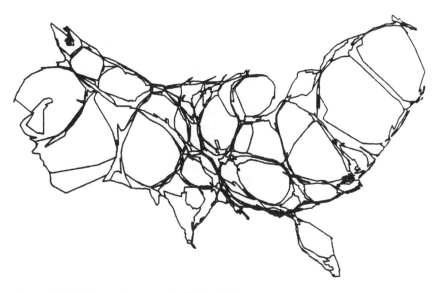

Figure 7.5 R&D employment, by BEA, 1985

services are 0.27 for legal services, 0.72 for security brokers, and 0.85 for research and development laboratories. The higher the coefficient of industrial concentration, the more skewed or distorted the cartogram becomes (see Beyers (1989) for more detail on this topic).

To describe the pattern of change in the distribution of individual producer services among the BEA economic areas, shift-share analysis was utilized. The employment data for individual producer service sectors shown in Table 7.3 for SMSA and non-SMSA territory were used in this shift-share analysis; Table 7.4 presents some summary results of these calculations. Net shifts amount to 1.4 million jobs, out of an overall change of 6.4 million. Net shifts are defined as differences between actual and expected change, where expected change is based on the aggregate rate of growth and the initial level of employment in the sector in each region (see Stevens and Moore (1980) for a detailed discussion of shift-share methodology). This result indicates that the amount of regional redistribution associated with the rapidly growing producer services was relatively small. Expressed alternatively, the analysis shows that aggregate growth of producer services employment was largely proportional to its distribution among the various subregions in 1974.

The shift components shown in Table 7.4 are divided into two parts, the industry mix and competitive components. The industry mix component is based on expected sectoral growth; thus, fast growing sectors have positive industry mix components, while slow growth sectors have negative industry mix components. The competitive component is the portion of the net shift not accounted for by industry-specific growth rates. The redistribution that

Table 7.4 Producer services shift-share summary, 1974–85

Big relative gains (SMSA) (% of total positive shift)		Big relative losses (SMSA) (% of total negative shift)	
Dallas	8.65	New York	30.6
Los Angeles	8.64	Chicago	15.4
San Francisco	7.25	Detroit	8.5
Washington, DC	5.55	Cleveland	5.7
Boston	4.56	Philadelphia	4.7
Houston	4.40	Pittsburgh	3.6
Tampa/St. Petersburg	3.99	Hartford	2.7
Orlando	3.8		
Phoenix	3.8		
Denver	3.61		
San Diego	3.27		
Atlanta	2.57		
Anchorage	2.44		
Austin	2.2		
Total	64.9	Total	71.3

Overall change	+ 6.4 million
Net shift	+/− 1.4 million
Industry mix shift	+/− 0.28 million
Competitive shift	+/− 1.24 million

Source: Computed from US Census Bureau, *County Business Patterns*

occurred was primarily associated with the competitive shift component, indicating that the 1974 distribution of individual producer services was not a good predictor of the regions having particularly fast or slow growth in given sectors.

Table 7.4 shows that 71 per cent of the negative shift was associated with New York, Chicago, Detroit, Philadelphia, Cleveland, Pittsburgh and Hartford, but this is a relative 'loss', for all of these cities had absolute growth in producer services employment. Their collective gain was one million jobs; New York alone had a gain of 550,000. Thus, these big cities with large prior (1974) endowments of producer service employment simply grew more slowly than some other cities over the past decade. The positive shifts shown in Table 7.4 are primarily in sunbelt and western cities, although notable north-eastern inclusions are Boston and Washington. Percentage changes in employment in producer services and manufacturing for the set of BEA economic areas shown in Table 7.4 are revealing. The regions with large positive shifts in producer services had aggregate producer services employment growth of 95 per cent, and manufacturing employment growth of 21 per cent. In contrast, the regions with large negative shifts in producer services had aggregate producer services employment growth of 35 per cent, and a loss of 21 per cent in manufacturing employment. The relationship between these varying experiences in manufacturing and growth rates in producer services will be discussed further in a later section of this chapter.

Patterns of net shifts for individual sectors with varying growth rates reveal interesting geographical patterns. Figures 7.6 to 7.9 show these patterns for four of the sectors included in the shift-share analysis. Two of these are rapidly growing sectors, personnel supply (Figure 7.6) and accounting (Figure 7.7), while advertising (Figure 7.8) had a moderate growth rate and banking (Figure 7.9) was a slow growth producer service. These figures show patterns of net shifts for SMSA and non-SMSA territory for each BEA economic area, as well as for the entire BEA economic area. The six-way classification shown in the legend includes three categories where overall BEA net shifts are negative or positive (labelled TOT), and then within each of these three categories the possible combinations of net shifts are identified: both SMSA and non-SMSA as negative or positive, and the mixed cases where either the SMSA was positive or negative, and the non-SMSA territory was negative or positive.

The two sectors with rapid growth rates shown in Figures 7.6 and 7.7 have quite different distributions, as shown in Table 7.3. Personnel supply agencies have only a modest presence in non-SMSA territory, as evidenced by their low location quotients, while accounting services are much more strongly represented in the industrial structure of non-SMSA territory. If both metropolitan and nonmetropolitan territory are sharing in the nationally rapid growth of these two sectors, then the patterns of net shifts should be predominantly in the three positive groups on Figures 7.6 and 7.7. In fact, these two figures show this to be the case.

Figure 7.8 shows a different picture, with many regions exhibiting negative shifts. Almost all of the metropolitan areas with large aggregate positive shifts shown in Table 7.4 exhibit positive shifts in Figure 7.8, while most of the regions experiencing aggregate negative shifts shown in Table 7.4 also had negative net shifts in advertising. However, there is a general tendency for BEA economic areas in the American industrial belt and in resource-orientated economies in the west to show negative net shifts.

The slowly growing banking sector shows a fairly uniform pattern of negative net shifts in Figure 7.9. However, the 'oil patch' in the south-central US shows up as an exception, reflecting the boom in financial activities that accompanied the prosperity of the oil industry during the study period.

Figures 7.6 to 7.9 are representative of spatial trends in net shifts for other producer services sectors. The rapidly growing sectors identified in Table 7.2 tend to have widely distributed patterns of positive net shifts, when the sectors are themselves dispersed among BEA economic areas (e.g. have relatively low values of coefficients of industrial concentration). Similarly, Figures 7.8 and 7.9 are typical of patterns for moderately and slowly growing sectors identified in Table 7.2.

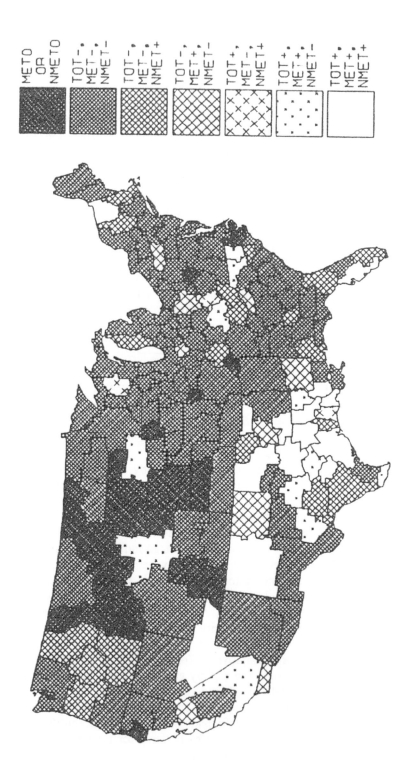

Figure 7.9 Net shift, banking, by BEA, 1985

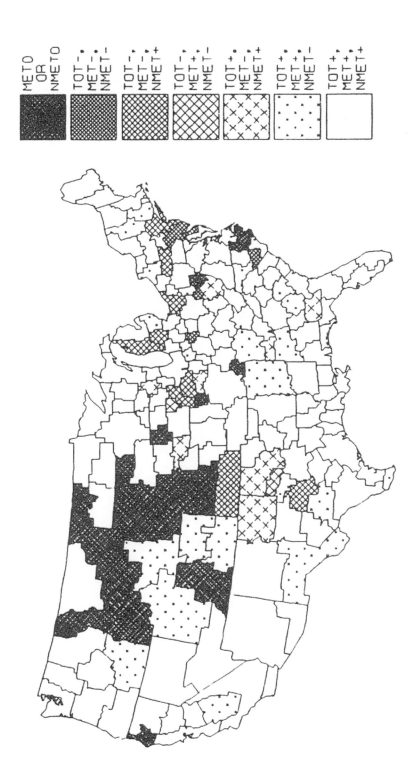

Figure 7.7 Net shift, accounting services, by BEA, 1985

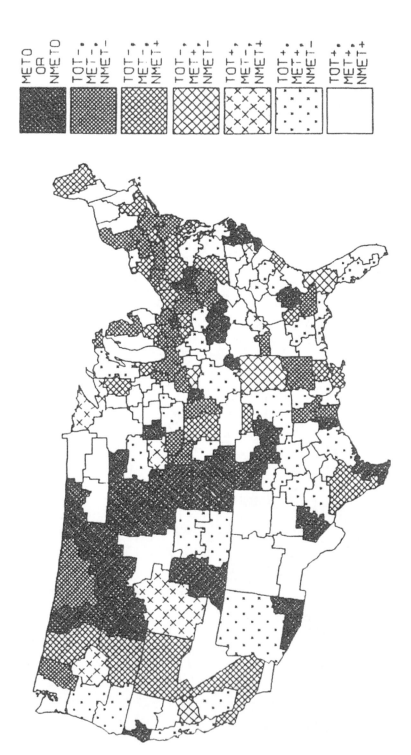

Figure 7.8 Net shift, advertising, by BEA, 1985

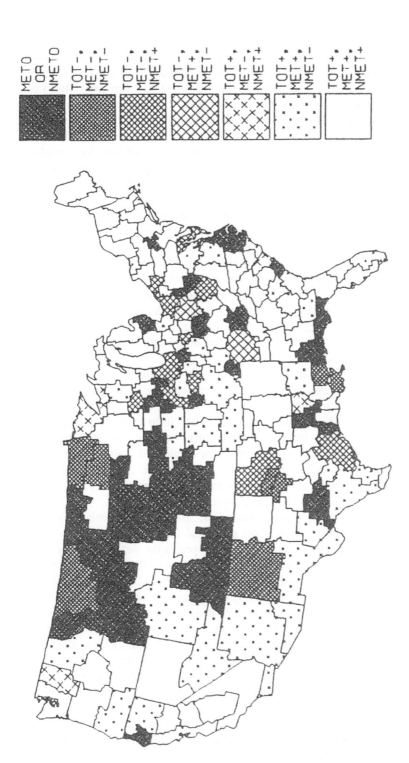

Figure 7.6 Net shift, personnel agencies, by BEA, 1985

Data presented above for the large contributors to the net shifts are suggestive of a relationship between the performance of a region's manufacturing sector and the growth rate in producer services. Statistical analyses of this relationship were undertaken, and it was found that there was a weak, but significant positive relationship between the changes in the two sectors (Beyers 1989). Earlier in this paper, reference was made to the presumed strong growth in demand for producer services by manufacturing. In considering this relationship between producer services and manufacturing, data on producer services markets in the US input-output tables were analysed (Beyers 1989). These analyses reveal that manufacturing is *not* an important market for most producer services. In fact, household consumers are very important customers for financial, insurance, real estate, and legal services; government is a very important client for accounting, labour supply, research and development, and consulting services. The various services are more important inter-industry markets for most producer services than are the goods producing sectors. Exceptions to this last generalization are found in the miscellaneous repair, equipment rental, architecture, and engineering sectors. Other recent research on the growth of the services has noted similar structural relationships (Petit 1986; Ochel and Wegner 1987).

These input-output data suggest that the relationship between growth rates in the producer services and a region's manufacturing sector is primarily an indirect one. In addition, it should be recognized that: (a) many regions in the United States have key industrial specialities other than manufacturing, including agriculture, mining, government, and services; and (b) the demand for producer services in a given region is not just intra-regional demand, but these services are also entering inter-regional and international trade to a growing extent (Beyers and Alvine 1985). Let us explore these various structural issues further.

In the case of regions whose industrial structure has a relatively weak manufacturing sector, there is no statistical basis for expecting regional growth in the producer services to be primarily determined by local manufacturing trends. However, we could expect that the fortunes of a region's other industrial specialities would contribute to the success of their local producer services. In a recently completed study, cluster analysis was used to classify the industrial structure of the BEA economic areas to help test this proposition (Beyers 1989). In an analysis of regions where net shifts in producer services moved opposite to net shifts in manufacturing (this was the case for fifty-five of 183 BEA economic areas), in two-thirds of these cases the regions were members of clusters with industrial specialities other than manufacturing. This finding supports the contention made above regarding diversity in the economic base of BEA economic areas.

No direct evidence on the magnitude of interregional or international exports of services from the BEA economic areas is available. To provide some estimate of the significance of these exports, a location quotient approach was taken to this question, using the detailed sectoring plan shown in Table 7.3 for each BEA economic area. Figures 7.10 and 7.11 show estimated absolute and percentage changes in export employment between 1974 and 1985. The level of export tied employment was estimated to increase from 2.7 to 4.0 million jobs over the study period; in 1985 the share of total producer services jobs linked to export markets was estimated to be 23.5 per cent. It is recognized that the location quotient approach produces conservative export estimates, so these figures are probably low (Tiebout 1962; Isserman 1980). Figure 7.10 shows the strong absolute concentration of the change in producer service export employment in the largest BEA economic areas (with the notable exception of Chicago, Detroit, and some other industrial belt centres). However, Figure 7.11 shows a very different pattern, with strong percentage growth in many of the smaller BEA economic areas. Figure 7.11 suggests that the producer services are becoming more important in the economic base of many regions in the United States, although their benchmark (1974) levels of export activity tended to be small.

These exports data and the evidence regarding the role of key sectors influencing the performance of the locally orientated producer services can be considered with the national input-output data to provide an overall framework for producer services demand.

The growth of producer services employment in a given region should be seen to be a function of the growth of producer services exports and exports of other key industrial sectors. These changing export demands in turn lead to local direct requirements for producer services and other inputs, and they directly create local income. These direct requirements lead to a chain of indirect effects, which have associated with them additional demands for producer services and other sectors, that also create additional income. Expenditures of locally created income leads through the consumption and government expenditure process to additional indirect and induced effects, including local final demands for producer services. Prior research suggests that the chain of multiplier relationships at the local level will be relatively powerful through the income earning and disposition-induced effects mechanism, which is primarily stimulatory to the services, including producer services (Beyers 1974). In regions with a strongly expanding export base, this set of linkages appears to have led to strong growth in producer services, and the opposite has been the case in regions with a declining or slowly growing export base. It is recognized that these linkage systems have leakages associated with them; imports from given regions are conversely components of the export demands of the supplying regions. Over time, the importance of these interregional connectivities may be increasing in the producer services as businesses are able to extend market areas through the use of advanced information processing and telecommunications technologies.

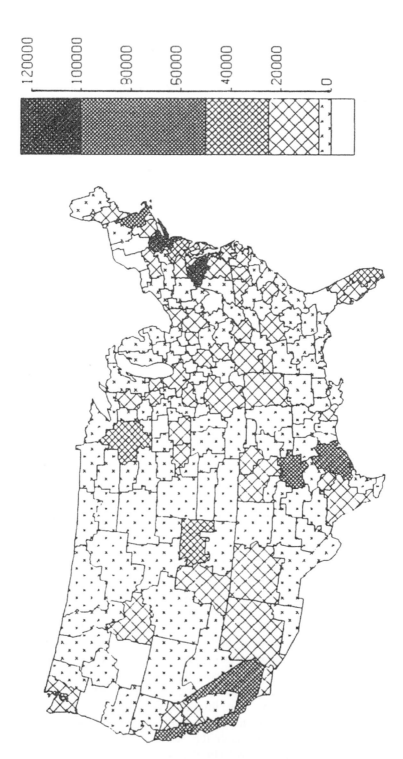

Figure 7.10 Producer services, change in export employment, by BEA, 1985

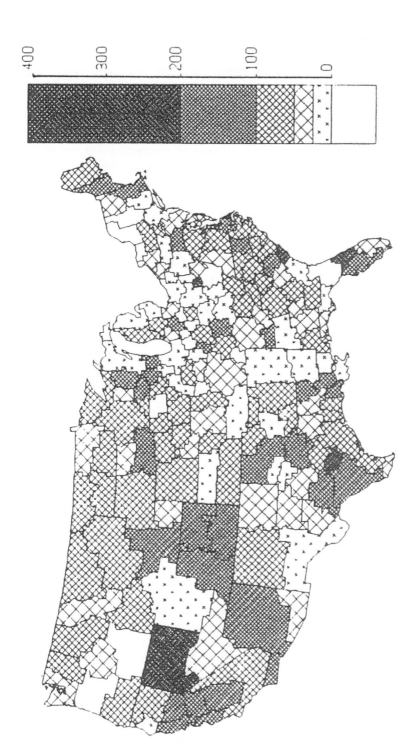

Figure 7.11 Producer services, percentage change in export employment, by BEA, 1985

Over time it also appears as though the demands for these services have been growing from their various clients, including household consumers, governments, and industries. Recent research suggests that the income elasticity of demand for producer services is above unity, that innovations in service provision have increased their use by households, and that lifestyle changes enhance their demand (Kravis *et al.* 1982; Ochel and Wegner 1987). Governments are apparently contracting out for more services (Tschettler 1987). Within the private sector, the intermediate demand for producer services per unit of output has also risen.

Research has focused on the bases of this rising use of producer services, exploring the degree to which this growth represents merely a shift of the locus of provision of producer services from within manufacturing or other industrial sectors to freestanding producer service firms. The evidence seems to suggest that this has not been a primary source of growth of the producer services (Petit 1986; Tschettler 1987). Rather, changes in the nature of work, changes in the types of producer service activity, and changes in the information processing and telecommunications technologies intimately associated with the provision of many lines of producer service activity appear to be important bases of expansion of producer services employment, both to supply localized and interregional or international markets (Collier 1983; Leveson 1985; McCrackin 1985; Petit 1986; Ochel and Wegner 1987; Ott 1987; Tschettler 1987). In addition, advances in information technologies have made it possible for market areas to be expanded geographically so that market areas for individual firms in the producer services have been able to expand and take advantage of scale economies, fuelling the increasing division of labour in this segment of the economy (Petit 1986; Ochel and Wegner 1987; Beyers 1989).

All of the factors just discussed help explain the growth of producer services even in regions experiencing declines in traditional sectors, as was the case for the major contributors to the negative shifts identified in Table 7.4.

CONCLUSIONS

Producer services appear to be growing rapidly in both metropolitan and nonmetropolitan America, in areas of decline as well as in areas of growth in manufacturing. While particular sectors have had differing trends, the amount of redistribution of employment in these industries has not been dramatic given their overall growth rates. Nonmetropolitan territory has a much lower concentration of producer services than metropolitan territory, and while nonmetropolitan growth has been strong as measured by percentage growth rates, the small base makes absolute growth in non-metropolitan regions pale in comparison to the expansion in metropolitan America. Hence, the producer services have exhibited negligible deconcentration from metropolitan areas over the past decade, but there has been modest redistribution of employment among areas. This has been a pattern of relative change, occurring largely through the relative decline of some of the nation's largest diversified urban centres, and the simultaneous relative growth of another group of large urban regions.

The results presented in this paper tend to support the observations of Noyelle and Stanback and O'hUallacháin regarding the lack of decentralization within the producer services. In the present analyses, the dominance of the share component in the shift-share analysis meant that the unequally distributed producer services had only modest redistribution. While growth rates were found to be rapid in both metropolitan and nonmetropolitan areas, confirming Kirn's observations about fast producer services growth rates in nonmetropolitan areas, in absolute terms the large metropolitan areas captured the majority of new producer service jobs over the study period. The shift component was found to be located primarily within a few of the larger metropolitan areas; this result cannot be directly compared against these other studies as they did not report upon redistributions in the same way.

This paper has briefly presented some aspects of trends in the producer services in the United States over the past decade. It has highlighted differences in sectoral and spatial trends, and touched upon the changing role of various sources of demand for these services in the process of regional development. However, the analyses which have been presented have been very broad-brush, and need to be supplemented with in-depth case studies of particular sectors and regions which would reveal more precisely the mechanisms at work in the development of this rapidly expanding component of the US economy.

ACKNOWLEDGEMENTS

The support of the US Economic Development Administration is gratefully acknowledged for this research; the findings presented here are the conclusions of the author, and should not be considered to be the position of this agency. The author would like to thank his research assistants, Ted Hull and Bill Wildprett, for their help in development of the data and graphics for this chapter.

REFERENCES

Bacharach, M. (1970) *Biproportional Matrices and Input-Output Change,* Cambridge: Cambridge University Press.

Baumol, W.J. (1985) 'Productivity policy and the service sector', in R.P. Inman (ed.) *Managing the Service Economy, Prospects and Problems,* Cambridge: Cambridge University Press.

Beyers, W.B. (1974) 'On geographical properties of growth center linkage systems', *Economic Geography,* 50 (3) July: 203–18.

Beyers, W.B. (1989) *The Producer Services and Economic Development in the United*

States: The Last Decade, Final Report to US Department of Commerce, Economic Development Administration, Technical Assistance and Research Division, January.

Beyers, W.B. and Alvine, M.J. (1985) 'Export services in postindustrial society', *Papers, Regional Science Association,* 57: 33–45.

Collier, D. (1983) 'The service sector revolution: the automation of services', *Long Range Planning,* 16 (6): 10–20.

Daniels, P.W. (1985) *Service Industries. A Geographical Appraisal,* London: Methuen.

Isserman, A. (1980) 'Estimating export activity in a regional economy: a theoretical and empirical analysis of alternative methods', *International Regional Science Review,* 5 (2): 155–84.

Kirn, T.J. (1987) 'Growth and change in the service sector of the US: a spatial perspective', *Annals of the Association of American Geographers,* 77, September: 353–72.

Kravis, I., Heston, A. and Summers, R. (1982) *World Product and Income: International Comparisons of Real Gross Product,* Baltimore: Johns Hopkins University Press.

Leveson, I. (1985) 'Services in the US Economy', in R. Inman (ed.) *Managing the Service Economy,* Cambridge: Cambridge University Press.

Marshall, N., Wood, P., Daniels, P.W., McKinnon, A., Bachtler, J., Damesick, P., Thrift, N., Gillespie, A., Green, A. and Leyshon, A. (1988) *Services and Uneven Development,* Oxford: Oxford University Press.

McCrackin, B. (1985) 'Why are business and professional services growing so rapidly?' *Federal Reserve Bank of Atlanta Economic Review,* August: 14–28.

Noyelle, T.J. and Stanback T.M., Jr. (1983) *The Economic Transformation of American Cities,* Totowa, NJ: Rowman & Allanheld.

Ochel, W. and Wegner, M. (1987) *Service Economies in Europe, Opportunities for Growth,* London: Pinter Publishers.

O'hUallacháin, B. (1987) 'Agglomeration of services in American cities,' unpublished paper presented at Association of American Geographers meetings, Phoenix, AZ, April.

Ott, M. (1987) 'The growing share of services in the US economy – degeneration or evolution?', *Review, Federal Reserve Bank of St. Louis,* June/July: 5–22.

Petit, P. (1986) *Slow Growth and the Service Economy,* London: Frances Pinter Publishers.

Piore, M. and Sabel, C. (1984) *The Second Industrial Divide,* New York: Basic Books.

Reifler, R. (1976) 'Implications of service industry growth for regional development strategies', *Annals of Regional Science,* 10: 88–104.

Rose, S.J. (1986) *The American Profile Poster,* New York: Pantheon Books.

Scott, A.J. (1988) *Metropolis, From the Division of Labor to Urban Form,* Berkeley: University of California Press.

Stevens, B. and Moore, C. (1980) 'A critical review of the literature on shift-share as a forecasting technique', *Journal of Regional Science,* 20 (4): 419–37.

Tiebout, C. (1962) *The Community Economic Base Study,* New York: Committee for Economic Development.

Tschettler, J. (1987) 'Producer services industries: why are they growing so rapidly?' *Monthly Labor Review,* December: 31–40.

US Bureau of Labor Statistics (monthly) *Employment and Earnings.*

US Bureau of Labor Statistics (1988) *Projections 2000.*

Wheat, L.F. (1986) 'The determinants of 1963–77 regional manufacturing growth: why the south and west grow', *Journal of Regional Science,* 26: 635–59.

Zysman, J. and Cohen, S.S. (1987) *Manufacturing Matters,* New York: Basic Books.

8 The service sector and metropolitan development in Canada

David F. Ley and Thomas A. Hutton

INTRODUCTION

In Canada, as in other advanced nations, the service sector has been a major element in the restructuring of the space economy since the mid-1960s, and nowhere has its transforming power been more evident than in the major cities. The downtowns of Toronto, Montreal, Vancouver, Calgary, Ottawa, and Halifax among others, have been virtually rebuilt since 1965. At the head of the urban hierarchy, the core area of the City of Toronto added 30 million square feet of office space between 1965 and 1985. Spillovers from this transformation have impacted the old industrial zone in transition and the ring of inner city neighbourhoods, producing an urban landscape with significant departures from the patterns of the past.

There is of course a legitimate argument linking the service sector to industrial production (and a related and more tendentious thesis that the service economy does not exist). We take this more conventional argument as given in this chapter, and will not pause to trace the relations between services on the one hand and manufacturing and resource extraction on the other. However, we shall comment on the diversity of the service sector, and how a generic title conceals a range of different activities and occupations. This attention to specificity is important not least because of the role of a subset of service industries, those defined as the corporate complex of head office functions and related producer services, in the remaking of the Canadian city.

We shall begin by discussing the place of the service sector in the national economy and international trade, including recent governmental awareness of its role and subsequent public policy directives. Next we contextualize the service sector geographically, emphasizing its unequal impact upon the urban hierarchy, and the disproportionate location of the corporate complex in Toronto, the primate city, and the inter-urban rivalries which are associated with this dominance. Third, we summarize the more familiar geographical literature on the location of the corporate complex within metropolitan areas, a pattern of spatial concentration which has so much to do with the restructuring of the urban landscape.

Finally, we trace some of the municipal responses to rapid, uneven growth of the service sector and the modest attempts at spatial strategies of growth management which have been declared within the largest cities.

SERVICES IN THE CANADIAN ECONOMY

The Canadian economy is conventionally described as a staple economy, and as such its typical geography is one of core and periphery, where central places exercise a trading function as brokers between the raw materials of the periphery and overseas markets. Consequently, service employment has been a persistent feature of the urban system. Indeed a review of the employment structure of seven leading industrial nations using the International Standard Industrial Classification, showed that only Canada had 30 per cent of its workforce engaged in social and personal services plus producer services as early as 1961 (Kellerman 1985), and, more generally, that North America continues to be at the forefront in the development of service economies among the advanced industrial nations. Recent American data bear this claim out (see also Beyers, Chapter 7, this volume). By 1986 one estimate places the service sector as accounting for 75 per cent of all jobs and contributing three trillion dollars or 71 per cent of the gross national product (Quinn *et al.* 1987). Workforce trends in the United States are continuing to favour the service sector.

The senior white collar occupations (administrative, professional and technical workers) all experienced growth of 45–60 per cent between 1973 and 1984, and were projected to continue their leadership in relative growth up to 1995. These trends are matched in Canada (Davis and Hutton 1989). Between 1951 and 1981 the service sector added 6 million jobs while the primary and secondary sectors added only 840,000 new positions. These trends have continued to the present. From 1980–7 the respective increments were +18 per cent for service industries and −0.5 per cent for goods producing industries, while in the most recent 12 month period (June 1987–June 1988) the service increment nationally was 229,000 jobs (2.8 per cent) while the remainder of the economy added 49,000 new positions (1.2 per cent). The lion's share of growth continues to be registered in the (primarily) senior white collar occupations of managerial, professional, technical and administrative positions (plus 144,000 jobs or plus 4.1 per cent, 1987–8), the strategic categories often associated with quaternary or, somewhat more ambiguously, new class jobs. For example, the number of lawyers and notaries in Canada more than doubled, 1971–81, and their proportional share of the labour force grew by 50 per cent (Gill 1988); consulting engineers and architects rose in number by 5.7 per cent a year through the same period, a rate well in excess of growth in the overall labour force (Hammes 1988).

While the rapid growth of services presents a fundamental commonality in Canada and the United States, some important points of contrast can also be identified. During the 1980s, employment in the business and financial services sector (which includes most of the producer services) in the United States expanded at nearly double the rate of such growth in Canada (Table 8.1). At the same time, growth in public and social services in Canada was much greater (by about 50 per cent) than in the United States over the same period. This may be seen broadly to reflect the greater role of government and public agencies in Canadian society *vis-à-vis* the United States; in many ways, the Canadian situation more closely parallels that of western European nations.

As numerous authors have shown, however, it is necessary to look beyond aggregate trends to the quality of jobs being created by the service sector. In the United States, data for 1975 showed that almost 70 per cent of jobs in consumer services and 44 per cent in retail services were placed in the lowest quintile of earnings; other categories including producer services and non-profit services were characterized by a dual labour market of well paid managers and professionals and poorly paid clerical staff (Stanback *et al.* 1981). Ten years later average wages in most service categories continue to lag behind earnings in other economic sectors (Quinn *et al.* 1987).

The transition in both numbers and earnings in the Canadian province of British Columbia is shown in Table 8.2. In this resource-based economy,

Table 8.1 Year-to-year percentage change in employment by broad industry groups, Canada and the United States, annual averages, 1980–7

	80–1	*81–2*	*82–3*	*83–4*	*84–5*	*85–6*	*86–7*	*80–7*
Goods producing sector								
Canada	2.0	−8.7	−1.6	3.1	1.2	1.9	2.0	−0.5
United States	−0.3	−5.6	−0.1	4.9	0.1	0.8	0.6	0.2
Service sector								
Canada	3.2	−0.4	2.2	2.1	3.5	3.2	3.1	18.1
United States	1.9	1.5	2.0	3.8	2.9	2.9	3.5	20.0
Public and social services								
Canada	4.2	1.6	4.1	0.9	2.1	2.0	3.6	20.0
United States	0.6	0.9	2.6	1.5	2.1	2.6	3.2	14.3
Business and financial services								
Canada	1.9	−1.0	−0.7	3.3	4.2	4.3	3.3	16.0
United States	2.7	2.6	1.7	4.7	4.7	4.2	4.8	28.3
Consumer and personal services								
Canada	3.6	−1.7	3.0	2.3	4.2	3.3	2.5	18.3
United States	2.2	1.2	1.7	5.1	2.1	2.0	2.4	17.9
Total								
Canada	2.8	−3.2	1.0	2.5	2.8	2.8	2.8	11.8
United States	1.1	−0.9	1.3	4.1	2.0	2.3	2.6	13.2

Source: Statistics Canada 71–001

Table 8.2 Selected aspects of the British Columbia labour force, by industry

Selected industry groups	Employees 1977 1986 change (thousands) (%)			Mean weekly earnings ($) 1986 increase index 1977–86 (%)		
Forestry	25	23	−8.0	636	107.2	143
Mining	13	19	46.2	758	111.8	171
Manufacturing	170	151	−11.2	569	92.9	128
Construction	80	68	−15.0	508	56.3	114
Transport, communi- cations, utilities	107	117	9.3	587	90.6	132
Trade	194	238	22.7	340	61.9	77
Finance, insurance, real estate	62	83	33.9	465	106.7	105
Services	300	450	50.0	354	59.5	80
All industry groups*	1,051	1,270	20.8	444	75.5	100

Note: *Includes agriculture, fishing, trapping and public administration sectors not shown in table
Sources: Statistics Canada. Planning and Statistics Division, B.C. Ministry of Finance and Corporate Relations

76 per cent of all employment in 1986 was in service industries. These jobs had grown by a third in the previous decade, whereas the goods producing sector had suffered a slight loss of jobs over the same period. By 1986 the largest single employment class was community and personal services. The reliance on the broadly based service sector during the deep recession of the early 1980s is notable. During the 1981–4 period in British Columbia, the advanced services, managerial and professional occupations, added 40,000 new jobs at a time when losses amounting to 90,000 jobs beset other occupational groupings (Daniels 1985; compare Quinn *et al.* 1987). However, growth in service industries is characterized to a considerable extent by the familiar proliferation of low paying jobs; while the goods sector enjoyed wages 30 per cent above the average, in the service sector they were 8 per cent below. American data have shown a considerable internal variation within services, and this trend is repeated in Canada. Notable is the weak earnings performance of community and personal services in Table 8.2, and particularly of food and accommodation. The concerted policy thrust by the provincial government towards tourism and hospitality is favouring a sector where weekly earnings fall well below half the average level.

It has frequently been contended that employment data alone exaggerate the economic performance of the service sector. Many writers have properly pointed to the linkages between manufacturing (and resource extraction) and services. Others have noted the supposedly modest export role of services and their lesser contribution to the national economy (though a pointed response is the Chrysler Corporation Chairman's remark that Chrysler buys more medical services per car than steel, indicating the basic role of services in manufacturing, including manufacturing for export). Productivity gains in services have lagged behind improvements in manufacturing or resource extraction, though the automated office is leading to rapid adjustments, particularly in finance and related industries. Commonly cited trends up to the early 1980s none the less indicate that output per person growth in services was half that of manufacturing and one third that of agriculture (Dobell *et al.* 1984). However while less impressive than its share of employment, the contribution of services to Canada's gross domestic product was 65 per cent and rising by 1983 (Table 8.3). Each of the goods categories lost ground relatively during the 1971–83 period, except for the utilities industry. Some interesting variations occurred among the services. The largest relative gain was registered against producer services, though distributive services (transportation, communications, trade) also showed healthy growth. Personal services were more stable, while non-market services provided by government experienced a relative decline in contribution to Canada's GDP, with the most substantial relative loss against education.

Table 8.3 Share of GDP in Canada by industry (constant 1971 dollars)

Industry	Percentage of total GDP	
	1971	1983
Goods		
Primary	8.0	6.1
Manufacturing	22.9	20.5
Construction	7.0	5.4
Utilities	2.7	3.4
Total goods	40.6	35.4
Market services		
Producer services	13.9	17.9
Distributive services	21.3	23.7
Personal services	5.1	5.4
Total market services	40.3	47.0
Non-market services		
Health, education, welfare	11.7	10.5
Public administration and defence	7.4	7.2
Total non-market services	19.1	17.7
Total services	59.4	64.7

Source: Ludwick and Associates (1987)

Services in Canadian international trade

Recent data also reveal the extent to which the trade in services is a
significant feature of international trade. Indeed Britain and the United
States (as well as many Third World nations pursuing tourist dollars)
increasingly look to services to reduce a serious balance of trade deficit.
Against an American negative trade balance in merchandise (which
reached $146 billion in 1986), a consistent positive balance has been
recorded for services; one careful assessment of trade figures suggested a
services surplus of the order of $15 billion in 1984 (Quinn *et al.* 1987). In
Canada, with its staple-led economy, the relations are dramatically
reversed. In 1984 service transactions (excluding investment income)
accounted for 10 per cent of Canada's international receipts and 14 per
cent of payments, leaving a negative trade gap in services of nearly $4.4
billion, compared with an overall positive trade balance of over $3 billion
(Statistics Canada 1986).

The sources of this deficit are of some interest and reflect the diversity of
the sector. Tourist dollars have shown an increasingly negative balance
(over $2 billion by 1984), but are subject to variation according to currency
fluctuations and special events. The 1988 Winter Olympics in Calgary led
to a substantial improvement in the international travel account for the first
quarter of the year, and was a significant contributor to a drop of $1.5
billion in the seasonally adjusted annual deficit in service trade. Business
services also generated a deficit of over $2 billion in 1984 and highlight the
dependent nature of a staple economy with considerable foreign investment. From sixteen classes of services, the two classes of management and
administrative services, and royalties, patents and trademarks amounted to
77 per cent of the negative 1984 trade balance in business services. These
losses were generated in particular by the Canadian affiliates of foreign
owned (mainly US) corporations purchasing services from their parent
companies. This trend was particularly clear for manufacturing corporations in Ontario, where there is a high degree of foreign (principally
American) ownership. The purchase of services by foreign controlled
subsidiaries from their international parent is substantial enough to
dominate the pattern of international trade in business services.

In contrast, in the category of consulting and professional services, a
maturing of the Canadian economy is revealed in the transition from the
status of net importer up to the mid-1970s to net exporter (Statistics
Canada 1986). By 1984, Canada enjoyed a net surplus of $0.7 billion in
this category, primarily associated with large overseas projects (many in the
developing world). A map of overseas contracts won by Montreal engineering consultants shows broad international coverage, with particular
success in Africa and Latin America; indeed the three leading Montreal
companies are among the ten largest in the world (Slack and Barlow 1981).
In Vancouver a survey of over 600 firms in producer services showed that

some 7 per cent of business by value originated outside Canada (Ley and Hutton 1987) but this figure varied greatly between firms. A third of geological and engineering consultants derived more than 10 per cent of their business from international clients; a similar level of transnational business was registered by 31 per cent of real estate companies and 27 per cent of firms in the securities and commodities markets (see Polèse and Stafford 1984). In contrast, no firms in personnel services or printing were seriously engaged in markets outside Canada. Larger firms were significantly more active than small ones in international sales of services. The most popular export markets for the Vancouver firms were in the United States, followed by Europe and then east and south-east Asia. However, in their future projections, after the United States, firms were much more orientated to the Pacific Rim (principally Asia) than to Europe as growth markets.

Services and government policy

Until comparatively recently, economic policy, regional development strategies, and trade policies and programmes, have been overwhelmingly orientated towards the resource, energy, and manufacturing sectors. Over the last several years, however, a very much heightened interest in service industries, notably in banking, finance, and the producer services, has been evident in the policy thrusts of all three levels of government, impelled both by development trends within Canada, as well as significant policy precedents and initiatives in other jurisdictions. Indeed, in their recent underwriting of massive tourist and leisure events like the Olympics, the Commonwealth Games, and world's fairs, the federal and provincial governments are declaring their endorsement of recreational services as an employment and investment multiplier (Ley and Olds 1988).

Some of these new policy initiatives involve important but relatively narrow adjustments, such as the increased eligibility of service firms for export marketing assistance under the Programme for Export Market Development (PEMD) framework. In other areas, however, the nature of change and innovation has been of a more fundamental nature, attracting in some cases very considerable controversy, with proponents and opponents split both on sectoral and on the familiar regional lines. The most profound area of change and controversy has been associated with the Canada–US Free Trade Agreement, signed by Prime Minister Mulroney and President Reagan and approved by the US Congress and the Canadian Parliament at the end of 1988. This major treaty is intended to facilitate bilateral trade in goods, resources and energy, but also addresses the liberalization of trade in services. Broadly speaking, the Agreement accepts that trade in services 'represents the frontier of international commercial policy in the 1980s' (External Affairs 1988), and sets out a modest number of changes in policies and regulations, while hoping to promote more substantial future innovations in bilateral and multilateral service trade. Immediate changes in bilateral trade as they relate to services include the following:

1. Extension of national treatment to regulatory provisions relating to commercial services (Article 1402), thereby lessening or prohibiting discriminatory treatment of American firms located in Canada, and to Canadian firms situated in the United States.
2. Clarification of obligations by both parties in the treatment of certain services, notably architecture, tourism and telecommunications and computer services.
3. Provision for the negotiation of additional sectoral agreements relating to services.
4. Facilitation of business travel, 'to ensure that business persons and enterprises will have the necessary access to each other's market in order to sell their goods and services and supply after sales service' (Chapter 15).
5. An agreement by both parties to continue liberalizing rules governing their respective financial markets and to extend the benefits of such liberalization to institutions controlled by the other party.

The essentially incremental nature of the provisions relating to services should be emphasized: each government will remain free to choose whether or not to negotiate, there is no obligation to harmonize policies and regulations, and the said provisions are for the most part 'prospective', i.e. not requiring either government to change any existing laws and practices.

However, despite immediate changes in bilateral trade in services of modest and incremental nature, some agencies have forecast substantial growth and employment generation among Canada's service industries as a result of free trade. The Economic Council, for example, has predicted that 182,805 jobs in services will be generated in Canada over the next decade, compared to only 37,454 in construction, 18,935 in manufacturing, and 13,107 in primary industries (*Globe and Mail* 1988). However, there is no unanimity on the scale of benefits flowing from the Agreement, or even on which regions will benefit most.

The Economic Council has identified western Canada and the Atlantic provinces as the principal beneficiaries, while the Federal Finance Department and the Informetrica Ltd. consulting group have indicated that central Canada (Ontario and Quebec) will fare better. This likelihood of differential benefits accruing to the various areas of Canada has led not unnaturally to a revival of regional tensions, with Ontario and Atlantic Canada being considered broadly in opposition to free trade and Quebec and the west largely in favour. Even in apparently supportive regions, however, there are strong concerns about how certain industries will fare: a western Canadian advocacy group has suggested that free trade 'offers considerable downside but little upside to western Canada's service industries' (Canada West Foundation 1986), although their view is challenged by other groups.

This inter-regional contrast in perspective recurs in a second federal government initiative in the area of international banking and finance. While acknowledging global trends towards deregulated financial markets, the federal government has responded not by promoting Toronto, Canada's pre-eminent financial centre, nor by broadening the scope for international banking across Canada generally but, rather, by designating Montreal and Vancouver as putative international banking centres (IBCs). Although the federal IBC legislation is quite limited in scope, the designation of Montreal and Vancouver as IBCs has drawn heated protests from Toronto interests on the grounds that it discriminates unfairly against Canada's pre-eminent banking centre, and may be prejudicial to Toronto's emergence as a major 'second rung' international financial centre. (This may have substance over the longer-term, as the federal and both relevant provincial governments have indicated that the IBC legislation may be broadened.) In response, the federal government has affirmed that the underlying strategy of the IBC legislation is to take advantage of Montreal's special and historical relationship with western Europe and an international francophone community (Polèse 1988a) and Vancouver's burgeoning trade and investment linkage with the Asia Pacific, and thus is not intended specifically to discriminate against Toronto.

Additional evidence of growing senior government appreciation of the role of services can be discerned in federal-provincial economic and regional agreements. Typically, they have tended to emphasize traditional target sectors such as resource development, energy, transportation, and manufacturing. Recently, however, this sectoral focus has included services, including producer services. In the case of the federal-provincial agreement on economic and regional development for British Columbia, for example, signed by the respective ministers in December 1986, the joint framework includes provisions not only for service industries and institutions in support of trade (e.g. financial and legal services), but also for encouraging direct trade in a range of producer services, such as architectural and engineering services (Governments of Canada and British Columbia 1986). Subsequently a management group was established under the rubric of the 'Asia-Pacific Initiative' (API) and including representatives not only of senior government but also the City of Vancouver and the private sector, to guide implementation over a three-year period, and focusing on the following areas: (a) international trade and finance, (b) transportation, (c) export of services, (d) tourism, and (e) cultural and social impact awareness. The new policy direction favouring services as embodied in the Canada–British Columbia agreement and the API may thus be seen as a significant innovation in the evolution of regional development policy in Canada.

In summary, the partial list of federal and provincial initiatives cited above indicate that while public policy usually lags economic trends, a more balanced approach inclusive of services as well as traditional staple and manufacturing industries characterizes contemporary economic development strategy in Canada. Broadly, this approach incorporates not only deregulation of markets and liberalization of trade, trends common to many of the OECD nations, but also substantial government support for service industry growth consistent with the recurrent themes of regional economic development strategy and trade policy.

THE CORPORATE COMPLEX AND THE CANADIAN URBAN SYSTEM

The preceding discussion established the significant role the service industries have come to assume within the economy of Canada, in terms of job generation, output and exports, and overall growth. The significance of service industries in Canadian employment growth also holds true for individual provinces, as shown in Table 8.4. While aggregate growth rates vary among the Canadian provinces, 'services' (including here the finance, insurance and real estate category, as well as the more broadly-defined services group) led growth in almost all cases over the 1975–87 period.

While this national perspective is important, it should be acknowledged that services are by no means dispersed evenly throughout Canada but are instead concentrated largely within urban areas, particularly in the metropolitan regions (Table 8.5). Here the most propulsive sector of the service economy is the interwoven network of head offices and producer services often called the corporate complex of economic activities. Indeed, consistent with trends and patterns observed in other OECD nations, the expansion of services and contemporary urban development in Canada are powerfully linked, and may in many ways be viewed as complementary phenomena (Ley and Hutton 1987). The range of influences associated with the expansion of service activities upon the development or 'metamorphosis' (after Gottman 1982) of urban areas in Canada, includes the following:

1. While staples development and goods production remain very important features of the Canadian economy, urban Canada is now largely dominated by service employment: services account for 70 to 80 per cent of total employment in most metropolitan centres (Table 8.5), proportions common also to the major American and western European urban centres (Polèse 1988b).
2. The expansion of service employment (and the concomitant relative and/or absolute contraction of goods employment) are primary agents in the fundamental 're-structuring' of Canada's cities.
3. Relative rates of growth in the service industries (especially with

Table 8.4 Growth of employment in selected industries for Canadian provinces, 1975–87

	Manufacturing		Transportation		Trade		Services		FIRE[a]		Public admin.	
	1987 ('000s)	change[b] (%)	1987 ('000s)	change (%)	1987 ('000s)	change (%)	1987 ('000s)	change (%)	1987 ('000s)	change (%)	1987 ('000s)	change (%)
Newfoundland	21	31.3	16	−20.0	35	9.4	58	45.0	5	0.0	18	50.0
Prince Edward Is.	4	n/a	n/a	n/a	9	12.5	17	41.6	n/a	n/a	5	n/a
Nova Scotia	45	2.3	27	−7.9	67	19.6	117	42.7	18	38.5	31	29.2
New Brunswick	37	5.7	24	0.0	54	22.7	92	58.6	12	50.0	24	41.2
Quebec	573	−2.2	232	0.4	521	30.3	984	45.9	169	43.2	209	21.5
Ontario	1,038	18.4	316	24.4	806	30.4	1,504	55.5	302	44.5	285	10.0
Manitoba	57	−12.3	47	0.0	89	11.2	161	47.7	29	52.6	40	42.9
Saskatchewan	24	14.3	33	10.0	80	35.6	141	65.9	21	61.5	32	33.3
Alberta	87	22.5	88	27.5	209	42.2	399	93.7	57	58.3	84	47.4
British Columbia	157	2.6	119	13.3	246	28.1	462	61.0	80	48.1	87	29.9
Canada	2,043	9.2	902	11.6	2,116	29.3	3,935	56.0	693	46.6	815	22.4

Note: (a) FIRE includes finance, insurance and real estate
       (b) 1975–87
Source: Statistics Canada

Table 8.5 Percentage of labour force employed in service occupations for selected Canadian census metropolitan areas, 1986

Census metropolitan area	Total employment (000s)	Service employment (000s)	Employed[a] in services (%)
Toronto	1,960.2	1,419.0	72.4
Montreal	1,467.3	1,073.9	73.3
Vancouver	724.5	544.6	75.2
Ottawa-Hull	454.2	372.0	81.9
Edmonton	432.1	314.2	72.7
Calgary	385.7	299.6	77.7
Winnipeg	331.5	240.0	72.4
Quebec	295.2	237.9	80.6
Hamilton	290.8	190.2	65.4
London	185.7	129.8	69.9
Halifax	160.1	125.8	78.6

Note: (a) Service employment includes the following occupations: managerial, administrative and related, natural sciences, engineering and mathematics, social sciences and related fields, occupations in religion, teaching and related occupations, medical and health occupations, artistic, literary, recreational and related occupations, clerical and related occupations, sales occupations, service occupations
Source: Census of Canada (1986)

respect to the complex of corporate activities) among Canada's metropolitan areas now also greatly influence the positioning of cities within the national urban hierarchy, explaining in large part the growing dominance of Toronto, and the emergence of Calgary, Ottawa and Vancouver as increasingly important centres of business and decision making.

4. The growth of services has had a major impact on the evolution of urban structure among many Canadian cities, characterized by the functional specialization of the CBD, in favour of the corporate complex (Gad 1985); processes of gentrification and displacement in the inner city (Ley 1988); and the multi-nucleation of the metropolitan region, marked by the dispersal of residentiary services, and the increasingly important role of commercial activity in the development of regional town centres (Hutton and Davis 1985).

5. While some evidence of divergence in the fortunes of urban centres and their respective hinterland regions can be discerned (Ley and Hutton 1987), the strength of regional linkages including head offices and business services remains a critical feature of well-established core-periphery relationships in Canada.

6. The growth of the corporate complex implicates the larger Canadian cities in the global network of the geography of services, as reflected in the international division of labour, the expansion of trade in services, and the emergence of what Gottman terms a 'neo-Alexandrine'

network of information-based world cities.
7. Finally, the expansion of the service sector among Canadian cities is strongly linked to a host of planning and policy issues, to be discussed later, including such problems as affordable housing and long-range commuting.

Service employment growth and the urban hierarchy

While it is beyond the scope of this paper to explore each of these themes in detail, we shall review some of the more salient features. Historically, analyses of the economic base and growth characteristics of metropolitan areas within the Canadian urban system have centred on resource trans-formation and goods production (see Boisvert 1978). In light of the staple-dominated structure of much of Canada's economy, this orientation is justified, but, increasingly, studies are also directed towards the role of services in urban development in Canada, and the growing significance of services in the urban export base (see Davis and Hutton 1981). Indeed, there is evidence to suggest not only that services are increasingly leading urban growth in Canada, but also that differential rates of growth in certain of the more specialized producer services influence the positioning of cities within the national urban hierarchy.

During the 1980s, overall employment growth among Canada's larger metropolitan centres has been led by service industries, including public and social services, and business and financial services, and by the more advanced, skill-intensive service occupations. More specifically, urban employment growth has been led by rapid expansion in the managerial, administrative and related category, which is closely associated with the development of producer services in urban areas. Table 8.6 demonstrates that for Canada's largest metropolitan areas, growth rates for this category far exceed overall employment growth rates, often by ratios of 2:1 or even 3:1. Close correlations between the expansion of the higher echelons of service employment and overall metropolitan growth can also be discerned. During the period 1981–6, for example, Ottawa-Hull led Canadian CMAs both in employment growth in the managerial, administrative and related category (52 per cent – see Table 8.6) and population growth (just over 10 per cent). Other cities which have experienced both rapid growth of managerial and administrative employment as well as high levels of population growth include Toronto, Vancouver and Calgary. As in the case of 'post-industrial' centres within other OECD nations, urban growth is seen as closely correlated with the expansion of producer service industries and the higher echelons of service employment.

These patterns can be amplified by reference to individual cities. In Vancouver CMA, for example, the 1970s were marked by rapid expansion in the more advanced service occupations, somewhat lesser but substantial

Table 8.6 Employment growth in managerial, administrative and related occupations for selected Canadian CMAs, 1981–6

	Employment in all occupations			Employment in managerial, admin. and related occupations		
	1981	*1986*	*change 1981–6 (%)*	*1981*	*1986*	*change 1981–6 (%)*
	(000s)			*(000s)*		
Toronto	1,668	1,960	17.5	191	263	37.7
Montreal	1,398	1,467	4.9	146	191	24.0
Vancouver	431	647	12.0	66	85	29.0
Ottawa-Hull	380	454	19.5	47	72	52.3
Edmonton	372	432	16.2	38	47	23.7
Calgary	347	386	11.1	41	48	19.0
Winnipeg	307	332	8.0	29	35	22.7
Quebec	273	295	8.3	29	40	36.4
Hamilton	275	290	5.6	23	29	26.7
London	151	186	23.1	13	20	47.6
Halifax	142	160	12.7	14	17	27.8

Source: Census of Canada (1981, 1986)

growth in clerical and sales employment, and moderate growth in direct production jobs. During the period 1981–6, however, although growth rates in higher-order service occupations continued at a high level (again, notably in the managerial and administrative category), the expansion of clerical employment virtually ceased and absolute reductions in many of the blue collar occupations were recorded (Table 8.7). These must be viewed against both the effects of the severe recession (which disproportionately impacted resource and manufacturing sectors and occupations), and of capital substitution processes which could account for some job-shedding among both the blue collar and clerical occupations. Also, the managerial and administrative category includes small firms with a high turnover rate. Even so, the very substantial and clearly growing role of producer services and higher-order service occupations in the growth of Vancouver and other Canadian cities must be acknowledged. This is supported by current data: employment in the 'services to business' industry group (incorporating many of the producer services) continues to lead employment growth in Vancouver CMA, accounting for over 8,000 new jobs in 1987.

The primacy of Toronto and inter-metropolitan rivalry

While growth in service employment, the producer services, and the advanced service occupations has been general among Canadian cities, special attention must be paid to the quite extraordinary development of

Table 8.7 Employment change in Vancouver CMA by occupation, 1971, 1981, 1986

Occupational category	Employment			Change (%)	
	1971	1981	1986	71–81	81–6
Managerial, administrative and related	21,030	52,805	85,020	151.1	61.0
Natural sciences, engineering, math.	14,215	25,190	26,220	77.2	4.0
Social sciences and related fields	5,440	12,720	17,530	133.8	37.8
Occupations in religion	795	1,040	1,455	30.8	39.9
Teaching and related occupations	16,135	25,260	25,915	56.6	2.6
Medical and health	19,495	32,920	37,960	68.9	15.3
Artistic, literary, recreational	5,315	10,370	14,770	95.1	42.4
Clerical and related occupations	88,700	143,290	146,780	61.5	2.4
Sales occupations	55,855	75,330	81,390	34.9	8.0
Service occupations	58,715	85,355	108,120	45.4	27.1
Farming, horticultural	7,760	9,600	12,295	23.7	28.1
Fishing, hunting, trapping	1,795	1,295	2,155	−27.9	66.4
Forestry and logging occupations	2,390	2,270	2,320	−5.0	2.2
Mining and quarry (incl. oil and gas)	1,130	935	790	−17.3	−16.5
Processing occupations	18,365	21,410	19,650	16.5	−16.6
Machining and related occupations	11,794	14,090	11,640	19.5	−17.4
Product fabricating, assembling	28,935	40,785	40,690	41.0	−0.3
Construction trades	32,140	41,705	40,980	30.0	−1.7
Transport equipment operating	19,170	25,285	27,510	31.9	8.8
Material handling and related, n.e.c.	15,865	17,340	15,410	9.3	−11.5
Other crafts and equipment operating	5,730	7,730	5,925	28.0	−19.2
	430,769	646,725	724,525	50.1	12.0

Source: Census of Canada (1971, 1981, 1986)

Toronto, and to its unchallenged and apparently increasing dominance of Canada's industrial, commercial and financial space economy. Table 8.6, for example, shows that the overall employment growth rate for Toronto CMA during the period 1981–6 was half again that for Vancouver CMA, and approximately three-and-a-half times that for Montreal CMA. More specific data relating to the corporate complex serve to further clarify the extent and depth of Toronto's dominance within Canada's urban system. By 1987, the City of Toronto was the head office site for 40 per cent of Canada's top 100 industrial institutions, for 45 per cent of the top 100 financial institutions, and for half the forty major insurance companies. Its control of business services was even greater, including the leading firms in investment dealing (73 per cent with Toronto head offices), accountancy (64 per cent) and management consultancy (90 per cent) (Financial Post 1988). Unlike the United States, where a more dispersed pattern of head office location is evident, the concentration of corporate control in Canada favouring Toronto is apparently increasing. As a result, apart from a few areas (such as consulting engineering and certain resource industries), Montreal and Vancouver have only secondary roles as national business centres, and are now largely focusing instead on regional and international development niches where they possess some element of comparative advantage.

Other measures of Toronto's dominance within Canada's urban hierarchy may be cited. Between 1970 and 1985 office space in Toronto's core increased 250 per cent, compared with 137 per cent in Vancouver and 68 per cent in Montreal; at the same time Toronto has experienced far more suburban office development than the other two centres. Growth among managerial and administrative occupations in Toronto CMA exceeded that in metropolitan Vancouver and Montreal both in absolute and relative terms (Table 8.6). By 1986, Toronto's workforce among the managerial and administrative groups almost equalled that for Montreal and Vancouver combined. Since 1986, Toronto's economic boom, and expansion as a corporate and producer services centre, has if anything intensified. A recent report noted that in 1987, some 52,800 office jobs were created in metropolitan Toronto, net office absorption in the downtown totalled 3.4 million square feet (annual averages of 0.6–0.75 million square feet for Vancouver), and office space in the 'pipeline' (projected for completion between 1988 and 1992) totalled almost eight million square feet (Colliers 1988). From 1985 to 1987, Toronto's office sector recorded some 7.6 million square feet of net absorption. The total for metropolitan Toronto now exceeds 110 million square feet, almost twice as much as for Montreal, and approximately three times as much as for each of the next largest Canadian office centres, Ottawa, Vancouver and Calgary (Royal Lepage 1988).

In part due to its pre-eminence within Canada's financial sector and corporate structure, Toronto is also emerging as an increasingly important centre of international banking and finance: 'Financiers in Toronto brag that their city, now the country's biggest, has a secure foothold alongside Frankfurt and Zurich on the second rung of the world's money markets ...' (*The Economist* 1988b). Toronto's stock exchange is now the seventh largest in the world, and ranks fourth in terms of capitalization (at over Cdn$250 billion), significantly ahead of Paris, Frankfurt, Zurich and Milan (*The Economist* 1988a). Of the fifty-eight international banking subsidiaries (termed 'Class "B"' banking institutions) in Canada, no fewer than forty-five have their Canadian head offices in Toronto, compared with eight for Montreal, and five for Vancouver (*Canadian Banker* 1988). If anything, trends in the international banking and financial sector are likely increasingly to favour Toronto over rival Canadian cities: a combination of such factors as financial deregulation and rapid improvements in business telecommunications means among other things that international banking and other business concerns can more easily direct their Canadian operations from Toronto, rather than having to establish branch offices elsewhere.

Other trends relating to corporate restructuring and international business may tend to further extend Toronto's control over commercial activity in Canada. In part the city has benefited from the corporate exodus from Montreal in the 1970s, an exodus driven by political and cultural as well as economic pressures. The movement of the Sun Life Assurance Company from Montreal to Toronto in 1977 removed the corporate assets of the largest insurance company in the country, valued at $8 billion in 1980 (Semple and Smith 1981; Semple and Green 1983). In part, too, a spate of corporate mergers and acquisitions has often involved large conglomerates based in Toronto acquiring major commercial and industrial concerns formerly headquartered in other Canadian cities.

During the 1970s, corporate relocations (whether through mergers and acquisitions or migration) led to a massive centralization of control in the urban system. The net gains and losses, measured in 1980 corporate revenue, from relocation contributed to major benefits for Toronto, major losses for Montreal and modest absolute changes for other cities with metropolitan populations of over a quarter of a million in 1981 (Table 8.8). This process has continued, indeed may well have accelerated, since 1980, though Montreal's losses do seem to have been checked. Recently, too, there has been some breaking down of provincial barriers in the form of regulations governing certain of the professions, which in Canada have traditionally been administered within provincial jurisdictions (McCarthy & McCarthy, one of Toronto's largest legal firms, has just merged with Shrum, Liddle & Hebenton of Vancouver, the first such merger in Canada). Some observers fear that this could lead in time to a weakening of the base of independent, locally-based producer services resident in each of Canada's metropolitan centres, as such mergers could in many cases result in the development of a head office-branch plant relationship

Table 8.8 Corporate headquarters relocations among major Canadian cities, by 1980 revenues ($ millions)

City	Resource, manufacturing utilities sector	Services, finance, real estate sectors	Net total
Toronto	11,467	19,974	31,441
Montreal	−5,209	−15,309	−20,518
Vancouver	572	1,177	1,749
Ottawa	112	−797	−685
Edmonton	−139	410	271
Calgary	390	924	1,314
Winnipeg	234	−2,293	−2,059
Quebec City	−497	1,845	1,348
Hamilton	294	−122	−172
St. Catherines-Niagara		−165	−165
Kitchener		−200	−200
London	16	52	78
Halifax		2,123	2,123

Source: Adapted from Semple and Green (1983)

between the merging firms, with the (likely larger) Toronto firm assuming overall control.

The recent rapid growth of Toronto's corporate complex has excited considerable competition and rivalry among the three major cities. By 1987, the Toronto Stock Exchange enjoyed a share volume four times greater than the Montreal Exchange, and Montreal traders complain that 'Toronto traders are too busy eyeing New York to consider what's happening in Montreal' (Robinson 1987). At the same time the Toronto exchange is actively raiding the Vancouver exchange which, although much smaller, has historically been North America's premier venture-capital market. Some 100 listings have moved from the VSE to the TSE since 1984, and, in the perception of Vancouver brokers, 'we are losing the $10 stocks and replacing them with 35 centers ... we run the risk of being the exchange of listings nobody wants' (*Macleans* 1988).

Toronto is fiercely protective of its financial primacy, and launched a lawsuit against the federal government to overturn the legislation desig-nating Vancouver and Montreal as international banking centres, to which an irate finance minister in British Columbia expressed his disappointment that 'central Canadians should be so paranoid about our relatively minu-scule attempts to preserve our own financial community' (*Vancouver Sun* 1988). In these everyday relations the pecking order on the urban hierarchy is won and lost. Primacy in the service sector also has a cultural face. Responding to jibes from Montreal and Vancouver, Toronto is now following their lead in building a modern in-town sports stadium. So too, following the 1976 Montreal Olympics and the 1986 Vancouver World's Fair, Toronto is an earnest bidder for the 1996 Olympics. Competition extends also to international film festivals and various other cultural events. To be legitimate, primacy must be expressed across the gamut of the service sectors, in consumption as well as production.

While it is apparent from the above discussion that Toronto has emerged as the undisputed centre of international finance and commerce in Canada, other Canadian cities have within their respective corporate complexes producer services with significant international dimensions. For example, there is a substantial and diversified presence of Asian business concerns in Vancouver (notably from Japan, China, Taiwan, Korea, Singapore, and Malaysia), and ties are especially strong between Vancouver and Hong Kong: the head office of the Hongkong Bank of Canada is located in Vancouver, as is the North American head office of Cathay Pacific airlines. The Vancouver chapter of the Hongkong–Canada Business Association has over 600 members, making it by far the biggest bilateral trade association in Vancouver, and one of the largest in Canada. Vancouver is second only to Toronto in attracting new immigrants from Hong Kong, swelling the over 100,000-strong ethnic Chinese community in Vancouver. Finally, Hong Kong interests own considerable commercial and residential property in Vancouver (a trend observed in other Pacific Rim business centres such as Sydney and Los Angeles), including the Concord Pacific site, the 200-acre setting of Expo '86 located on the periphery of the downtown, and to be one of the largest urban redevelopment projects in North America undertaken over the next twelve to fifteen years (Keast 1988).

In addition to the international and national dimensions cited above, inter-urban and intra-regional service flows represent a vitally important element of producer services within the Canadian urban system. This is especially so in areas of Canada outside Toronto, in light of the enormous expanse of hinterland regions, the general lack of intermediate-size cities, and the resulting near-total dominance of primate cities within their respective regions. Thus, while some element of competition among cities within Canada's urban system can be discerned with respect to the marketing of producer services, we find by and large that the larger metropolitan centres play specialized service roles within their respective hinterlands, rather than directly competing in these markets. Thus, Montreal functions as a specialized business service centre for much of Quebec (see Polèse 1982; Coffey and Polèse 1983), Edmonton provides producer services for much of the Canadian northlands, Vancouver imposes almost total control on head office and producer service flows within British Columbia (Ley and Hutton 1987), Halifax is the central place for Atlantic Canada, and so on. In some cases too these specialized regional service roles also have a distinct sectoral bias, reflecting the economic structure of the various hinterland regions in Canada. Thus,

Calgary, for instance (which has a commercial office complex on the same scale as that of Vancouver's, despite a population little more than one-half of metropolitan Vancouver's) specializes in producer and financial services for Alberta's petroleum industry (Barr and Szplett 1984).

While most analysis has been directed towards the larger metropolitan centres in Canada, there is evidence that producer service flows may be assuming importance even for smaller, metropolitan centres situated within staple-dominated regions. A recent study has indicated that Saskatchewan, whose largest city has a population of under 200,000, has experienced rates of growth in service exports comparable to that for Alberta and British Columbia, which have much larger urban centres, suggesting that 'not only are service exports unrelated to size of firm ... but they may also be unrelated to the size of the metropolitan area in which the firms are situated' (Stabler and Howe 1988). While this specific and potentially profound hypothesis has yet to be verified, the general observation of the Saskatchewan experience serves to reinforce the universality of the importance of producer services within the Canadian urban system as a whole.

THE CORPORATE COMPLEX WITHIN THE CANADIAN CITY

Analysis by location quotient shows that the highest concentration of employment in producer services and finance, insurance and real estate occurs in the largest Canadian cities. Significantly, these were the categories with the strongest growth rate in employment across the family of service industries between 1971 and 1981 (Coffey 1988; Coffey and Polèse 1988). But the corporate complex remains heavily concentrated not only in major cities but also within their core areas. A prosaic illustration from business services are the 1987 addresses of members of the Architectural Institute of British Columbia. With some 15 per cent of the provincial population, the City of Vancouver accounts for 57 per cent of members' addresses, while 28 per cent are listed in five postal districts in the downtown peninsula. A broader review of the location of business services in the Vancouver region indicated the high level of downtown concentration and its continued dominance over time (Table 8.9). The explosive growth of some groups has been remarkable, including an increment of 1,200 legal firms in the urban core and high rates of relative growth for most categories. Over the entire period only some 10–20 per cent of firms have been located in the suburban municipalities, with some 60–80 per cent in the CBD and its margins. Indeed distinct patterns exist within the urban core for particular services; there are over 300 legal firms clustered in a single downtown block.

This spatial bias holds across most Canadian cities, which as a group display higher levels of office centralization than American cities (Hutton and Ley 1987; Goldberg and Mercer 1986). Thus while some 43 per cent

Table 8.9 The growth of producer service firms in metropolitan Vancouver, 1960–83

	Number of firms		Percentage in CBD	
	1960	*1983*	*1960*	*1983*
Advertising agencies	26	103	72	68
Architects	60	215	53	44
Business consultants	14	89	71	62
Data processing	7[a]	124	71[a]	50
Lawyers	696	1,935	89	82
Management consultants	20	213	75	59
Market research	5	43	60	53

Note: (a) in 1966 – no entry in 1960
Source: Ley and Hutton (1987)

of office space in large American metropolitan areas was in the central city in 1980 (Leitner 1988), the figure in all large Canadian cities was very much higher: 68 per cent in Toronto's 'central district', comprising a large part of the City of Toronto (Gad 1985), 79 per cent for the central City of Vancouver (Ley and Hutton 1984) , and 68 per cent for Montreal's downtown and Old Town (Polèse 1988a). The dynamics of office location have been examined most thoroughly in Toronto, at the top of the Canadian urban hierarchy, particularly for advanced services; in the 1983–7 period a quarter of all new jobs in finance, insurance and real estate in Canada were generated in the Toronto census metropolitan area (City of Toronto 1987b). For the first half of the decade office-related employment was accounting for some 64 per cent of new jobs in the central city (City of Toronto 1987a). At the same time, while central city office space continues to grow apace, both relative and absolute growth is greater in the suburbs; from 1976 to 1984 average annual absorption of 125,000 square metres of office space in the central core compared with over 200,000 square metres in the suburbs. Gad (1985, 1986) has examined the geography of the office sector in some detail and detected important city–suburban differences, by industry, by nationality and by size. Very few of the principal offices of financial institutions have left the city; among business services the pattern is more varied, ranging from 74 per cent of advertising agencies to only 30 per cent of engineering consultants with a location in the central district. Nationality is a significant factor with head offices of foreign companies far more likely to decentralize than Canadian head offices. In terms of size, the larger business services retain a central location with smaller suburban offices catering primarily to a local market; thus in firms with more than ten professionals, only three out of sixty-five legal firms are in the suburbs, only four out of fifty-four advertising agencies and none of the top seventeen accounting offices (Gad 1985).

Back office relocation, or the dispersion of certain, primarily record-keeping functions, out of the downtown head office has frequently been referred to in the American literature. Associated with it is a sense of the importance of a pool of suburban secretarial labour which can be employed in a data processing capacity. While the fragmentation and partial decentralization of head office activities has been noted in Canada, notably in Toronto, the labour and gender constitution of back offices seems to be far more varied than the American profile has suggested. Both census tabulations and case studies in metropolitan Toronto indicate a contrary pattern: an increase in the female share of central area office employment, and a sustained plurality of male office staff in the suburbs (Huang and Gad 1988). These unexpected results are attributed to growing female participation in higher status jobs downtown, and the significant male employment associated with suburban data centres.

Conventional explanations for the concentrated location of the corporate complex have emphasized the significance of linkages, particularly the imparting of specialized information in face-to-face transactions and the same argument has been demonstrated in Canada (Gad 1979; Code 1983). Surveys of both head offices and producer services in Vancouver confirm the significance of centrality and specialized transactions in the location of the corporate complex (Hutton and Ley 1987). Some 70 per cent of both groups identified access to business contacts, notably clients, as the principal advantage of their location. Among head offices a range of secondary considerations included the quality and availability of space, prestige and core area amenities. Among the producer services, the only other advantage of downtown mentioned by a significant minority was the prestige of a central address. Linkages and business contacts monopolized the locational decision making of producer service firms in the sample; 80 per cent accounted for their location in terms of proximity to clients or business services, and 15 per cent identified a more general notion of centrality. Location and proximity are dominant concerns in an economic sector characterized by tight linkages and vertical disintegration.

For the corporate complex, separation from the interaction peak of centre city can be costly (Code 1983). In Vancouver downtown head offices were significantly more likely to report integration into a contact-rich business environment than their surburban counterparts. A detailed study of a large head office which had relocated 10 kilometres to an inner suburb revealed considerable disapproval by managers in departments with extensive external linkages (such as marketing and finance) even five years after relocation had occurred (Ley 1985a). Some managers alluded to the need to make frequent trips downtown, and the lack of local prestige hotels or related services was mentioned by others as a problem aggravated when entertaining business guests. One middle manager noted that he missed downtown business contacts, while senior management missed the business clubs. 'Downtown business trips are made,' he added, 'but they are time-consuming and it is easier to choose not to make them at all.' This ominous remark offers support to Code's (1979) fears that decentralization may lead to a decrease of firm productivity due to extended travel time and that potentially favourable business meetings may be forgone because of the cost of separation from face-to-face contact. As a Greater Vancouver planning survey noted of large head offices which had decentralized from downtown:

> They frequently had to travel downtown to meet with their lawyers and bankers, to use the law courts or to meet with major clients. In some cases this caused considerable inconvenience and was cited as a significant disadvantage of locating outside the core.
>
> (GVRD 1982: 13)

These misgivings raise the question of why relocation from the centre city should occur at all. In Vancouver, the principal locational irritants for the downtown corporate complex were cited by senior management to be high rental costs and various forms of traffic problems, though with rental costs accounting for only 11 per cent of head office costs, considerable locational inertia was anticipated, particularly in the sluggish commercial real estate market of the mid-1980s. In terms of the relation between urban size and congestion costs, it is salient to note the suggestion of a size threshold below which office decentralization is unlikely to occur; no Australian cities, it was concluded, reached this threshold (Daniels 1986). In this context, an interpretation might be offered for the rapid suburban office growth which occurred in Toronto in the late 1970s as its population nudged 3 million and which coincided with both a decision to refrain from major new transportation initiatives and a rapid escalation of downtown office rents, partly in response to growth management policies. Public policy at that time irritated worries about both rental and congestion costs; perceived constraints on the supply of downtown office space at a time of unusual demand, and the prospect of worsening congestion costs in the future, helped create a more vibrant demand for suburban office space.

At the same time a widely recognized specialization of office functions is occurring over space. In recent years two-thirds of downtown office development in Vancouver has consisted of high prestige (AAA) buildings, which are not being constructed elsewhere in the metropolitan area. So too an increase in occupied space per office worker has been noted. The interpretation of these changes is a growing specialization of the Canadian downtown in high status managerial and professional activities. Between 1975 and 1985 floor space per worker ratios in Toronto's central core grew almost 15 per cent; associated with this was a doubling of managerial and professional employees in the central area between 1971 and 1981, while all office workers increased by only 40 per cent (City of Toronto 1986a).

There is evidence here of social as well as functional polarization. The thesis of the dual labour market in the service sector has posited the juxtaposition of well paid, primarily male, managerial and professional workers in the upper echelons of the corporate complex compared with modestly paid, primarily female, clerical and service workers. Vancouver data show that the gender bias of this dichotomy is being blurred at the margins (Hutton and Ley 1987) while evidence from Toronto indicates more rapid movement of women into advanced services (Huang and Gad 1988). But at the same time, geographical polarization is increasing. The high status jobs of the corporate complex are prompting the embourgeoisement not only of downtown retailing and leisure outlets but also of inner city neighbourhoods. The restructuring of the downtown labour market and the gentrification of the inner city housing market are two sides of the same coin (Ley 1988). The fifteen year downtown office boom has coincided with a period of acute problems of housing availability and affordability in the inner city. The growing spatial disequilibrium of work place and home place driven by the office boom has attracted the attention of civic politicians and policy makers.

METROPOLITAN POLICY INITIATIVES

In Canada the response to the challenges of office location has devolved, by default, to urban municipalities. The major exception to this rule is the special case of Ottawa where the federal government under Prime Minister Trudeau engaged in an extensive policy to disperse government offices, particularly to the Quebec side of the Ottawa River. Several provinces, including Alberta, have discussed deconcentration of office staff away from the provincial capitals, but achievements to date have been meagre.

It is in Toronto and Vancouver that dispersal policies have been attempted, notably as a result of community pressures in the mid-1970s. Toronto's burgeoning downtown office growth was already, by the early 1970s, exerting development pressures upon inner city neighbourhoods and the City's reform council responded in 1976 with its Central Area Plan which introduced a containment policy on downtown office development within the constraints of existing transportation capacity. In an attempt to trade off the regional economic function of downtown against local social and environmental impacts, local objectives set the growth limits of permitted development (City of Toronto 1986b). Office deconcentration was a significant component of the Plan, intended to lead to a multi-centred metropolitan structure served by transit. Simultaneously, a strikingly similar plan was conceived by the Greater Vancouver Regional District. *The Livable Region 1976–1986* (GVRD 1975) also reflected social objectives and resulted from extensive community participation. The objective of the plan was to minimize the journey to work, and improve the regional distribution of services and amenities. The solution was to be a deconcentration of office space and other uses into regional town centres.

We noted earlier that since the mid-1970s, considerable growth of suburban office space – 55 per cent of 1976–85 metropolitan development – has indeed occurred in Toronto. However, the identification and promotion of suburban city centres has taken place somewhat belatedly, so that by the end of 1985 offices had dispersed to a range of suburban locations, none of them containing as much as 5 per cent of regional office space (City of Toronto 1986c; Matthew 1986). Major growth has occurred in suburban office parks not served by major transit lines, though since 1980 planned sub-centres in close proximity to rapid transit have gained momentum. In metropolitan Vancouver, less than half the size of metropolitan Toronto, office decentralization has been slower. While the downtown share of regional office space fell from 63 per cent to 56 per cent, 1974–82, only 20 per cent was located outside the city (Ley and Hutton 1984). Aside from Burnaby-Metrotown, on the region's rapid transit line, the pattern of suburban office growth has disregarded the distribution of regional town centres designated in the Livable Region Plan (Hutton and Davis 1985).

The welfare of employees in terms of the journey to work, while a priority of municipal planners, is a low order objective for firms themselves (GVRD 1982). None the less there is evidence that decentralization can indeed reduce the length of the journey to work and provide access to cheaper, suburban housing, particularly for clerical staff. Clerical workers at the head office of a utility corporation in suburban Burnaby reported a saving of 25 minutes in their journey to work over a comparable group in a downtown head office (Ley 1985b). For managers the saving, though more modest (16 minutes) remained statistically significant. In addition, in the rapidly inflating housing markets of Vancouver and Toronto since 1971, suburban relocation can project employees towards more affordable housing opportunities.

Yet this evidence of improved welfare for office workers is in many respects a pyrrhic victory for municipal planners. First, other than relatively inflexible zoning powers and moral suasion municipal governments have limited powers to redirect either private or public sector development, particularly in the face of the sometimes competing agendas of separate jurisdictions. Second, decentralization policies have not succeeded in deflecting the economic pressure on central city land. The complex interplay of commercial, residential and environmental considerations have led to some unanticipated consequences. The warning that deconcentration policies in Toronto would contribute to spiralling downtown rents has been proven true, by association if not by causality (Code 1979). The perception of future shortages, in a period of keen demand, may well have fuelled the rapid rise of commercial rentals in downtown Toronto in the late 1970s, and as a result, the trend towards 'the executive city', the concentration of high status white collar jobs downtown. Perhaps more decisively in the inner city housing market the economic (and thus social) costs of central city conservation policies have been self-evident. The association between downtown office development and inner city gentrification has been demonstrated in Canada as elsewhere (Ley 1986), and in such service dominated cities as Vancouver, Toronto, Ottawa and Halifax, the conservation politics of inner city professional households led to residential down zonings in many neighbourhoods. The environmental gains made in these districts have led to predictable upmarket pressures, with the renovation of rooming houses to one- and two-family dwellings, or redevelopment of affordable rentals to luxury condominiums. In the City of Toronto, for example, housing deconversion led to the loss of over 18,000 units between 1976 and 1985 (Howell 1986).

Inflating residential markets have their own pernicious consequences for the cost of public services. The increasing price of residential land both removes affordable housing and raises the public cost of providing it. At the same time the forced separation between employees seeking cheaper suburban housing and their downtown workplace adds to the pressure for expensive new transportation corridors. In major Canadian cities the spatial disequilibrium between jobs, housing and transportation is more acute in the late 1980s than at any other time. Simultaneously, government action has become far more piecemeal, leaving final solutions to the market place. Discernible, however, is a more permissive environment for downtown office development, and attempts at the incremental densification of inner city neighbourhoods. Ironically, it may be argued, under present allocation mechanisms in urban Canada, the relaxation of environmental standards promises a more democratic urban structure.

The service sector and economic development planning

In addition to the range of planning approaches outlined above, most of Canada's major urban centres now explicitly address service industries within the framework of economic policy, reflecting the increasingly dominant role of services within urban areas. Typically, municipal and regional economic development approaches for the service sector involve marketing efforts aimed at attracting head offices, banking and finance, and other producer services; information services; promotional task forces or 'partnerships' with business organizations to encourage business services to establish; and lobbying senior government to enact or amend pertinent regulations, and to upgrade critical infrastructural elements such as international airports.

All three of the largest metropolitan centres are currently pursuing economic policies for the service sector although, interestingly, Toronto is presently assigning a higher priority to manufacturing, despite a recent study warning of complacency over the future prospects of Toronto's office employment base (City of Toronto 1988). After the disappointments of the 1970s, Montreal, on the other hand, has undertaken a vigorous and proactive strategy with respect to finance and the producer services generally, emphasizing in particular international banking and finance, and export-orientated producer services such as engineering. Montreal, too (like Vancouver), has an active collaboration with local universities in the area of research on the producer services, as an input to economic policy development. Lobbying pressure from the City of Montreal and its business community was in large part responsible for Montreal (with Vancouver) being designated as an international banking centre by the federal government.

In Vancouver, services are at the forefront of economic policy both at the municipal and regional levels. The City's initial economic strategy, approved by Council in 1983, emphasized the role of banking and finance and producer services, especially export services. The current (draft) strategy for the City of Vancouver builds upon this effort, and includes as well measures for the 'design services', such as architecture, interior design, graphic and commercial arts, and fashion design. Policies and programmes in the area of the design services include, for example, annual exhibitions, juried competitions, industry development, education and training, and export marketing (City of Vancouver 1988). The new regional economic strategy, developed for the metropolitan area as a whole, also highlights the importance of export-orientated services, implicitly acknowledging that the expansion of producer services has now to some extent at least spread beyond the corporate complex of the CBD, to include some of the metropolitan sub-centres among the suburbs (GVRD 1988).

CONCLUSION

In this essay we have demonstrated the substantial role played by the different sectors of the service economy in Canada. The most privileged segment of the service economy, the corporate complex of head offices and producer services is highly concentrated, not simply in cities, but in the largest cities, and within these a further concentration occurs in the downtown area, a degree of clustering ironic for the sector of the economy most impregnated by the information revolution, a revolution which in some respects has the capacity to annihilate space and the friction of distance altogether.

As usual there has been a lag between economic realities and government response, and only within the past five years have the employment and, increasingly, the export potential of services drawn them into the discourse of public policy concerning trade and regional development. But for Canada, at least, services offer no easy panacea to economic planning. As a branch plant economy there is a significant overall trade deficit in services; while, spatially, the service sectors with the best jobs are highly segregated in areas already enjoying a relatively buoyant economy. Within metropolitan areas, the uneven distribution of the rapidly expanding corporate complex in the downtown core contributes to the major metropolitan problems of the current decade: rapidly inflating land prices, problems of housing affordability, the challenge (and the political will) to balance rapid growth against environmental quality, and the increasing costs of separation from work for commuters. The fact that cities like Toronto and Vancouver (in particular) as world cities (if of the second and third rank) are popular investment havens for global capital in the ever-freer private market magnifies these tendencies significantly. The need not simply to understand but more fundamentally to manage the rapid changes to the space economy of the service sector is a pressing priority.

ACKNOWLEDGEMENTS

We gratefully acknowledge the helpful comments of Trevor Barnes, Roslyn Kunin, Mario Polèse and Keith Semple to an earlier version of this paper.

REFERENCES

Barr, B. and Szplett, E. (1984) 'Resource-based energy industries: related head office and corporate linkages in Canada's energy capital', in B. Barr and N. Waters (eds) *Regional Diversification and Structural Change*, Vancouver: Tantalus.

Boisvert, M. (1978) *The Correspondence between Urban System and the Economic Base of Canada's Regions*, Ottawa: Economic Council of Canada.

British Columbia Ministry of Economic Development (1987) *British Columbia: Earnings and Employment Trends*, Victoria BC: Central Statistics Bureau, Government of British Columbia.

Canada West Foundation (1986) *Putting the Cards on the Table: Free Trade and Western Canadian Industries*, Calgary, Alberta.

Canadian Banker (1988) 'The chartered banks of Canada and foreign banking subsidiaries', mimeo: Toronto.

City of Toronto (1986a) *Trends in the Utilization of Office Space*, Quinquennial Review Background Paper No. 5, City of Toronto Planning and Development Department.

City of Toronto (1986b) *Overview Report*, Quinquennial Review, City of Toronto Planning and Development Department.

City of Toronto (1986c) *Changes in the Distribution and Demand for Office Space in the Toronto Region*, Quinquennial Review Background Paper No. 1, City of Toronto Planning and Development Department.

City of Toronto (1987a) *Trends in Employment*, Quinquennial Review Background Paper No. 6, City of Toronto Planning and Development Department.

City of Toronto (1987b) *Toronto Economic Trends*, City of Toronto Planning and Development Department.

City of Toronto (1988) *Downtown Employment Study*, Final Report of the Downtown Employment Study Steering Committee.

City of Vancouver (1988) *A Strategy for Vancouver's Economic Development in the 1990s*, Vancouver Economic Advisory Commission.

Code, W. (1979) *The Planned Decentralization of Offices in Toronto: A Dissenting View*, Geographical Papers No. 42, University of Western Ontario.

Code, W. (1983) 'The strength of the centre: downtown offices, and metropolitan decentralization policy in Toronto', *Environment and Planning A* 15: 1361–80.

Coffey, W. (1988) 'Service industries and regional development: patterns, theory and policy', in J. McRae and M. Desbois (eds) *Traded and Non-Traded Services,* Halifax: Institute for Research on Public Policy.

Coffey, W. and Polèse, M. (1983) 'Towards a theory of the inter-urban location of head office functions', paper presented to the European Congress of the Regional Science Association, Poitiers, France.

Coffey, W. and Polèse, M. (1988) 'Locational shifts in Canadian employment, 1971–1981', *Canadian Geographer* 32: 248–56.

Colliers (1988) *Downtown Toronto: Market Survey,* Toronto.

Daniels, P.A. (1985) 'A geography of unemployment in Vancouver CMA', unpublished MA thesis, Department of Geography, University of British Columbia.

Daniels, P.W. (1986) 'Office location in Australian metropolitan areas: centralization or dispersal', *Australian Geographical Studies* 24: 27–40.

Davis, C. and Hutton, T. (1981) 'Some planning implications of the expansion of the urban service sector', *Plan Canada* 21: 15–23.

Davis, C. and Hutton, T. (1989) 'The two economies of British Columbia', *BC Studies*: forthcoming.

Dobell, R., McRae, J., and Desbois, M. (1984) *The Service Sector in the Canadian Economy: Government Policies for Future Development,* Victoria, BC: Institute for Research on Public Policy.

Economist, The (1988a) 'Waiting for Taiwan to burst', 13–20 August: 70.

Economist, The (1988b) 'Canada: bleeding-heart Conservatives', 8–14 October: 54.

External Affairs (1988) *The Canada–US Free Trade Agreement,* Ottawa.

Financial Post, The (1988) *The Financial Post 500,* Toronto.

Gad, G. (1979) 'Face-to face linkages and office decentralization potentials: a study of Toronto', in P.W. Daniels (ed.) *Spatial Patterns of Office Growth and Location,* London: Wiley.

Gad, G. (1985) 'Office location dynamics in Toronto: suburbanization and central district specialization', *Urban Geography* 6: 331–51.

Gad, G. (1986) 'The paper metropolis: office growth in downtown and suburban Toronto', *City Planning,* Fall issue: 22–6.

Gill, D. (1988) 'Legal services', in J. McRae and M. Desbois (eds) *Traded and Non-Traded Services: Problems of Theory, Measurement and Policy,* Halifax: Institute for Research on Public Policy.

Globe and Mail (1988) 'Economic Council of Canada becomes free-trade oracle', 20 October.

Goldberg, M. and Mercer, J. (1986) *The Myth of the North American City,* Vancouver: University of British Columbia Press.

Gottman, J. (1982) 'The metamorphosis of the modern metropolis', *Ekistics* 49: 7–11.

Governments of Canada and British Columbia (1986) *Canada British Columbia Memorandum of Understanding: British Columbia as a Pacific Centre for Trade, Commerce and Travel.*

GVRD (Greater Vancouver Regional District) (1975) *The Livable Region 1976–1986,* Vancouver.

GVRD (1982) *Suburban Office Development in Greater Vancouver: A Survey of Existing Firms,* Vancouver.

GVRD (1988) *Achieving Greater Vancouver's Potential: an Economic Vision and Action Plan for the Livable Region,* Burnaby.

Hammes, D. (1988) 'Architects and engineers: an overview', in J. McRae and M.

202   *Services and metropolitan development*

Desbois (eds) *Traded and Non-Traded Services*, Halifax: Institute for Research on Public Policy.

Howell, L. (1986) 'The affordable housing crisis', *City Magazine* 9(1): 25–9.

Huang, S. and Gad, G. (1988) 'Office decentralization and the gender composition of central area and suburban office employment', paper presented to the Canadian Association of Geographers, Halifax.

Hutton, T. and Davis, C. (1985) 'The role of office location in regional town centre planning and metropolitan multinucleation', *Canadian Journal of Regional Science* 8: 17–34.

Hutton, T. and Ley, D. (1987) 'Location, linkages and labor: the downtown complex of corporate activities in a medium size city, Vancouver, British Columbia', *Economic Geography* 63: 126–41.

Keast, G. (1988) 'The world's longest commute', *Equity* 6(1): 26–9, 38.

Kellerman, A. (1985) 'The evolution of service economies', *Professional Geographer* 37: 133–43.

Leitner, H. (1988) 'Urban politics and downtown redevelopment: the case of six American cities', unpublished paper, Department of Geography, University of Minnesota.

Ley, D. (1985a) 'Downtown or the suburbs? A comparative study of two Vancouver head offices', *Canadian Geographer* 29: 30–43.

Ley, D. (1985b) 'Work-residence relations for head office employees in an inflating housing market', *Urban Studies* 22: 21–38.

Ley, D. (1986) 'Alternative explanations for inner city gentrification: a Canadian assessment', *Annals, Association of American Geographers* 76: 521–35.

Ley, D. (1988) 'Social upgrading in six Canadian inner cities', *Canadian Geographer* 32: 31–45.

Ley, D. and Hutton, T. (1984) 'Office decentralization and public policy in Greater Vancouver', in B. Barr and N. Waters (eds) *Regional Diversification and Structural Change*, Vancouver: Tantalus.

Ley, D. and Hutton, T. (1987) 'Vancouver's corporate complex and producer services sector', *Regional Studies* 21: 413–24.

Ley, D. and Olds, K. (1988) 'Landscape as spectacle: world's fairs and the culture of heroic consumption', *Society and Space* 6: 191–212.

Ludwick, E. and Associates (1987) *The Canadian Transportation Industry in a Deregulated and Free Trade Environment*, Halifax: Institute for Research on Public Policy.

Macleans (1988) 'The lure of western gold', 7 March: 36–7.

Matthew, M. (1986) 'Decentralization of Toronto's office space', *City Planning*, Fall issue: 27–9, 57.

Polèse, M. (1982) 'Regional demand for business services and inter-regional service flows in a small Canadian region', *Papers, Proceedings Regional Science Association*, 50: 151–63.

Polèse, M. (1988a) *Les Activités de Bureau à Montréal: Structure, Evolution et Perspective d'Avenir*, Montréal, Institut National de la Recherche Scientifique-Urbanisation.

Polèse, M. (1988b) 'La transformation des économies urbaines: tertiarisation, délocalisation et croissance économique', *Cahiers de Recherche Sociologique* 6(2): 13–25.

Polèse, M. and Stafford (1984) 'Le rôle de Montréal comme centre de services: une analyse pour certains services aux entreprises', *L'Actualité Economique* 60: 39–57.

Quinn, J., Baruch, J., and Paquette, P. (1987) 'Technology in services', *Scientific American* 257(6): 50–8.

Robinson, A. (1987) 'Montreal-Toronto conflict flares on exchange floors', *Globe and Mail*, 1 September: B1,4.

Royal Lepage (1988) *The Royal LePage Market Survey*, Toronto.

Semple, R.K. and Green, M. (1983) 'Interurban corporate headquarters relocation in Canada', *Cahiers de Géographie du Quebec* 27: 389–406.

Semple, R.K. and Smith, W.R. (1981) 'Metropolitan dominance and foreign ownership in the Canadian urban system', *Canadian Geographer* 25: 4–26.

Slack, B. and Barlow, M. (1981) 'Montreal – the international city', in D. Frost (ed.) *Montreal: Geographical Essays*, Department of Geography, Concordia University, Montreal.

Stabler, J. and Howe, E. (1988) 'Service exports and regional growth in the postindustrial era', *Journal of Regional Science* 28: 303–16.

Stanback, T., Bearse, P., Noyelle, T., and Karasek, R. (1981) *Services: The New Economy*, Totowa, NJ: Allanheld and Osmun.

Statistics Canada (1986) *Canada's International Trade in Services*, Catalogue 67–510, Ottawa.

Vancouver Sun (1988) 'Banking centre challenge launched', 28 June: El.

9 Producer services and metropolitan development in Australia

Kevin O'Connor and David Edgington

THE AUSTRALIAN URBAN SYSTEM

The Australian urban system is small, simple and dispersed. Although only a handful of cities have more than 250,000 people, two metropolitan areas rank among the top fifty in the world in population terms. The coastal nature of Australian settlement, reflecting the low density use of the central part of the continent, has meant the urban system fringes the country. This can be seen in the pattern of daily direct air services between metropolitan areas (Figure 9.1). The key nodes in the network are Sydney and Melbourne (3 million), Perth, Brisbane and Adelaide (1 million); other places on the daily air network, due to their political role, are Canberra and Hobart, with populations in the range 150,000 to 250,000. The pattern in Figure 9.1 excludes some major industrial cities, and regional cities in Queensland, that also have populations in the 150,000 to 250,000 range. A broadly based introduction to the urban system is provided by Logan *et al.* (1981).

FACTORS IN METROPOLITAN DEVELOPMENT IN AUSTRALIA

A simple way to understand the growth and development of an urban system such as that of Australia is to recognize that metropolitan areas develop in response to demand. Such demand can emerge from different sources at different times. To develop this idea it is convenient to think in terms of demand at an international, national and regional scale. This means metropolitan areas develop activities to serve demand from overseas, throughout their country, and within their hinterland (which in the Australian case is usually their state). To understand the present structure of, and change in, the Australian urban system there are three important aspects to be taken into account. The first is that national level demand has become much more important; previously regional demand was the main source of growth for the Australian metropolitan areas. Second, the impact of national demand has been geographically more selective than in the past with the result that the Australian urban system has entered an era of

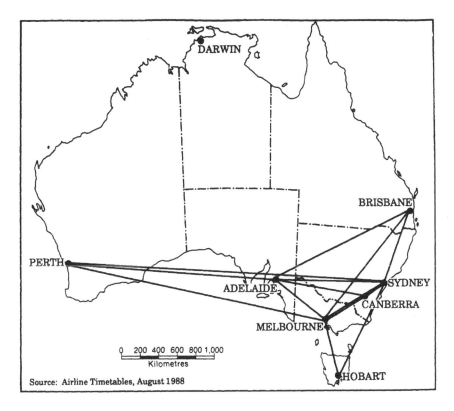

Figure 9.1 Daily inter-metropolitan air links, Australia, 1988

strong inter-city competition, which was not a characteristic of its past growth and change. Third, international demand is now much more important and tends to reinforce the national pattern. This fundamental shift in the source of metropolitan growth in the Australian urban system can be illustrated by the case of producer services which will be the main focus of attention in this paper.

As background to these ideas, one can simplify the links between Australian economic history and Australian metropolitan areas into three stages. The first was one where international development was the main source of growth and change – indeed the growth of Australian metropolitan areas can be seen as the result of international development of the world economy in the 1700s. The initial activity of Australian cities was the shipping and banking services in the areas near the docks; these services dealt with the world demand for our original resources. In this situation Australian metropolitan areas were gateways, providing openings to world demand.

A natural next step in this growth was the emergence of activity that served the regional market, and this was the foundation of the economy of

Australian metropolitan areas well beyond the period when Australia ceased to be a system of colonies and was established as a single political unit. The regional demand was so significant because each state had its own rail system (each with a different gauge), education system, road building authority and actively sought economic development in its own right. In a sense the country was only superficially a national unit, and the regional sub-economies were the main source of metropolitan development. Indeed, even today each state still maintains promotion offices in major world capitals. When this regional perspective was prominent, the potential for growth of each metropolitan area was set largely in terms of the level of development of its state. Those cities in intensively settled states with large resource deposits, such as Victoria and New South Wales became very large, while the metropolitan areas of South Australia, Western Australia, Tasmania and Queensland were smaller, primarily because their states had fewer resources, and were settled less intensively.

National market based activity began to emerge, but was less important than the regional focus in the period perhaps even up to 1960. Prior to this time interstate transport of goods and people remained a difficult task, so few activities spanned the nation (O'Connor 1987). It is the contention of this paper that national level demand is now the most important source of growth for metropolitan areas because national markets have emerged for most commodities. This shift has ushered in an era of inter-city competition where growth is not based on the level of state resources, but on the capacity to service national demand. When metropolitan areas were dependent mainly on their regional markets, they were in a sense insulated from one another; once they started competing for a unified national market the situation was quite different, and growth and change in one can be at the expense of others.

Responding to national scale demand can often require a concentration of activity in a few places, and this shift can have an important impact upon an urban system. The study of producer service location is a most appropriate means to illustrate these changes, as these activities have grown largely in association with the development of national and international markets.

PRODUCER SERVICES AND URBAN SYSTEMS

Producer services involve activities that include a reaction to, and influence over, corporate development and change; in turn at a national or world scale, they are involved with decisions that influence the direction of the national economy which has important consequences for development and change in a broad social and cultural sense. In short they involve creative people, individuals who devise solutions to production problems. This characteristic means their choice of location can be important to the fortunes of regions and nations. Their impact is reinforced through their leadership role, as they act as a magnet for other types of activity which are dependent on decisions and new directions created by these activities. Consequently, both their function and their locational concentration make them a major component of the vitality of certain metropolitan economies. In effect, these services create the conditions for one or other city's leadership in national activity, ideas and fashion.

Australia provides an unusual context for this competition as it has two almost equal-sized major cities. The location of producer services – or the choice in location between these two places – can determine the relative role they play in the national economy and society generally. In this way producer service location can redirect the pattern of growth of the Australian urban system.

Producer services have certain locational needs, often only met in a small number of places, which means they often cluster in only one or two places within a nation. Their geography, and in particular their location in certain cities, has attracted recent study. Cohen (1981) pointed out that the top end of the US national hierarchy, and only a few world cities, have attracted the majority of firms in these activities. Noyelle (1983) has plotted the spatial distribution of producer services in the US cities, and found that New York, Chicago, Los Angeles and San Francisco form a group of 'national advanced service centres'. That observation is consistent with results of Pred's (1977) earlier work on the US urban system, and fits with analyses of head office location by other researchers. Certain sectors, like international banking, are even more concentrated, and in the US cluster in New York and Los Angeles. The tendency to centralize, and the need for face-to-face contact, has been closely related to the uncertainty of information and its non-routine nature, ideas developed in Goddard's (1973) original work on office location. Sassen-Koob (1985) has shown these services do not need to be close to their consumers but do need to be close to one another.

These ideas, however, seem inadequate to explain the powerful concentration of producer services in a few cities. What is needed is an understanding of how the internationalization of producer service activity has special locational needs, which in turn has produced a few places throughout the world as centres of activity; in turn these places dominate their own national systems. A step towards this understanding has been provided by research on the location of production facilities of transnational corporations in foreign countries; to date, however, much of this work has been upon manufacturing plants and concerned with regional policy. Reviewing this work, Dunning (1981) indicates there are no easily predictable outcomes for the location of the subsidiaries of multinational enterprises. In research directly upon the office activities of transnational corporations, by Dunning and Norman (1983) and Heenan (1977), the attractions of large markets, good communications (especially airports) and 'access to specialized services' figured prominently in location decisions.

The latter factor indicates there is a circular and cumulative causation in this process – new offices locate where others established previously – and so increase the probability that other offices will select the same centre.

Heenan's (1977) work on headquarters location indicated that the regionalization of the world market, which is a common form of administration for transnational corporations, has generated regional headquarters (and hence clusters of producer services) in just a few cities. The International Labour Organization (ILO) and Food and Agriculture Organization (FAO), for example, have headquarters or regional operations in certain cities, which strengthens those places within the international network. These two forces together – regional administration and international organizations – have together created opportunities for just a few cities to play an international role and attract producer service type activities. Naturally this elevates them over competitors within their own country. Heenan (1977: 87) concludes that 'an adequate financial infrastructure, favourable regulatory environment, freedom from confiscatory measures and a strong historical role in commerce' are essential characteristics for a city to attract regional headquarters functions. In addition, he quotes Johnston's (1976) work on Panama showing a strong complementarity between a city's tourist industry and its role in international business. Summarizing much of this early work, Barlow and Slack have identified 'international cities' that

> perform three main functions: they are the loci of specialised activity serving the global community; they are nodes in international transportation and communication networks; and they are points of contact and interaction between national and international systems.
>
> (Slack 1985: 337)

These international cities are the most likely centres to host producer services, and the services develop 'global control capability' to be exercised from these few locations. The latter are usually the very big cities of the world (Sassen-Koob 1985: 238). Within any nation, the extent of international contact, or the potential for international linkages will be the factors that differentiate between each city's ability to attract these services. These two ideas together lie behind a revitalization of Hall's (1966) idea of 'World Cities'; Friedmann and Wolff (1982) show how the organization of world capitalism has generated a few 'control centres', which are a new form of world city. Thrift (1984) has applied that idea to a world property market, with special attention to Hong Kong as a focus for Pacific region activity.

From this work, two aspects stand out as critical for producer service location. The first is airline networks and airports, because of the accessibility they give to international markets. A place on the international air network ensures centrality within a region of the world and boosts a centre's attraction for producer service development. The second aspect is the characteristics of the local service sector. Generally large metropolitan areas can supply the necessary support services (the hotels, office space, local professional connections and the like) and so minimize uncertainty and create opportunities for firms to grow. Large metropolitan areas are more likely to have airports and other transport linkages so this factor is linked with the first. A crude measure of this aspect could be total population, but a more sensitive measure would be the amount of office space and recent trends in its construction. Other things being equal, producer services will locate in the bigger, busier office markets, and their presence will be made obvious by the number of new office buildings. In addition, hotel construction could be an indicator, in that hotels cater for international business travellers. There is the possibility that hotel construction could be for tourism rather than business travel; a check could be the coincidence of office and hotel construction. The latter would then be a good guide to producer service location.

The coincidence between producer service location and nodes in the world air transport system emerged because the latter places act as gateway cities to nations, and in some cases to groups of nations, and so provide producer services with good access to markets. This is especially relevant for services with world markets. The forces that influence gateway functions, and how cities compete one with another for hegemony in this role, emerges as a basic geographic dimension in understanding producer service development. Competition between centres as gateways lies behind important new developments in some national urban systems. The powerful resurgence of the New York metropolitan economy, reflected in recent office building and a strengthened banking role (even in the face of some headquarters relocations) is an important development in the US urban system (Gerard 1984). Soja *et al.* (1983) have analysed the competition between Los Angeles and San Francisco, and identified Los Angeles' dominance. In West Germany, the structural shifts in the national economy away from steel, mining and energy towards research and innovation-based industries has led to the rise of Frankfurt, Stuttgart and Munich as national headquarters centres; the former has a major international airport (Olbrich 1984). In Canada, 'Montreal has lost its status as Canada's national metropolis ... to Toronto' (Barlow and Slack 1985: 338); significantly activity at Toronto Airport has begun to surpass that at Montreal.

This background suggests that producer service development can have a major impact on urban system change, and will be more advanced in cities that have the following characteristics – international air traffic, regional headquarters for world organizations, and well-developed business service infrastructure, reflected in the volume and growth in office space and hotel construction.

PRODUCER SERVICES IN AUSTRALIAN METROPOLITAN AREAS

In the following section, data on a variety of measures of producer services, and information on the key characteristics of metropolitan areas associated with producer service location will be analysed, to provide a detailed understanding of these activities in Australia at the present time.

To provide an initial perspective, data on employment in a narrow range of occupation groups that involve management and control were assembled for metropolitan areas in 1986. The following occupations were selected:

> General manager
> Specialist manager
> Managing supervisor
>> Sales and service
>> Other business
> Business professionals
> Miscellaneous professionals

These were taken from a recently established Australian Standard Classification of Occupations and represented a good measure of producer service type jobs, because they referred to key decision makers. It was not possible to look at the data separately by industry so it is likely that some jobs included in this list are not dealing in the traditional core of producer services which is finance and banking. However the data is a good first measure of the activity under scrutiny (see Table 9.1, Column 1).

The information makes clear the current management role that Sydney plays in the Australian urban system. Even though it is roughly the same size as Melbourne, it has some 20 per cent more jobs in this category. This concentration is similar to that which has been found in other countries. A second important piece of information is that the contrast between the largest two cities and the rest is very obvious; these types of jobs are disproportionately clustered in the big cities, and smaller places do not experience anywhere near the same level of growth as big cities.

Table 9.1 also displays a range of measures of city development that relates to producer service location. The second column shows the total amount of office space available for commercial use. Ideally, the data should refer just to offices for selected producer service activities but that information is not available; what is available is information on the use of office space as surveyed by Jones Lang Wootton. This data shows even more graphically the dominance of Sydney – the latter has more than twice the space that is found in Melbourne.

The data confirm results in a property industry survey carried out by Adrian and Stimson (1986: 29). This showed the Sydney market led all others in investor perceptions of potential for property development. The total number of rooms available in Australian metropolitan areas in 1986 is

Table 9.1 Measures of producer service location: capital cities, Australia

City	Management jobs (1)	Office space (2) ('000 m²)	Hotel rooms (3) ('000)	Hotel occupancy rates (3) (%)	Head offices (4)	International activities (5)	International airport traffic inbound aircraft passengers 1985 (6)	International airport traffic inbound aircraft movements 1985	Population 1986 (7)
Sydney	182,471	2,650	14,871	73	50	205	1,395,454	10,093	3,364,858
Melbourne	148,977	1,200	8,815	64	36	156	594,984	5,738	2,832,893
Brisbane	49,115	1,030	6,155	57	4	66	251,372	2,190	1,149,401
Perth	46,686	790	6,160	56	2	79	256,350	1,477	994,472
Adelaide	43,371	460	3,932	69	7	67	55,353	443	977,721
Hobart	7,484		na	na	nil	19	7,911	97	175,082

Sources:
1. Group of occupations selected from Australian Standard Classification of Occupations, Census of Australia 1986
2. Jones Lang Wootton (1987) *Property 1987*, Pages 11–21 for cumulative absorption in each market as per analysis
3. Jones Lang Wootton (1987) *Property 1987*, Page 63
4. Number of firms in top 100 companies in Australia (Edgington 1983)
5. Entries in telephone book labelled 'International'
6. Department of Aviation (1985) *International Air Transport*

shown in Table 9.1, Column 3, and indicates that Sydney has 70 per cent more rooms available than in Melbourne, and twice that in other cities. The Sydney hotels led the nation, too, in their occupancy rate at 73 per cent, some four percentage points busier than its nearest rival in this respect, Adelaide. Previous research by Taylor and Thrift (1981) has established that Sydney houses a predominant share of the top 100 largest companies in the nation. The data in Table 9.1, Column 4 is an update of that information, and shows that there has been no change to the pattern found in the earlier research. Head office location is very concentrated in Australia's two largest cities. For example, the two largest cities have two-thirds of the office space in all metropolitan areas (roughly their share of metropolitan population) but together have eighty-six of the headquarters of the top 100 companies.

As indicated in the preceding section, the location of international organizations is an important indicator of producer service activity within a nation. An effort was made to identify some key international activities – organizations that need a base in a country as part of an international network. These could be the regional offices of international professional or trade organizations, offices of United Nations departments or other international regulatory bodies. In the Australian case it was expected that some organizations would use an Australian city as a base for a South Pacific or perhaps a South East Asian regional office. Details of this activity was difficult to find. As a crude estimate, the research relied on a simple summation of all telephone book entries that began with the word 'International'. Data on this crude indicator (in Table 9.1, Column 5) show a surprising congruence with patterns in other data, with a major cluster of activities in Sydney.

The distribution of international air traffic between the cities of a nation is another indicator of producer service location. Two simple indicators here are the number of passengers that travel through an airport, and the number of air traffic movements. Here Sydney's role as the international gateway to the country is apparent – more than 50 per cent of passengers travelling into Australia in 1985 came through Sydney, even though all capitals have international airports. It had twice the aircraft movements of Melbourne, and five and ten times that of other places. This in part reflects the choice made by international airline operators; many serve Sydney as their only Australian port of call, allowing domestic operators to move passengers to and from their Sydney terminal. The concentration of international air traffic in the two major cities is indicated by the fact that Melbourne and Sydney accounted for 80 per cent of international air traffic movements and international passengers travelling into Australia in 1985.

The data in Table 9.1 shows that producer service activity in Australia has two characteristics. First, it is concentrated in the two largest metro-politan areas, and second, Sydney is the dominant producer service centre.

Although the first characteristic does not appear new within Australia's development – the distribution of most activity has favoured the two largest cities for most of the country's history – there have been subtle shifts in producer service location that have given these two cities an even greater influence than their population share would suggest. The second characteristic is a new feature in Australia. Economic history has ascribed to Melbourne the role of 'commercial capital' and its political power has traditionally matched that of its big corporations (Blainey 1984). As an example Taylor and Thrift (1981) (in data to be discussed below) found that in 1950, Melbourne housed half of the headquarters of Australia's top 100 companies. Today, however, Sydney is the key international and national centre while Melbourne ranks as a second nationally significant metropolitan area. This differentiation has emerged due to the concentration of particular producer services in Sydney rather than Melbourne. This shift in emphasis is the focus of the following analysis, which explores recent shifts in producer service location, one part of a re-alignment of the geography of economic activity which is taking place within the nation, with significant consequences for the urban system.

CHANGES IN PRODUCER SERVICE LOCATION IN AUSTRALIA

As a first measure of change, data on increases in professional and technical jobs within industry groups were distilled from the 1971 and 1981 census. It was not possible to bring this up to 1986, as the classification of occupations has changed. These are the activities most likely to be involved in processing information, making decisions and providing leadership and direction, so they provide a good measure of producer service jobs. To reduce a substantial amount of data to a manageable form, industries were grouped into a few categories, loosely in accord with the approach adopted by Stanback *et al.* (1981). This involved aggregating industry groups with trade, transport, communication, finance and banking graphed into one category called producer services. Public utilities, public administration and community services made up a second group, called government. For purposes of discussion, agriculture and mining industries were grouped together, and manufacturing left as a single category. This four way classification was designed to show the cities where employment change in producer services had taken place, and second, to show the relative size of that growth compared to professional and technical job development in other industries.

Table 9.2 assembles that information for Australian metropolitan areas. The important insight in this data is that over the 1971–81 period, Sydney attracted many more new producer service jobs than any other city; considering it is only a little larger in population terms it was surprising to find it with such a substantial lead over Melbourne. Perth, too, did surprisingly well, with more new jobs than Brisbane which has a larger

Table 9.2 Job change in professional and technical occupations within industry groups, 1971–81

Industry groups	Sydney	Melbourne	Brisbane	Perth	Adelaide	Hobart
Producer services	21,337	15,682	5,534	5,827	3,365	447
Manufacturing	791	1,530	265	879	−129	−85
Agriculture, mining	178	−448	220	579	320	108
Public utility, administration, community services	54,247	47,747	17,258	19,025	19,827	11,783
Total	76,553	64,511	23,277	26,310	23,383	12,253
Share in producer services (%)	29.1	25.9	25.8	27.6	14.3	3.6

Source: ABS Census (1971, 1981)

population. To put that in perspective though, it attracted only a third of Melbourne's new jobs and a quarter of those in Sydney.

A second important insight in the table is that the share of new professional and technical jobs in the producer services group was larger in Sydney and Perth than the other cities. Perth's prominence is due in large part to its key role in mining industry growth, as well as some strong links with South-east Asia. The other metropolitan areas were less involved in leadership and control activities, and their economies were generating professional jobs in activities largely involved in local or regional markets – especially administration and community services. The significance of this distinction is reinforced by the data for Adelaide and Hobart, two small and slower growth metropolitan areas. In these two cities the share of producer service jobs in overall professional job development is very low. Perhaps the point is made most clearly by comparing Adelaide and Brisbane. Though Adelaide had more new jobs in total over the time period, Brisbane had many more in the producer service activities; Brisbane has been the faster growing of the two metropolitan areas in this period, in part because it attracted more of these leader activities.

A third important piece of information is the location of new professional and technical jobs in manufacturing. These are concentrated in Melbourne, with twice as many as in Sydney, confirming Melbourne's role as the major manufacturing centre in the nation. The latter has been illustrated in terms of production or employment (Rich 1987); the data in Table 9.2 supports that claim from the perspective of management and innovation. This points to a shift towards greater complementarity between Melbourne and Sydney, with the latter attracting the new service sectors, while the former is consolidating an industrial role by developing management and control functions.

When the producer service category in Table 9.2 is disaggregated (see Table 9.3), the predominance of the finance sector within the overall group is obvious, and it is in this sector that Sydney has the greatest leadership in new jobs. It is here too that the international focus is the strongest. As expected Sydney is a leader here. Perth's better performance is obviously related to growth in the finance category; in that group Perth outdoes Brisbane by quite a margin. Producer service development is currently led by the finance industry, and in Australia between 1971 and 1981, that sector has generated many more jobs in Sydney than in any other city – in fact more jobs than both Melbourne and Perth combined.

Looking at a longer time period Taylor and Thrift (1981) used conventional approaches to the location of head offices of the top 100 firms to show that the number of offices in Melbourne and Sydney has almost exactly see-sawed between 1953 and 1983, in favour of Sydney (Table 9.4). Apart from numbers of companies, it is important to consider their

Table 9.3 Job change in professional and technical occupations within producer service industry groups

Industry groups	Sydney	Melbourne	Brisbane	Perth	Adelaide	Hobart
Trade	2,818	1,424	438	566	276	92
Transport	1,424	832	429	384	172	20
Communication	2,542	2,733	729	411	468	48
Finance	14,553	9,269	3,938	4,466	2,449	287

Source: ABS Census (1971, 1981)

Table 9.4 Location of top 100 company headquarters by number of firms, Australia

City	1953	1963	1973	1978	1983
Melbourne	50	47	39	38	36
Sydney	37	44	50	52	50
Brisbane	4	2	3	4	4
Adelaide	6	3	6	5	7
Perth	1	2	1	1	2
Boyer (Tas)	1	1	1	–	–
Newcastle	–	1	–	–	–
Mt Morgan (Qld)	1	–	–	–	–
Bundaberg (Qld)	–	–	–	–	1
Total	100	100	100	100	100

Source: 1953, 1963, 1973 and 1978, Taylor and Thrift (1981); 1983, derived from Edgington (1983)

size and influence. One measure is the declared profits of organizations headquartered in a city (Table 9.5). This information shows just how much Sydney has 'caught up' with Melbourne.

One other way to track shifts in producer service location is to follow the office space markets of metropolitan areas. Table 9.6 draws on an analysis of building activity in Australian states in the early 1960s (Linge 1966). Assuming the majority of each state's building took place in its metro-politan area, the leadership of Sydney in the category of buildings called 'offices and banks' was apparent even at that time. It is extremely difficult to trace the pattern of construction any further back than the 1960s.

Looking at the current situation, Figure 9.2 shows the level of prime office rentals in Australian metropolitan areas since 1976. Apart from Sydney's dominance, this figure is important as it shows the strong rise of Perth to be the fourth ranked market in 1987, up from the bottom place in 1976. Adelaide, a place identified earlier as having few producer services,

Table 9.5 Share of profits declared by top 100 companies in Australia

City	1953	1963	1973	1978	1983
Melbourne	60.0	53.1	59.3	51.6	45.8
Sydney	27.7	40.0	33.7	40.2	45.9
Brisbane	7.9	5.3	3.9	4.9	2.8
Adelaide	2.8	1.6	2.4	2.1	3.8
Perth	0.4	0.6	0.5	1.1	0.1
Boyer (Tas)	0.4	0.7	0.2	–	–
Newcastle	–	0.8	–	–	–
Mt Morgan (Qld)	0.9	–	–	–	–
Bundaberg (Qld)	–	–	–	–	–
Total	100	100	100	100	100

Source: 1953, 1963, 1973 and 1978, Taylor and Thrift (1981); 1983, derived from Edgington (1983)

Table 9.6 Share of building in category offices and banks, percentage of national total in each state

State	1961–2	1962–3	1963–4	1964–5	1965–6
New South Wales	51.90	56.85	54.29	48.13	43.63
Victoria	21.47	22.61	26.74	26.66	33.25
Queensland	7.97	5.63	7.33	8.06	10.10
South Australia	10.72	7.77	6.09	4.80	5.50
Western Australia	5.50	4.84	3.92	6.31	5.94
Tasmania	2.39	2.28	1.60	1.00	1.55
Total	100	100	100	100	100

Source: Data reproduced in Linge (1966)

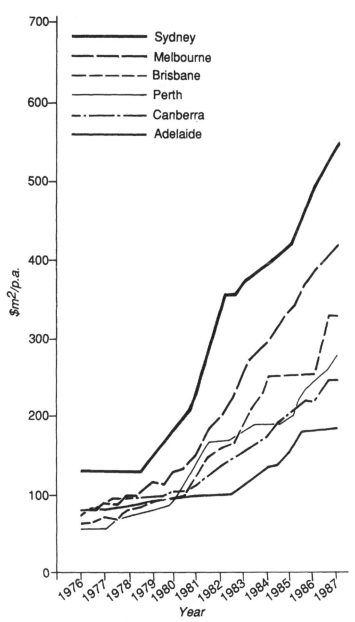

Notes: 1. Prime Office Rents are expressed at gross occupancy cost, excluding cleaning
2. Figures used in graphs represent the top end of the rental range achieved over the period
3. All years assessed at June

Figure 9.2 Prime office rents at June 1987

has fared poorly in this competition, beginning ahead of Brisbane, but falling well behind. This information indicates that the concentration of producer services in favour of Sydney probably dates from the early 1960s and the latter's leadership in this area is now virtually unchallenged. In addition, the particular locational needs of these activities means that some cities have attracted them (e.g. Perth) but others have not (e.g. Adelaide), a change that will eventually mould a different geography of metropolitan growth in Australia.

Change in the urban system along these lines is likely to be enhanced by recent deregulation of the finance market in Australia. New banking licences were made available in 1985, and the Australian dollar was allowed to float. Although new foreign banks can be found in all of the metropolitan areas, Sydney attracted the most new firms, and the largest operators. Foreign exchange dealings, a rapidly expanded activity, are predominantly centred in Sydney. In this way, changes in producer service location have contributed to the differentiation in role of each metropolitan area. The international emphasis, together with corporate restructuring and reorganization has produced a very strong headquarters role for Sydney, while manufacturing activity has tended to concentrate in Melbourne. Perth, too, has asserted a more prominent role than its equal sized neighbour, Adelaide, due to its finance sector, while Brisbane seems to have developed in Sydney's shadow. Within this change and reorientation the respective roles of Melbourne and Sydney have probably experienced the greatest change. Attention now turns to that aspect.

SHIFTS IN PRODUCER SERVICES IN AUSTRALIA: MELBOURNE VERSUS SYDNEY

Logan *et al.* (1981) have shown that the emergence and development of the system of settlement in Australia can be understood in terms of gateway cities – ports that link hinterlands and overseas markets (Burghardt 1971). The proposition explored in this section is that Sydney has become the major gateway for Australia. That shift, related to the concentration of producer service development, identified in the two previous sections, has realigned the two largest places within the national urban system.

It has long been recognized that initial points of contact for international trade often lead a local economy. Maintenance of this leadership is not assured, however, and depends upon the activity in the surrounding area, as well as the direction and technology of travel. The service industry growth will be strongest in cities that face incoming traffic, or are most likely to be first point of call. A gateway's vitality depends on the technology of travel, as a change in method of transport could favour a new location, for technical or other reasons. Many by-passed ports in the world illustrate the impact that changing technology can have on the vitality of a gateway city economy. Direction and technology can be interrelated; a new type of transport could involve a new approach to a national market. As will be discussed below, that change happened when air travel took over from sea transport as the principal form of travel to Australia.

Activity within the respective hinterlands of Melbourne and Sydney was comparable, except for the greater richness and length of mining of the Victorian goldfields, which was followed by a major role in the management and control of the world's richest silver deposits at Broken Hill (Blainey 1963). Though the latter were located in New South Wales, they were more closely linked to Melbourne than to Sydney; even the Broken Hill water supply company was based in Melbourne. This major difference was crucial to financial sector growth. The Broken Hill wealth came hard on the heels of a long period of successful corporate gold mining that had used Melbourne's financial institutions as a conduit for capital raising and profit distribution. By 1890, it had the headquarters of the majority of banks, which were generally subsidiaries of British banks. In addition it provided a financial foundation for the development of Australia's largest industrial corporation – Broken Hill Proprietary Limited – which developed the nation's iron and steel industry involving New South Wales coal, South Australian iron ore, and Victorian management, from a Melbourne headquarters (Blainey 1963).

Looking at early travel and trade patterns, strong links with the UK and Europe were apparent, and the most common form of travel was by ship. In that era, Melbourne had a strong gateway role, being the first major city for eastbound shipping from Europe. Over time these patterns, as well as the activity in each hinterland, changed to favour Pacific basin countries. The long-run effect was to change the gateway roles of Melbourne and Sydney.

To illustrate and test these ideas, an attempt was made to measure the gateway functions of Melbourne and Sydney from 1950 to 1980. The lack of readily available data in comparable form to measure external contacts made this a difficult task. Trade or financial information was sought, but was not in the form needed. However, data on the state nominated as a residence by international migrants was available from a single source, without any changes in definition or terminology since 1950. In addition, it gave an outsider's view of the most appropriate location to begin settlement in Australia; it is not known whether that decision was made in the country of origin, or upon arrival in Australia. Over a period of years in the 1950s, roughly half of all assisted settlers nominated Victoria as a state of intended residence, and would have entered the country through Melbourne (Figure 9.3). During the 1960s Victoria's share fell to level out at around 25 per cent of the national intake. At the same time there was a shift in transport, with air transport used for assisted migration, rather than sea transport, which was normal in the earlier period. Sydney at the time

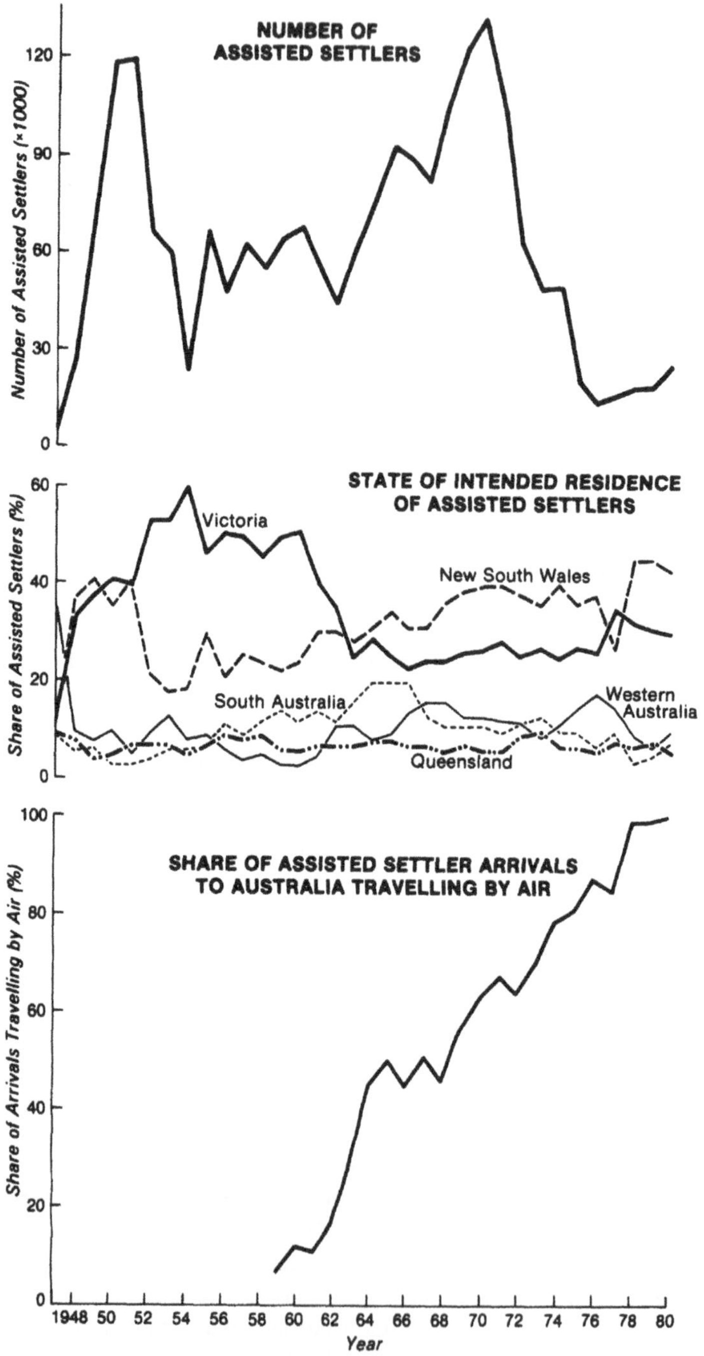

Figure 9.3 Assisted migration to Australia, place of residence and means of transport, 1948

had the main international airport, and the best level of service (Blainey 1984) so that it was in effect the first port of call.

Though there is probably no link between a migrant's residential decisions and producer services, patterns in the material at hand are confirmed by some data on foreign company headquarters location. For example, Edgington (1984a) has traced the establishment and location of major foreign companies in Australia from 1900, and notes a shift in favour of Sydney brought about largely by US and Japanese companies after 1960. In particular US manufacturing companies showed a preference for Sydney with 60 per cent of the new manufacturing companies (and 71 per cent of the new capital) locating there in the period 1965–75. Japanese companies in particular have shown a market preference for locating their head offices in Sydney (Edgington 1984b). The material reviewed by Edgington recognizes that links with the US, and then later with Japan dominated Australia's trade from 1960. In this world, a position on the Pacific Ocean was much more fortuitous than an outlet to the Southern Ocean, and so Sydney assumed the role as Australia's international gateway. Consistent with that interpretation Friedmann (1986) nominated Sydney in a network of 'world cities'; Melbourne is not marked on Friedmann's map. It is in this way that the internationalization of the Australian economy has been felt at a metropolitan scale. This trend, which has a particularly strong impact upon the location of producer services, reinforces the changes taking place at the national level, and differentiates cities one from another. Sydney has become the international finance centre for Australia, while Melbourne is the nation's major manufacturing centre.

The material discussed above has emphasized a narrow part of economic activity, but producer services can have important links with many activities, especially those involving creative endeavour. These links arise directly from the skills needed in management and control, and indirectly, through connections with sectors like advertising and finance. The latter can assist the growth in the motion picture and television drama production, for example, because of a coincidence of skills and a capacity to raise money for risky projects. In the creative world, there has also been debate on changes in the respective roles of Melbourne and Sydney; the changes parallel those discussed in this paper, with Sydney taking over leadership from Melbourne (Davidson 1986). In politics, too, Sydney, and the NSW Labor Party, have been seen to have a stronger role in national politics in recent years (Australian Bulletin of Labour 1984: 60) especially in setting the national agenda for the federal government. Hence there has been a major redirection in the operation of the Australian urban system. The final section will explore the consequences of that change.

CONCLUSION AND IMPLICATIONS

Material analysed in this paper has two broad consequences for research into service sector development. First, measures of producer service activity were used to differentiate types of metropolitan areas in Australia. This information discriminated Sydney from Melbourne, for example, even though the two are roughly similar in population, and also distinguished between Perth and Brisbane on the one hand and Adelaide and Hobart on the other. It also showed the absolute dominance of the two largest cities in the economic life of Australia.

Second, an attempt was made to reinterpret the idea of the gateway city in the circumstances of recent development in Australia. This reinterpretation identified Sydney as the international gateway, because of its links to Australia's Pacific and Asian neighbours, where much of the impetus for development has come in the immediate past. This interpretation sharpens an understanding of the divergent paths followed by the roughly equal sized metropolitan areas of Melbourne and Sydney in the last ten years, which is relevant to the economic dimension of government urban policy in Melbourne and Sydney.

Governments responsible for both cities have looked to boost growth in recent years. In interpreting these actions, it is important to recognize what Ross *et al.* (1980) have called the 'global constraints of local planners'. In the present context, these global constraints operate through the factors that affect the international gateway role of a metropolitan area – airline networks, international hotels and national headquarter functions – most of which are beyond the influence of local planners. In reviewing this situation Ross *et al.* (1980) believe that some structural initiatives (changing the mix of activity in a region) and some competitive initiatives (improving the local context for economic activity) can provide opportunities for places to expand their present role in this global competition. Working along similar lines, the OECD (1986) have identified five programme directions for urban economies, including improving the local urban environment and diversifying business activity. Recent policy initiatives in Sydney and Melbourne provide examples of these actions.

Sydney has recently moved to capitalize upon the trends identified in the earlier parts of the paper, and based considerable promotional effort on very similar data. This involved a claim to be the Financial Centre of the Pacific (NSW Government 1983), which was followed by government sponsored redevelopment of old dockland, for hotel, convention centre and office space, a strategy that has been adopted in several metropolitan areas in the world, like Baltimore and Boston for example.

Efforts to improve Melbourne's commercial role are based on changes in local administrative arrangements to take advantage of benefits expected to flow from the floating of the Australian dollar, and other deregulation steps in the finance market generally. These also involve encouragement to new financial institutions and activities to locate in Melbourne. These include listing a secondary market in government securities and also for mortgages expanding a domestic rating service and requesting the national government to upgrade powers of the Reserve Bank office in Melbourne. In addition, the Victorian government has made repeated submissions to Qantas to improve Melbourne's direct air service connections to the US and Europe. It is also active in planning for extra capacity at Melbourne Airport. Spatial strategies that back these initiatives include government sponsorship of new office buildings on vacant CBD and CBD fringe land. Efforts designed to bring this land on to the market are backed by a considerable variety of short-term actions designed to improve the amenity of the central city.

Whether these actions in Melbourne can overcome the problems implied in the earlier analysis remains to be seen. A difficulty is the head start that Sydney obviously has with respect to producer services. Melbourne does have advantages and assets associated with space around the docks, the airport and its position at the junction of highways and rail links to other capital cities. These advantages provide a solid base for Melbourne to capitalize on its role as the manufacturing and freight centre of the nation. The latter is the prosaic base of the commercial world, while Sydney seems to have the sectors with the charisma in modern capitalism, which are so captivating for governments. Perhaps separate roles, leading to integration rather than competition, could be a resolution of current difference? Whatever the outcome it is likely that the respective roles of these places will remain central to the evolution and change of the Australian urban system.

At the level of the smaller cities many of these issues have been addressed and policies implemented, with the objective of improving an international and national role. Brisbane, for example, has rebuilt its airport, with special emphasis upon the international terminus, and provided direct freeway access to tourist resorts to the north and south. This has coincided with the staging of Expo in that city. Adelaide competed aggressively to capture the Australian Grand Prix, and has undertaken extensive inner area redevelopment in association with a casino. Perth has adopted a similar approach, with special attention to central city redevelopment.

Even allowing for these changes, however, the gap between the activity in the second level cities, and that taking place in Melbourne and Sydney has probably widened, especially in the area of producer services. The latter reflects some major changes that are taking place within the Australian economy, which can be seen in a range of activities. Detailed data assembly on these changes is under way, and work will involve investigating the broad patterns established by this paper to individual sectors such as research and development, advertising, media and publishing. This will produce a broadly based understanding of how

Australia's urban system will be shaped during the early twenty-first century.

REFERENCES

Adrian, C. and Stimson, R. (1986) *Australian Property Industry Survey*, AIUS Publication 126, Canberra AIUS.

Australian Bulletin of Labour (1984) 'The dialectical crunch', *ABL*, 10(2), 60–1, Flinders University Institute of Labour Studies.

Barlow, M. and Slack, B. (1985) 'International cities: some geographical considerations and a case study of Montreal', *Geoforum*, 16, 333–45.

Blainey, G. (1963) *The Rush that Never Ended*, Melbourne: MUP.

Blainey, G. (1984) *Our Side of the Country: The Story of Victoria*, North Ryde, Methuen Haynes.

Burghardt, A.F. (1971) 'A hypothesis about gateway cities', *Annals*, Association of American Geographers, 61, 269–85.

Cohen, R.B. (1981) 'The new international division of labour, multinational corporations and urban hierarchy', in M. Dear and A.J. Scott (eds) *Urbanization and Urban Planning in Capitalist Society*, London: Methuen.

Davidson, J. (ed.) (1986) *The Sydney-Melbourne Book*, Sydney: Allen and Unwin.

Dunning, J. (1981) *International Production and the Multinational Enterprise*, London: Allen and Unwin.

Dunning, J. and Norman, G. (1983) 'The theory of multi-national enterprise: an application to multi-national office location', *Environment and Planning A*, 15, 675–92.

Edgington, D. (1983) *Central Melbourne in its National Context*, Background Working Paper, Central Area Task Force, Melbourne: Ministry for Planning and Environment.

Edgington, D. (1984a) 'Australian cities and foreign multinationals', *Papers, Australian and New Zealand Section Regional Science Association*, 9th Meeting, Melbourne, 88–116.

Edgington, D. (1984b) 'The location of Japanese transnational activity in Australia', *Environment and Planning A*, 16, 1021–40.

Friedmann, J. (1986) 'The world city hypothesis', *Development and Change*, 17, 69–83.

Friedmann, J. and Wolff, G. (1982) 'World city formation: an agenda for research and action', *International Journal of Urban and Regional Research*, 6, 309–44.

Gerard, K. (1984) 'New York city's economy: a decade of change', *New York Affairs*, 8, 6–18.

Goddard, J. (1973) 'Office linkages and location', *Progress in Planning*, 109–232.

Hall, P. (1966) *World Cities*, London: Wiedenfeld and Nicolson.

Heenan, D.A. (1977) 'Global cities of tomorrow', *Harvard Business Review*, 55, 79–92.

Johnston, H. (1976) 'Panama as a regional financial center', *Economic Development and Cultural Change*, 24, 261–86.

Linge, G.J.R. (1966) 'Building activity in Australian metropolitan areas: a statistical background', in P.N. Troy (ed.) *Urban Redevelopment in Australia*, Canberra: ANU Research School of Social Sciences.

Logan, M.I., Whitelaw, J. and McKay, J. (1981) *Urbanization: The Australian Experience*, Melbourne: Shillington House.

Noyelle (1983) 'The rise of advanced services', *American Planning Association Journal*, 49, 280–90.

NSW Government: Department of Industrial Development and Decentralization

(1983) *Sydney, Financial Growth Center for the Pacific,* Sydney: Peat Marwick and Mitchell for the Department.

O'Connor, K. (1987) 'The restructuring process under constraints: a study of recent economic change in Australia', *The Australian Journal of Regional Studies,* 1, 23–36.

OECD (1986) *Revitalising Urban Economies,* Paris: OECD.

Olbrich, J. (1984) 'Regional policy and management jobs: the locational behaviour of corporate headquarters in West Germany', *Environment and Planning C,* 2, 219–38.

Pred, A. (1977) *City Systems in Advanced Economies,* New York: Halstead Press.

Rich, D.C. (1987) *The Industrial Geography of Australia,* London: Croom Helm.

Ross, R., Shakow, D.M. and Susman, P. (1980) 'Local planners – global constraints', *Policy Sciences,* 12, 1–25.

Sassen-Koob, S. (1985) 'Capital mobility and labour migration', in Timberlake, M. (ed.) *Urbanisation in the World System,* New York: Academic Press.

Soja, E., Morales, R. and Wolff, G. (1983) 'Urban restructuring: an analysis of social and spatial change in Los Angeles', *Economic Geography,* 59, 195–230.

Stanback, T., Bearse, P., Noyelle, T. and Karasek, R. (1981) *Services: The New Economy,* Totowa, NJ: Allanheld, Osmun and Co.

Taylor, M.J. and Thrift, N. (1981) 'Spatial variations in Australian enterprise: the case of large firms headquartered in Melbourne and Sydney', *Environment and Planning A,* 13, 137–46.

Thrift, N. (1984) *World Cities and the World City Property Market: The Case of South East Asian Investment in Australia,* unpublished manuscript.

Victorian Government (1984) *Victoria: The Next Step. Economic Initiatives and Opportunities for the 1980s,* Melbourne: State Government Printer.

10 Advanced office activities in the Randstad-Holland metropolitan region: location, complex formation and international orientation

Pieter P. Tordoir

INTRODUCTION

According to recent theory, metropolitan economic fortunes are related to the diversity of advanced producer services supporting internationally orientated business. This is examined in this chapter with reference to the results of an empirical study of demand and supply networks of high level professional service activities located in Amsterdam, Rotterdam, The Hague and Utrecht. These four cities are the major centres within the Randstad metropolitan region (the 'Rimcity') in the western part of the Netherlands, which accommodates five million inhabitants. The main question addressed is whether these four urban centres together form an integrated and diversified complex of advanced office activities. The results of the study indicate functional specialization at the level of individual cities within the Randstad region and complementarity and diversity at the level of the region as a whole.

The rise of advanced office activities is a major aspect of the structural transformation of cities in advanced economies. The concept of advanced office activities used here includes specialized producer services (intermediary financial services, insurance, brokerage, and professional business services) and corporate staff services. The adjective 'advanced' denotes the knowledge- and information-intensive character of the activities regarded here.[1]

Most urban economies in the developed countries show marked changes in industrial and professional structure, due both to structural transformation of advanced economies as a whole and to transformation of spatial divisions of labour. The process of de-industrialization and the rise of advanced office activities has had dramatic implications for the physical and spatial structure and the use of factors of production in cities, including labour and infrastructure in its widest sense.[2] In the Netherlands, the rise of advanced office activities in the Randstad metropolitan region has recently induced a range of policy questions both on the metropolitan level and on the national level. There is a general feeling among Dutch policy makers that these activities are strategic for general economic development and should be stimulated. At the metropolitan level, agencies for economic development and physical planning acknowledge the increasing competition regarding the attraction and development of advanced office activities, among cities and regions within the Netherlands and, more important, among metropolitan regions at the European level. At the national level, the National Physical Planning Agency is concerned for the provision of adequate infrastructure and optimal spatial allocation of development resources in order to support the international orientation of advanced office complexes in general and the international competitiveness of the Randstad region in particular. The Dutch government acknowledges the particular function in this regard of the largest and most diversified cities. For this and other reasons, these cities were recently given a special development status.[3]

The Randstad region (see Figures 10.1 and 10.3) lacks a dominant central place and accommodates a multitude of small and medium-sized office centres. Given its concern with the development of an environment for advanced offices which can meet international standards, the National Physical Planning Agency is interested in eventual economic hierarchies, functional differentiations and complementarities (complex formation) among these centres. Apart from this practical interest, there are theoretical issues involved in the analysis of spatial and functional relations among advanced offices and these are discussed in the next section of this chapter. These practical and theoretical issues provide the framework of an empirical survey of 600 firms located in the main office centres in the Randstad region. A third section contains a presentation and analysis of the main results of this survey. The chapter concludes with a feedback on theory and a discussion concerning further research.

THEORY ON PROFESSIONALIZATION AND METROPOLITAN DEVELOPMENT

Two related processes play a central role in the development of advanced industrial capitalism in the twentieth century. First, producers and institutional agents devote increasing efforts to control processes in economic and technological systems, at the level of individual firms, markets, resources and general institutions. Second, the labour involved with these efforts is professionalizing, with implications for the division of labour, the development of institutions and the development of practical and theoretical knowledge. These processes have resulted in a growth of specialized professional work complementary to the production of goods and services: a rise of advanced office activities. The sociological and the managerial aspects of these processes are brilliantly analysed by, respectively, Bell (1973) and Chandler (1977). The economical logic underlying these processes is partly clarified by Williamson (1980).

The production and circulation of goods and services evokes three essential kinds of costs: the costs of production itself, the costs of transactions and the costs of adjustment. Transaction costs are incurred in matching supply and demand in ever increasing markets and circuitous production systems. Such costs include the costs of transportation and communication, credit- and debt-circulation, measurement, co-ordination and control, and risk and uncertainty. Adjustment costs are incurred in matching the quantity and quality of resources for production (in the widest sense) which have long gestation periods to the needs of an ever-changing production system. Examples of such resources are skilled labour, knowledge and technology, infrastructure, and organizational and institutional structures.

The continuing extension and change of markets and the increasing pace of change and divisions of labour all increase economic complexity. The relative shares of transaction costs and adjustment costs in total costs grow more than proportionally with this increasing complexity. The amount of discrete operations induced by transactions and adjustment grows disproportionally with total growth. Moreover, the labour involved in these operations shows on the average a relatively low level of labour productivity and is therefore relatively expensive.[4] This argument, coupled with the professionalization argument given above, explains the vast shift in the advanced economies from direct productive labour towards managerial, professional and administrative labour.

The rise of these 'professional overheads', including advanced office activities, could be seen as having a 'draining' effect on general production. Most often, however, the labour involved in overhead inputs results in increasing efficiency and productivity in direct production. As long as the resulting economies from developing transactional and adjustment labour are more than compensating for its rising costs, the whole economy is better off with a further shift to the advanced office activities involved.[5]

Transactional and adjustment costs enter into the production function of firms as professional overheads. The surge in the relative share of these overheads is presenting a major challenge for all producers of goods and services. The efforts to improve the productivity and the quality of the involved labour have important implications for organizational structure at the micro- and the meso-economic level, and consequently for the spatial structure of advanced economies.

An initial step producers can take in reducing transaction and adjustment costs is to develop specialized staff units for professional work. The organization of large corporations into divisions is partly induced by their efforts to extend the internal market for professional staff units in order to reduce their relative costs and enhance their leverage. Professional overhead costs are further reduced by locating corporate and divisional staff units in places where the resources needed are best equipped. Such resource requirements are, on the average, best met in large and diversified service agglomerations where the costs of accurate managerial information are lowest (much depends on the character and the routinization of the functions involved, however). This results in a spatial division of labour within firms.

A second step in reducing transaction and adjustment costs is to externalize corporate staff functions to specialized service suppliers. Such a policy is becoming more and more appropriate in countries with a highly developed business service sector, including the Netherlands.[6] This externalization movement is currently reinforced by the rise of flexibilization strategies in the corporate world. Increasingly, large firms try to flexibilize their professional overheads in order to gain reactive power in an environment typified by discontinuities.

Recent empirical research shows that such flexibilization strategies including professional functions are only efficient if firms can choose the best matched suppliers among an extensive array of professional service firms close at hand. The relation between the user and the supplier of the services involved must in general allow for both flexibility and intensive personal interaction.[7] It is supposed here that such conditions are in general best met in the largest as well as most diversified spatial concentrations of advanced professional services.

The above argument leads to an expectation of cumulative causation and 'filtering up' of advanced office activities in urban systems, reinforced by recent externalization strategies. Spatial concentration and increasing diversification of these activities result in lower transaction costs and enhanced opportunities for further specialization and diversification, and therefore in growth. Cities and regions lacking advanced office activities will have small chances to attract these. Cities accommodating a relatively well-developed sector of advanced offices could attract further growth on the one hand but could lose on the other hand specialized high level activities to larger and more diversified metropolitan centres (see Stanback and Noyelle 1984). Following from the above argument, the development of complexes of advanced office activities would stimulate a multi-tier system of cities whereby the relations among high ranking agglomerations would gain more strength and induce more growth than relations among lower ranking agglomerations and relations between high ranking and lower ranking agglomerations.

On the other hand we expect some differentiated downward spatial filtering or spatial division of labour due to differences in the level of routinization of office activities, to differences in sensitivity to spatially fixed resources such as infrastructure, and due to differences in sensitivity to eventual linkages with other kinds of activities such as manufacturing and transportation.

It is presumed here that, on the national scale, spatial office concentration (reinforced by the observed increase in the interwovenness among office activities, by the recent demand for flexibility, by the increasing sensitivity of advanced offices to high level resources and finally reinforced by the ongoing shift of employment towards the most advanced segments within all office activities) will more than offset forces of spatial deconcentration, even if these latter forces are stimulated by strong advances in telecommunication technology or rising spatial differentiations in operating costs as a consequence of agglomeration diseconomies.

The above theory partly explains the current vigorous development of advanced offices in some of the largest western metropolises. At the intra-metropolitan level, patterns of advanced office development seem however to be extremely complicated and less easily explained. There is a wide range of spatio-economic changes at the intra-metropolitan scale: growth in CBDs but also office suburbanization, office dispersal into very wide 'urban fields', axis-development, strong spatial divisions of office labour but also vigorous competition among different office locations. The research results presented in the next section unravel some parts of this complexity in the case of the Randstad metropolitan region.

SPATIAL STRUCTURES IN ADVANCED OFFICE ACTIVITY: EMPIRICAL RESULTS FOR THE RANDSTAD METROPOLITAN REGION

The spatial distribution of advanced office activities

The sensitivity of Dutch advanced office activities to urban proximity is indicated by a comparison of regional location quotients for advanced office activities employing 100 or more employees per establishment[8] (Figure 10.1) with the pattern of 'urban proximity' (Figure 10.2), using an agglomeration index based on three variables: distance to the nearest city, size of the nearest city, and inter-urban influence.

The similarity of the office pattern and the proximity pattern weakens if less advanced office activities and smaller office establishments are incorporated in the selection of activities.[9] The first order office centres and other main concentrations of advanced office activity in the Randstad region are shown in Figure 10.3

Four of these centres are traditional downtown CBDs in the four largest cities (Amsterdam, Rotterdam, The Hague and Utrecht). The other six office concentrations emerged during the 1970s and 1980s near major traffic arterials. Together, these areas account for approximately 50 per cent of all office space in the Randstad region. The indicated first order centres accommodate however at least 75 per cent of all advanced office activities in the region.

Although the new first order office centres are all located in the direct sphere of each of the four main cities in the region, their location near the ring of main traffic arterials connecting these cities indicates a trend towards economic integration at the higher level of the Randstad region as a whole. The result is an 'inward suburbanization' of offices at the Randstad

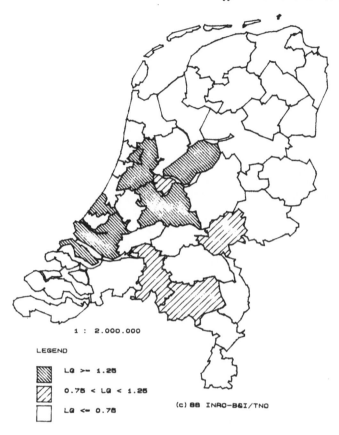

1 : 2.000.000

LEGEND

LQ >= 1.25

0.75 < LQ < 1.25

LQ <= 0.75

(c) 88 INRO-BGI/TNO

Figure 10.1 Regional location quotients for advanced office activities in the Netherlands

regional level, checked only by the political decision to leave the large central area, or the 'Green Heart', open.

Specifications of the survey

After determining the general distribution of advanced office activities within the Randstad region (Figure 10.3), the research focused on the office establishments located in the ten main centres indicated above. Two of the most recently developed new office centres, one on the northern fringe of Rotterdam and another on the western fringe of The Hague, were excluded from the survey because they are still in a phase of rapid development with the number of advanced office establishments insufficient for statistical analysis. The two main office concentrations in Utrecht are combined in the survey in order to achieve the threshold numbers; the new centre in Utrecht is in fact in a direct line with the downtown centre and

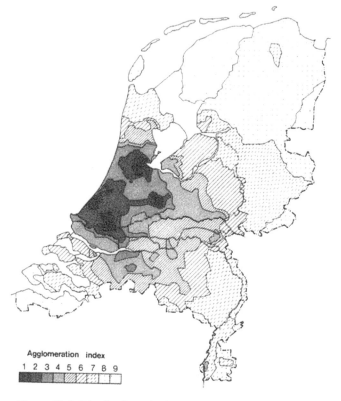

Figure 10.2 Distribution of urban proximity
Source: Dieperink and Nijkamp (1986)

forms an extension of the latter. The data for analysis therefore covers seven centres: Amsterdam Downtown, Amsterdam South, Amsterdam South-east, Rotterdam Downtown, The Hague Downtown, The Hague South-east (Rijswijk), Utrecht (Downtown/Southern extension 'Kanaleneiland').

The surveyed population of advanced office activities composed the following:

(a) headquarter establishments (corporate staff and divisional staff) with fifty or more personnel, of large companies in mining, manufacturing, trade and distribution;
(b) all establishments in investment banking, insurance, and real estate with twenty or more personnel; and
(c) a selection of knowledge-intensive activities within the group of professional business services – establishments with five or more personnel.

The distribution of all establishments in these three categories for each centre is shown in Table 10.1.

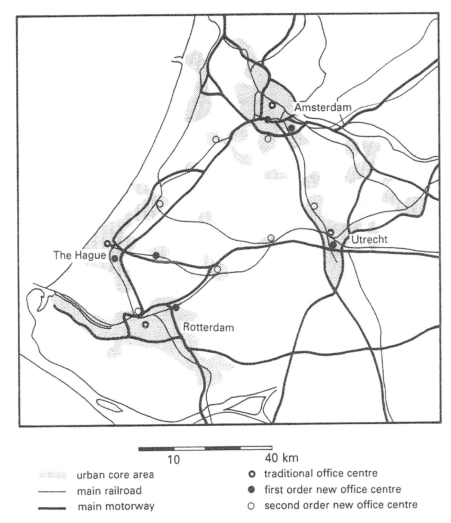

Figure 10.3 Distribution of main centres of advanced office activities within the Randstad metropolitan region
Source: Based on de Smidt *et al.* (1987)

A functional typology of advanced office establishments and locations

Except for the establishments in Group C (advanced business services), the nature and function of the investigated establishments do not so much relate to the industrial heading under which they are classified as to the status and function they have within the corporate structure. Our question-naire contained various variables concerning status and function, which were combined into three composite variables: the level of the managerial

Table 10.1 Distribution of advanced office establishments in survey centres, by number and industry

	Amsterdam Downtown	Amsterdam South	Amsterdam South-east	Rotterdam Downtown	The Hague Downtown	The Hague South-east	Utrecht	Total
Group A	133	96	58	141	61	41	76	
Group B	153	70	33	117	81	11	42	
Group C	185	127	33	162	147	42	77	
Total	471	293	124	420	289	94	195	1,886
Of which:	(%)	(%)	(%)	(%)	(%)	(%)	(%)	(%)
Mining						1.7		0.3
Manufacturing	9.6	4.0	16.1	3.3	6.6	19.1	11.8	8.3
Construction	3.1	3.2	4.8	2.1	4.2	4.3	8.7	3.5
Trade	11.5	7.2	23.4	10.5	8.3	17.0	16.4	11.9
Transport	4.0	0.4	1.6	12.6	0.3	3.2	1.5	4.4
FIRE[a]	32.5	31.4	26.6	30.0	28.0	11.7	21.5	28.4
Bus. serv.	39.3	57.0	26.6	41.5	50.9	44.7	39.5	43.3

Note: (a) Finance, insurance, real estate
Source: Dwarkasing et al. (1988)

functions performed, the level of the specialist functions performed, and the international structure of the organizational environment. Statistical grouping of these composite variables resulted in the following typology into five functional types of advanced office establishments:

Type 1 Headquarter establishments of Dutch corporations with a largely international structure and market environment.

Type 2 Headquarter establishments of Dutch corporations with a largely national structure and environment.

Type 3 Divisional, branch or daughter-firm office establishments owned by Dutch corporations.

Type 4 Divisional and branch office establishments of foreign corporations.

Type 5 Office establishments selling advanced business services on a commercial basis (90 per cent of these are members of Group A).

The differentiations in functional structure over office centres are marked and stronger than the industrial differentiation (see Table 10.1). Downtown Amsterdam has a diversified functional structure. Amsterdam South is characterized by specialist offices (in particular management consultancy) and branch offices, which correlates with the lack of large office buildings and the high quality of small buildings in the area. Amsterdam South-east is the principal domain of branch offices of foreign corporations. Foreign owned offices are also over-represented in downtown Rotterdam, but do not dominate there. Downtown Rotterdam shows a fairly even distribution of types, with over-representation of offices with an international orientation. Foreign offices are under-represented in The Hague Downtown. The downtown office centre of The Hague is characterized by headquarters of Dutch corporations (which might relate to the seat of the national government) and independent business services (in particular engineering consultancy). The suburban office centre The Hague South-east is dominated by the seats of Dutch multinational corporations. The Utrecht office centre has the least advanced complex of office functions, with over-representations of Types 2 and 3, and marked under-representations of Types 1 and 5 (software houses dominate Type 5).

Spatial structure of economic linkages in general

The spatial structure of the linkages between the surveyed establishments and their economic environment (consumer markets, factor markets and suppliers) and internal organizational environment (branch offices, plants) is investigated at four spatial scales: the city in which the office establishment is located, the rest of the Randstad region, the rest of the Netherlands, and foreign countries. The offices were asked to give the distribution of all contacts over these four levels using a five-point scale (Table 10.3).

Table 10.2 Distribution (%) of advanced office establishments by location and functional type

	Amsterdam Downtown (%)	Amsterdam South (%)	Amsterdam South-east (%)	Rotterdam Downtown (%)	The Hague Downtown (%)	The Hague South-east (%)	Utrecht (%)	Total (%)
Type 1	7.8	0.0	7.7	9.3	10.0	7.6	3.6	7.6
Type 2	5.9	3.1	7.7	3.7	10.0	5.9	10.7	6.2
Type 3	21.6	18.8	0.0	7.4	10.0	11.8	28.6	15.1
Type 4	19.6	28.1	46.2	35.2	20.0	29.4	25.0	27.6
Type 5	45.1	50.0	38.5	44.4	50.0	35.3	31.2	43.6

Source: Dwarkasing *et al.* (1988)

Table 10.3 Intensity of external linkages: by spatial scale and functional type

| | External linkages | | | |
	Within city of location	Within rest of Randstad	Within rest of country	With foreign countries
Type 1	2.3[a]	3.3	3.4	2.9
Type 2	2.8	3.4	2.8	2.2
Type 3	2.9	3.7	3.1	1.9
Type 4	2.6	2.9	3.3	2.4
Type 5	3.4	3.4	3.1	1.8

Note: (a) Values and averages on a five-point scale, running from 1 (0 per cent of all business relations per establishment) to 5 (more than 50 per cent of all relations per establishment)
Source: Dwarkasing *et al.* (1988)

Table 10.3 underpins the expectation that Groups 1 and 4 have geographically extended business contacts, whereas Group 5 is more orientated towards suppliers and clients in the adjacent urban environment. The most important result however is that the average business network of the establishments is quite evenly distributed over all geographical ranges. Excepting rare cases, external linkages are focused on a specific spatial scale. On average, the Randstad as a whole is the most important arena for linkages, although the differences between the various spatial scales are strikingly small in this respect. In general, relations of the offices extend over all ranges; none is dominant. These results point to the highly complex nature of spatial networks among advanced office activities.

It is interesting to see whether the various office centres differentiate regarding office linkage structures (Table 10.4). With the exception of the Utrecht office centre, the differences in spatial linkage structures between the office centres are quite marginal. Amsterdam South-east accommodates the most internationally orientated office population, but does not have a prime position in this regard. The Utrecht office community linkage structure is deviant because of the paramount orientation to the Randstad region.

Differentiations in the structure of office establishments by type (Tables 10.2 and 10.3) explain most of the differences in average linkage structure between office centres. For example, branch office establishments explain the relatively strong international orientation of the Amsterdam South-east centre, whereas branch offices of domestic firms explain the Randstad orientation of Utrecht. By the same token, the relatively strong local orientation of offices in Amsterdam South is explained by the dominance of Type 5 (business services) who find a large market in the nearby centres of Amsterdam Downtown and South-east.

A well-known hypothesis suggests that nearby contacts are more often face-to-face than extended contacts (Goddard and Morris 1976). This

Table 10.4 Intensity of external linkages: by spatial scale and office location

	External linkages			
	Within city of location	*Within rest of Randstad*	*Within rest of country*	*With foreign countries*
Amsterdam Downtown	3.0[a]	3.2	3.0	1.9
Amsterdam South	3.4	3.2	3.1	2.2
Amsterdam South-east	3.2	2.9	3.1	2.4
Rotterdam Downtown	3.0	3.3	3.4	2.2
The Hague Downtown	3.2	3.2	3.0	2.0
The Hague South-east	2.3	3.4	3.6	2.3
Utrecht	2.8	4.1	2.9	1.8
Total	3.0	3.3	3.1	2.1

Note: (a) Values and averages on a five-point scale, running from 1 (0 per cent of all business relations per establishment) to 5 (more than 50 per cent of all relations per establishment)
Source: Dwarkasing *et al.* (1988)

hypothesis is tested by comparing Tables 10.2 and 10.3 with the results of an analysis of the spatial distribution of face-to-face contacts. The comparison shows that the proportion of face-to-face contacts in all contacts does not seem to diminish as the geographical range of linkages increases. The spatial elasticity of personal contacts is almost zero. The advanced and non-routine nature of many transactions for the offices investigated here might explain this result. This tends to contradict existing theory on the substitution of telecommunications for face-to-face contacts with increasing distance between actors.

Spatial structure of relations with suppliers of professional business services

One of the main questions addressed by the survey concerned the role of professional business services in the formation of office complexes within metropolitan economies. The respondents were asked to indicate the frequency of use of external business services and the geographical origin of the services used by the office establishment (Table 10.5). Financial services in the widest sense (banking, insurance and accountancy) are used most frequently. These services are most often acquired locally (city of location), although insurance and accountancy services do show important relations at the level of the Randstad region. Specialization of the services involved might play a role in this regard. Relations with simple services, including administrative services are also most often local, although these latter services are not so often externally acquired when compared with banking, insurance and accountancy services.

Specialized knowledge-intensive services, including consultancy, R&D,

Table 10.5 Use of external business services by office establishments, by geographical origin of the service supplier

Service	Externally supplied		Distribution by geographical origin of supplier			
	Frequent	Infrequent	City of location	Rest of Randstad	Rest of country	Foreign country
Management consultancy	6.9[a]	20.3	26.5	57.7	29.8	3.3
Market research and information	10.0	18.6	33.2	49.0	28.3	11.2
Technical R&D and consultancy	8.2	14.7	33.6	45.9	33.6	6.1
Information techn. serv.	17.7	21.6	30.5	44.5	36.4	9.4
Legal services	20.8	42.4	66.9	30.2	9.5	4.4
Financial services	45.5	12.6	66.4	20.3	6.8	3.4
Insurance services	33.3	21.6	55.0	33.5	15.5	4.4
Advertisement and PR services	24.2	19.9	44.4	51.7	9.3	2.0
Administrative services	11.7	5.2	63.3	29.0	13.0	0.0
Accountancy	44.6	19.9	60.9	36.7	5.7	2.8
Human resources	7.4	18.2	31.6	54.7	14.1	7.0
Congresses and fairs	4.8	16.9	24.9	39.6	27.2	16.6
Simple services (travel etc.)	28.1	14.7	79.0	21.3	7.5	1.2
Average	20.2	18.9	47.4	39.3	18.2	5.5

Note: (a) % of all respondents
Source: Dwarkasing *et al.* (1988)

market research, and information technology services, are most often acquired at the higher level of the Randstad region, with the exception of legal services. The role of foreign based service suppliers is only marked in the case of market research, IT services and the organization of congresses and fairs.

There is an increasing geographical extension of the area for services acquisition with diminishing intensity of demand (Table 10.5). More precise hypotheses on the factors explaining the spatial structure of service acquisitions could be formulated by disaggregation of the figures for each of the locations studied. Space does not permit the presentation of more disaggregated figures, but these reveal that the following important factors share the spatial networks between demand and supply of business services:

(a) the localization and diversification of demand for, and supply of, professional producer services;
(b) the specialization and complexity of the process (labour) and the product of the professional service involved, in other words the level of routinization of the service;
(c) the mode of transfer of the service and the intensity of interaction between supplier and consumer;
(d) the ability of professional service firms to branch spatially, either based on a division of labour (front- and back-offices) or by cloning (dispersal of identical offices within the same firm); and
(e) the sensitivity of interactions to the kind and quality of infrastructure for personal transport and communication.

Extensive local complex formation, measured by the amount to which service demands, including high level services, are met locally, appears only in the Amsterdam area (which accommodates three major office locations). This area has the most diversified structure of advanced office establishments, so that in relation to demand and supply of specialized producer services, Amsterdam is relatively self-supporting. The least extensive complex formation appears in Utrecht, smallest and least diversified of the four urban centres; demand as well as supply of professional services in Utrecht is quite strongly orientated towards the other cities in the Randstad region.

The diversity of office activities in Rotterdam and The Hague is comparable with that of Amsterdam, but local complex formation is less marked. The conclusion which can be drawn is that the four cities in the Randstad region together constitute a fairly integrated complex of supply and demand of professional services. Three hierarchical levels can be distinguished: an upper tier diversified in demand as well as supply (Amsterdam), a middle tier diversified in demand but less so in supply (Rotterdam, The Hague) and a lower tier less diversified in demand and rather specialized in the supply of services (Utrecht).

CONCLUSIONS

Although there has been a marked development of office activities in general in cities outside the Randstad region, documented by Van Dinteren (1987) and Buursink (1986), the most advanced segment of office activities (that is, corporate management and internal and external professional services) is still to a large extent concentrated and over-represented in the four major cities in the Randstad. In other words, there is a marked positive correlation between the level of office activities and their agglomeration at the national scale.

According to the hypotheses discussed in the theoretical section of this chapter, the advanced metropolitan economy is characterized by both diversity and integration of advanced office activities. This applies for the Randstad economy as a whole: the interweaving among advanced office activities is most marked at the regional Randstad level, and is less developed at either the level of the separate cities within the Randstad and at the national level. At the intra-Randstad level, only the city of Amsterdam has a relatively high degree of diversity and local interweaving of advanced office activities. For the other Randstad cities, we could state that their integration into the larger context of the Randstad economy compensates for the lack of local diversity and complex formation. The whole is more than the sum of the parts.

From the practical point of view of planning and economic development, the question could be asked whether efficient network development among advanced offices at the Randstad scale is hampered by the polycentric nature of the region. According to the surveyed office establishments, the geographical situation of the establishment within the Randstad region and the access to the Randstad intercity motorway network are, apart from the prominence and physical quality of office buildings and their immediate neighbourhoods, prime location factors. The rising importance of the infrastructural ring connecting the four major urban centres and the airports (see Figure 10.3) is evident from the successful development of new advanced office centres at the intersections of radial urban motorway rings and the larger Randstad motorway ring. These new centres, efficiently situated from the perspective of the Randstad as a whole, are in most cases highly competitive *vis-à-vis* the traditional downtown CBDs. The observations presented here are underpinning the relevance of intercity infrastructural connections within the Randstad region, the upgrading of which is currently a major issue in Dutch politics. Of course, an alternative way to facilitate efficient business networks at the metropolitan scale would be the concentration of high level offices at a very small number of large and diversified locations. The National Planning Agency proposes, partly based on the observations presented here, to adopt an intermediate policy: stimulation of advanced office concentrations in the three largest and most diversified centres (Amsterdam, The

Hague, Rotterdam) coupled with modernization of Randstad intercity connections. This policy would leave the typical polycentric nature of the region partly intact and would render it more efficient.

The polycentric complexity of the spatial pattern of advanced office development in the Randstad is also theoretically significant. On the one hand, the surveyed office locations, scattered over the four cities, are to a certain degree functionally differentiated. Multivariant analysis indicated five functional groups of activities for which the locations show the most marked differentiation: top managerial services within Dutch multinational corporations (prime location: The Hague SE), top managerial services within Dutch nationally orientated firms (prime locations: downtown Utrecht and downtown The Hague), lower order managerial services in Dutch corporations (prime location: Utrecht), managerial services within foreign based multinational corporations (prime location: Amsterdam SE) and finally specialist suppliers of advanced business services (prime locations: downtown CBDs of the three largest cities, and Amsterdam South). This spatially differentiated functional structure explains much of the differences between locations in network development, including inter-national orientation. It explains partly why complexes are most often formatted at the Randstad level, and much less so at the urban level (except in the Amsterdam case which comprises three advanced office locations that are complementary in functional structure). On the other hand there is also a hierarchical structure, in which Amsterdam takes the upper tier, Rotterdam and The Hague the middle tier and Utrecht the lower tier.

Regarding the general subject of this volume, some significant con-clusions can be drawn from the observations concerning the use of external professional producer services. The extent of the market area (or the acquisition area) of external business services is positively correlated with the complexity of the service involved (cognitive level, organization of interaction between supplier and client). Complex formation regarding the most advanced services is most extended at the inter-city scale, less so at the intra-city scale, and almost not at all at individual office locations. Except for some particular services, international exchanges do not weigh much regarding total exchanges in business services.

The results presented here suggest that the body of geographical theory on urban development and the role of transactional activities, shaped among others by Gottman (1983) and Pred (1977), needs to be extended and improved. New theories and facts on the economic, functional and spatial development of advanced professional activities, engines of the competitive urban economy, should be accounted for in the formation of geographical theory which can then deal better with the rising complexity of urban service economies.

NOTES

1. The adjective 'advanced', introduced by Noyelle (1983), may lead to confusion because of its resemblance with 'progressive' in the Baumolean sense. Baumol (1984) uses the concept of 'progressive services' for those services showing high increases in productivity. It may be clear that 'advanced' here means quite a different thing. In fact, most 'advanced services' in the present sense are among the least 'progressive services' in the Baumolean sense.
2. Evidence on these processes and their implications is particularly rich in the United States. De-industrialization and its impact on US cities is documented by Bluestone and Harrison (1982); the rise of advanced office activities in US cities is documented by Stanback and Noyelle (1984).
3. See the recent White Paper on Spatial Planning by the Dutch National Government *(Vierde Nota over de Ruimtelijke Ordening*, 1988).
4. This latter argument, defended by Baumol (1984) and many of his fellow economists, is however not unchallenged (see, for example, Delauney and Gadrey 1988).
5. The functionality of many new professional services is becoming a matter of increasing debate, however. See P.P. Tordoir, 'The professional economy', thesis forthcoming in 1990.
6. For a comparison of all OECD countries regarding the development of specialized business services see Elfring (1988).
7. This is a main conclusion of research on the corporate demand for advanced professional services (see ERMES 1988).
8. The basis of this map is a careful selection of four-digit headings within the Dutch Standard Industrial Classification, including only the most specialized financial and professional service establishments and corporate staff establishments. The selection totals 998 establishments for the whole country (INRO-TNO 1988).
9. This information on the spatial distribution of office activities in the Randstad region resulted from two preliminary studies (Tordoir and De Haan 1987; De Smidt *et al.* 1987).

REFERENCES

Baumol, W. (1984) *Productivity and the Service Sector*, Discussion Paper Series No.1, Fishman-Davidson Center, University of Pennsylvania.

Bell, D. (1973) *The Coming of Post-Industrial Society*, New York: Basic Books.

Bluestone, B. and Harrison, B. (1982) *The Deindustrialization of America*, New York: Basic Books.

Buursink, J. (1986) *Urbanization in the Dutch Service Economy*, Research Paper no.21, Institute of Geography and Planning, Catholic University of Nijmegen.

Chandler, A.D. (1977) *The Visible Hand*, Cambridge (Mass): Harvard University Press.

Dieperink, H. and Nijkamp, P. (1986) 'De Agglomeratie-index: een ruimtelijke indeling op grond van agglomeraieverschijnselen', *Planning, methodiek en toepassing*, No.27.

Dwarkasing, W., Hanemaayer, D., de Smidt, M. and Tordoir P.P. (1988) *Ruimte voor Hoogwaardige Diensten*, Netherlands Geographical Studies, no. 71, Royal Dutch Geographical Society, Amsterdam.

Elfring, T. (1988) *Service Employment in Advanced Economies*, PhD Thesis, Groningen University, Groningen.

ERMES 'Etudes et Recherches sur les Mutations Economiques des Services' (1988) *La Demande de Services Complexes des Firmes Multinationales et L'Offre Correspondante*, Groupe de Travail LAST-CLERSE (CNRS), UFR de Sciences

244 Services and metropolitan development

Economiques et Sociales, Université de Lille I, Villeneuve D'Ascq.

Goddard, J.B. and Morris, D. (1976) *The Communications Factor in Office Decentralisation*, Progress in Planning vol. 6, 1, Oxford: Pergamon Press.

Gottman, J. (1983) *The Coming of the Transactional City*, University of Maryland: Institute for Urban Studies.

INRO-TNO (1988) *Economisch-Technologische Vernieuwing en Ruimtelijke Ordening*, Delft: Institute of Spatial Organization TNO.

Noyelle, T. J. (1983) 'The rise of advanced services', *Journal of the American Planning Association*, summer.

Pred, A. (1977) *Cities in Advanced Economies*, London: Hutchinson.

RPD (1987) *Notitie Ruimtelijke perspectieven*, Staatsuitgeverij, Den Haag.

Smidt, M. de, van Dijck, B. and Hanemaayer, D. (1987) *Eindrapport vooronderzoek Toplocaties voor Commerciele Dienstverlening*, Geographical Institute, State University of Utrecht.

Stanback, T.M. and Noyelle, T.J. (1984) *The Economic Transformation of American Cities*, Totowa NJ: Rowman & Allenheld.

Tordoir, P.P. and de Haan, M.A. (1987) *De Economische en Ruimtelijke Ontwikkeling van Internatiuonaal Georienteerde Kantooractiviteiten. Trends en Implicaties voor de Randstad*, Delft: Institute of Spatial Organization TNO.

Van Dinteren, J.H.J. (1987) 'The role of business services in the economy of medium-sized cities', *Environment and Planning A*, 19 (5), 669–86.

Vierde Nota over de Ruimtelijke Ordening (1988) Tweede Kamer, vergaderjaar 1987–1988, 20490, nrs. 1–2, SDU Uitgeverij, Den Haag.

Williamson, O. (1980) 'Transaction cost economics: the governance of contractual relations', *Journal of Law and Economics*, March.

11 The internationalization of New York services

Barney Warf

Throughout much of the mid-1970s, the New York metropolitan region demonstrated a textbook series of symptoms of urban decay: high unemployment, exacerbated by a mass evacuation of manufacturing firms; net out-migration; declining tax revenues, forcing local governments to the brink of bankruptcy; a steady exodus of corporate headquarters; a collapse of the real estate markets; dropping per capita income; and mounting consternation in the business community (Shefter 1985; Starr 1985). The crisis, however, also generated the preconditions for a renaissance and ushered in a powerful restructuring process currently reshaping every facet of the region's economic and social landscape. In the 1980s, the trans-Hudson river New York metropolitan region has experienced a remarkable resurgence to become one of the healthiest parts of the United States. Its unemployment rate is well below average; its real estate market has boomed, unleashing an enormous wave of new construction; it has compensated for its earlier population losses; and a variety of new firms, both foreign and domestic, have been attracted to the region.

The reasons for this change of affairs lie largely in the region's growing international orientation. Always the most international city in the United States, New York has emerged, along with Tokyo and London, as part of the tripartite axis that dominates the global geography of finance in the 1980s. This chapter examines the New York region's recovery in the light of the internationalization of the service economy, focusing on the role played by financial and business services. Because data on international services trade are poor even at the national level and are even worse at the regional level, the analysis employs a variety of sources to demonstrate the progressive increase in linkages between the New York region and the international economy. It argues that this process is largely responsible for the restructuring of the region's office and labour markets, as well as the changes in its residential spaces.

The New York metropolis, of course, has never been a stranger to the international economy; indeed, it has always been an important financial and media centre, a focus of manufacturing and a magnet for waves of immigrants (most recently from the Caribbean). The historical development of the city for centuries was associated with port activities, which facilitated the export and import of large volumes of commodities and sustained an enormous complex of service firms, including importers, banks, wholesalers, insurance companies (Danielson and Doig 1982).

Today, the New York region is one of the planet's premier 'world cities' (Moss 1987; Thrift 1987). It is home to the largest conglomerate of service sector firms in the world, including roughly 70 per cent of all US advertising companies, 40 per cent of the securities industry, a major conglomeration of insurance firms (Young 1979), and large shares of the nation's accounting, legal and other business services (Stanback 1981). It stands at the apex of the national and international banking hierarchy (Wheeler 1986) and serves as the headquarters of numerous *Fortune* 500 firms. Service firms are especially attracted to Manhattan because of the potential for 'face to face' interaction, the ability to minimize forward and backward linkages with clients and suppliers that is central to the location decisions of many firms in the service sector (Scott 1983). In addition, New York offers large numbers of specialized ancillary services, a diverse and relatively skilled labour pool, an enormous consumer market, a well-developed infrastructure (including telecommunications systems), the Stock Exchange, and the bond market available through the New York Federal Reserve Bank. While Manhattan's advantages are by no means unique, the size and diversity of its office functions give the New York region an advantage in financial and business services not easily duplicated elsewhere.

It would be erroneous, however, to attribute the region's recovery entirely to the service sector. High-tech firms, for example, have boomed on Long Island and in New Jersey, in part sustained by large military contracts. Even today, despite three decades of industrial loss, New York City has more manufacturing jobs than Baltimore, Denver, New Orleans, Philadelphia, and San Francisco combined (Gurwitz 1983). None the less, it is the financial and business services, particularly the most internationalized ones, which have been primarily responsible for the region's recovery since the mid-1970s, and are the focus of attention here.

NEW YORK IN THE AGE OF GLOBAL CAPITAL

More than any other part of the US economy, the New York metropolitan area has been bound up with the international sector, and as a result, has disproportionately benefited from the global changes in the 1980s. The relationship has become sufficiently obvious that it raises the question as to whether the region is not now more sensitive to international fluctuations than national business cycles. As Quinn *et al.* (1987) argue, services tend to be less sensitive to cyclical fluctuations than manufacturing, a trend that has accentuated the differences between the New York metropolitan economy and the nation's. In 1982, for example, when the United States faced its deepest recession since the Great Depression of the 1930s, New York suffered only marginally (Gurwitz and Rappaport 1984). In subsequent years, conversely, even the hint of default by Third World borrowers or a reduction in foreign purchases of government securities produces shocks felt far more severely in New York than nationally.

The contributions of international services trade to the US economy have received astonishingly little attention in the literature given that 70 per cent of the US labour force is engaged in service sector work. Unlike the trade in merchandise, in which the United States routinely registers enormous deficits, trade in services is a net revenue generator for the United States. Figure 11.1

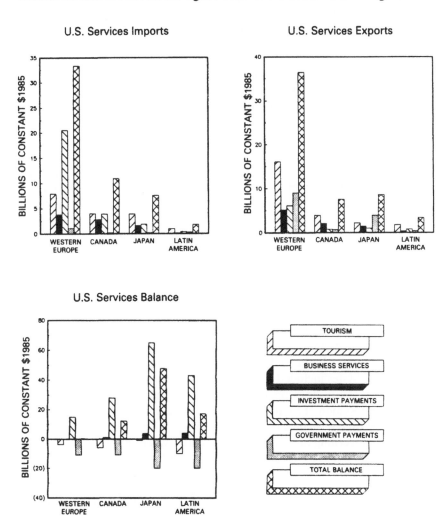

Figure 11.1 Composition and distribution of US services trade, 1985
Source: Survey of Current Business (1986)

portrays the types and magnitude of the US international services trade since 1960, including net government payments and repatriated profits from foreign investments. It is notable that both the level and composition of US services trade have changed. If one controls for repatriated profits (the largest single component), the United States exhibits a net deficit in services, in large part due to the large net government payments overseas that sustain US military bases. The geographic distribution of services points to the dominance of western Europe, a pattern that has favoured New York over other areas in the United States.

At the local scale, the contributions of service sector exports to regional growth are exceedingly difficult to assess empirically. As Beyers and Alvine (1985) have shown, many urban areas export services to markets throughout the American and, increasingly, global economy. Service exports from the New York region are diverse, including the extension of loans, bond underwriting, brokerage services, franchising, leases, royalties, intellectual property rights, entertainment and tourism, and trade in advertising, consulting, accounting, legal services, research and development, architecture and engineering, and so forth.

Unquestionably the most potent force driving the New York economy is its banks; banking is to New York as the automobile industry is to Detroit or steel was to Pittsburgh. The rapid growth in banking and other financial activities is primarily attributable to the conjunction of several factors, including the wave of deregulation that began in the 1970s, the oil crisis and recycling of petrodollars, the pre-eminence of Japan and the transformation of the United States into the world's largest debtor, the explosion of global securities trading, and the growth of secondary markets such as foreign exchange transactions and the Euromarket (Ingo 1988).

The deregulation of the global financial industry contributed enormously to New York's status in the 1980s. Deregulation may be traced to the 1973 collapse of the Bretton-Woods agreement and the shift to floating exchange rates, which permitted unprecedented transnational capital flows that circumvented national regulatory controls designed to protect the currencies of individual countries. Other deregulatory measures instituted in the United States included the lifting of controls on foreign direct investment, raising the ceilings on interest rates, the acceleration of interstate banking, the abolition of fixed commissions on the Stock Exchange, the creation of a host of new financial instruments (e.g. 'junk bonds') and the relaxation of rules governing the use of pension funds, which created an enormous pool of capital for investment, much of which found its way into the New York securities market. Deregulation also created investment opportunities overseas, such as with London's 'Big Bang' in 1986. Ultimately, deregulation raised competition in the financial sector, liberated large quantities of new funds for investment, heightened the volatility of the capital markets, encouraged large institutional investors (e.g. pension funds) to enter the field, and facilitated a wave of corporate mergers, takeovers, and

leveraged buyouts. In the United States all of these forces disproportionately favoured the New York area.

The recycling of petrodollars following the 'oil shocks' of the 1970s had
a dual set of effects on New York. On the one hand, as a centre of
international banking, New York became the repository for vast quantities
of OPEC funds. On the other hand, this trend firmly tied New York to the
Third World debt crisis as many banks heavily increased their debt
exposure to Latin American nations (Cline 1983). Despite recurrent fears
of default, however, Third World debt has proved less than catastrophic as
banks have written off loans, arranged debt-equity swaps and sold others
on the secondary debt market. In 1988, the repayment of past-due interest
bolstered the earnings of banks such as Citicorp and J.P. Morgan (Quint
1989). The debt issue is particularly important to New York because it
disproportionately affects the regional economy. After Mexico's near-
default in 1982, for example, the growth in banking employment in New
York City came to a virtual halt. Indeed, were it not for the rising number
of jobs in foreign banks, New York's total banking employment would
have declined in the 1980s.

New York's status as a hub of international banking activity is also
symbolized by the presence of foreign banks. Figure 11.2 indicates that
while New York lags behind London in terms of the number of foreign
banks represented there, it is by far the leading centre of foreign banking in
the United States, where Los Angeles remains a distant second heavily
orientated towards Pacific Rim financial circuits. Foreign banking has
become increasingly important to the New York labour market (Table
11.1): in 1979, 244 foreign bank branches operated in the New York
metropolitan region, employing 19,000 persons; by 1987, there were 351,
with combined assets of $329 billion (half of all foreign bank assets in the

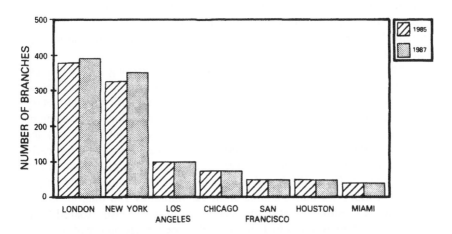

Figure 11.2 Foreign banks in London and selected US cities
Source: *The Banker*, 1987 (March)

Table 11.1 Foreign banking in Manhattan, 1979–87

	Number of foreign nations	Number of foreign banks	Foreign bank employment	Total bank employment	Foreign as % of total
1979	48	244	18,900	125,151	15.1
1980	52	253	20,600	131,802	15.6
1981	52	255	22,500	141,454	15.9
1982	61	285	25,300	147,208	17.2
1983	62	294	26,336	157,292	16.7
1984	63	307	27,300	152,845	17.9
1985	64	326	27,718	151,723	18.3
1986	65	356	34,875	152,301	22.9
1987	66	351	33,717	147,891	22.8

Source: *The Banker* (1979–88)

nation), employing more than 33,700 workers, or 23 per cent of all banking jobs in Manhattan.

New York's largest foreign banking presence is unquestionably Japan. Banking and investment giants such as Dai Ichi Kangyo (with $250 billion in assets in 1987, the largest firm in the world) are but one expression of Japan's increasingly formidable economic status throughout the world. Buoyed by a high national savings rate, the strong yen, and relatively few domestic investment opportunities, Japan has become the largest creditor nation in the world, with net external assets greater than those of OPEC at its apogee (Vogel 1985). By purchasing more than 40 per cent of US Treasury Department securities, for example, Japanese investors finance much of the American federal budget deficit. In 1987, Japanese banks employed more than 4,500 New Yorkers and were increasing faster than those of any other nation (Warf 1988); in the 1980s, most of the new entrants have been small banks that have entered the international market relatively late.

Despite the importance of banking, the fastest growing sector in New York in the 1980s is the securities industry, now the region's second largest employer. Since 1982, the US Federal Reserve's expansionary monetary policy has raised excess liquidity, which increased the flow of investment funds into the stock market and, along with foreign investors and institutional players, contributed to the enormous bull market of the early 1980s. The Dow Jones Industrial Average, for example, rose from 750 in 1976 to over 2,400 in 1987, and the number of shares traded daily on the New York Stock Exchange soared from 12 million to 150 million shares. Concomitantly, employment in the region's securities industry soared from 60,000 in 1975 to 160,000 in 1987, growth closely correlated with the rise in the Dow Jones (Figure 11.3). Not surprisingly, average annual incomes

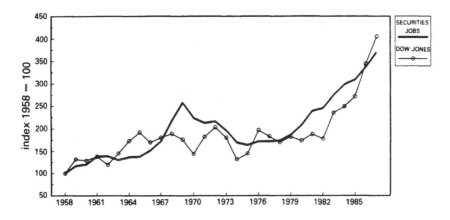

Figure 11.3 Relative growth of New York securities employment and Dow Jones Industrial Average, 1958–86

in the securities industry rose to the highest in the nation, from $41,000 in 1977 to $66,000 in 1987.

As with banking, the securities industry has become increasingly internationalized. Between 1980 and 1985, the number of foreign securities firms in New York increased from twenty-nine to forty-one, representing eleven nations (Table 11.2). As with banking, the Japanese firms are particularly important, notably the 'Big Four' of Nomura, Yamaichi, Daiwa and Nikko, all of which have become major players in the financial markets (Warf 1988). The Big Four account for half of all stocks on the Tokyo Exchange (currently the world's largest), 20 per cent of all US federal bond sales, and 10 per cent of all stock transfers on the New York Exchange (*Business Week* 1987). Similarly, several New York investment banks, such as Merrill Lynch, Morgan Stanley, and Saloman Brothers, have opened foreign offices, notably in Tokyo, the site of the world's largest stock market.

The degree of internationalization in the securities market became painfully evident on 19 October 1987, when the Dow Jones plunged by 508 points, or 22 per cent. The crash brought immediate, if less drastic, declines on other markets throughout the world (Table 11.3). Despite the resulting fall in investor confidence, however, the long-term effects of 'Black Monday' are probably negligible. The Dow Jones has stabilized, and begun to creep towards its former maximum. Layoffs by New York securities firms, totalling about 11,000 persons, have been far less than originally expected. While the crash dampened the escalation in the region's real estate prices, it demonstrated most clearly the enormous volatility of the financial markets in the age of global capital.

One indication of the reliance New York's financial and business services sector has upon global markets may be seen through a simple time-

Table 11.2 Foreign securities firms in Manhattan, 1980 and 1985

	1980	1985
Canada	11	9
Japan	10	12
Switzerland	4	5
Belgium	1	0
Australia	0	3
West Germany	0	2
United Kingdom	0	2
Bahamas	0	1
France	0	1
Israel	1	1
Hong Kong	0	1
Consortia	1	4
Total	28	41

Source: *Crain's New York Business*, June (1986)

Table 11.3 Percentage decline in indexes for major world stock markets, 19 October 1987

New York	22.6
Tokyo	2.3
London	10.8
Mexico City	16.5
Paris	6.1
Milan	6.3
Amsterdam	7.8
Frankfurt	7.5
Toronto	9.1
Hong Kong	9.3

series model in which the New York region's FIRE employment is regressed against total capital flows into and out of the United States. Figure 11.4 indicates that the model performs very well over the 1969–87 period ($r^2 = 0.92$), although the relationship is less stable in the highly volatile markets of the 1980s, a period in which New York finance became considerably more orientated to international rather than domestic sources of investment.

THE GLOBALIZATION OF NEW YORK BUSINESS SERVICES

Business (or producer) services, which comprise a rapidly growing part of the New York service economy, include a variety of advertising, public relations, legal services, architecture and engineering, and accounting firms. Business services employment totalled 2 million in the metropolitan area in

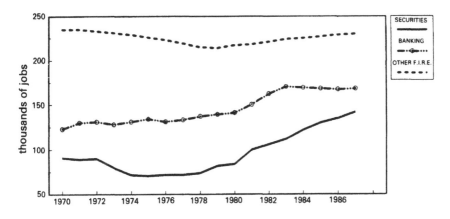

Figure 11.4 New York city FIRE employment, 1970–86

1987, and increased by 6 per cent annually during New York's recovery from the fiscal crises of the 1970s. Table 11.4 illustrates the degree of export orientation for four business services from the New York region, indicating that they are less reliant upon extraregional revenues than their counterparts elsewhere in the United States.

The rapid growth of business services has occurred as they have become a significant aspect of global services trade (Kakabadse 1987). Table 11.5 illustrates the relative importance of foreign revenues to six categories of US business services. At the regional scale, anecdotal evidence indicates that several New York engineering and construction companies worked upon the design of Japan's new Kansai airport in Osaka Bay, a $7 billion project. The growth of international business services trade has become increasingly important in the light of the Uruguay round of the GATT negotiations, which focused heavily on the barriers to service trade.

Accounting firms tend to be among the most internationally orientated of business services. In the United States accounting is dominated by the 'Big Eight' firms (see Table 11.5), most of which are headquartered in New York and have extensive international operations. The chargeable hours of the Big Eight firms, for example, are distributed as follows: North America, 54 per cent; Europe, 22 per cent; Asia, 15 per cent; South America, 5 per cent; Africa and Middle East, 4 per cent. US operations generally account for a relatively small proportion of their total worldwide employment (Table 11.6).

The advertising industry is another significant business service; the New York region, with 291,000 persons employed in this sector, contains roughly one-third of the total advertising employment in the United States, including seven of the ten largest firms. Advertising offers a notable example of global services trade: the largest forty-three advertising agencies in the United States derive 40 per cent of their gross revenues from

Table 11.4 Selected service exports[a] from the New York SCSA, 1982

SIC	Sector	Exports as % of total receipts	
		US	New York
737	Computer data processing	13.1	9.9
7392	Consulting and public relations	22.4	26.8
7394	Equipment rental	34.6	7.0
891	Engineering and architecture	21.2	17.8

Note: (a) National and international
Source: Computed from *Census of Service Industries* (1982)

Table 11.5 Percentage of gross revenues derived from foreign sources for selected US business services, 1985

Telecommunications	2.5
Licensing	4.0
Advertising	15.0
Engineering	21.0
Motion pictures	22.6
Accounting	25.0

Source: Office of Technology Assessment (1986)

Table 11.6 Distribution between US and global offices and employment of 'Big Eight' US accounting firms, 1988

	US offices employees ('000s)		Worldwide offices employees ('000s)	
Arthur Anderson	64	14.3	221	25
Arthur Young	94	9.0	385	20
Coopers & Lybrand	98	9.8	510	36
Deloitte Haskins Sells	100	6.9	330	19
Ernst & Whinney	115	11.0	235	21
Peat Marwick Main	86	3.0	400	44
Price Waterhouse	113	10.7	400	33
Touche Ross	100	11.0	490	21
Total	770	75.7	2,971	219

Source: Individual company reports (1988)

foreign sources. Table 11.7 indicates that international clients accounted for as much as 77 per cent of the gross revenues of large New York advertising firms in 1987, or $2.3 billion in revenues. Numerous foreign firms, particularly British companies such as Saatchi and Saatchi, have also entered the regional advertising market.

New York is a premier site for the practice of corporate law, accounting for one-third of the largest 100 US legal firms (Mollenkopf 1984). The region's law firms have built a flourishing business overseas, where they service financial transactions, underwrite securities, and assist in corporate takeovers (Farnsworth 1987). Table 11.8 illustrates the distribution of attorneys employed in the largest 100 New York law firms in 1978 and 1987, revealing very rapid growth in the United States (especially Los Angeles) and a decline in the proportion stationed abroad. The foreign branches, however, remain sizeable. For example, two Manhattan firms – Davis, Polk and Wardwell, and Cleary, Gottlieb, Steen and Hamilton – have been engaged on both sides of an Italian industrialist's attempts to take over the Belgian conglomerate Société Générale de Belgique. Table 11.9 offers other examples. Although foreign restrictions to legal services have been relaxed, they still create severe obstacles to this sector: for example, foreigners still cannot council Japanese clients about Japanese law. The shifting US macroeconomic position has also affected the nature of international law: whereas previously American law firms advised

Table 11.7 Foreign revenue sources of largest fifteen New York advertising firms, 1986

	Non-US revenues ($ million)	*Non-US revenues as % of total*
HCM	40.9	77.2
TBWA Advertising	50.0	76.1
SSC&B Lintas	168.0	70.9
McCann-Erickson	280.9	65.7
DYR	16.0	61.8
Saatchi and Saatchi	295.0	60.3
Ogilvy and Mather	209.9	45.7
J. Walter Thompson	211.3	44.9
Young and Rubicam	270.3	43.0
D'Arcy Masius Benton Bowles	144.1	42.8
Ted Bates	203.4	41.9
Greers Gross	12.0	34.3
Grey Advertising	105.5	34.1
DDB Needham	124.6	33.2
BBDO	138.1	31.0
Average		50.9

Source: *Advertising Age*, 26 March (1987)

Table 11.8 Distribution of attorneys employed outside of New York in the largest 100 New York legal firms, 1978 and 1982

Location	1978	1982	Change (%)
Washington, DC	170	498	192.9
Los Angeles	17	160	841.2
Other US	72	165	129.1
Total US[a]	259	823	217.7
Paris	106	134	26.4
London	33	49	48.5
Brussels	30	30	0.0
East Asia	30	41	36.6
Latin America	11	1	−90.9
Middle East	3	4	25.0
Total Non-US	213	259	16.9
Grand Total	472	1,082	129.2
Non-US as % of Total	45.1	23.0	

Note: (a) Not including New York region
Source: Mollenkopf (1984: 17)

Table 11.9 Examples of New York law firms and foreign clients, 1987

Firm	Client	Nation	Type of business
White and Case	Teletas	Turkey	Underwrite securities
Wachtell, Lipton, Rosen and Katz	Bridgestone of Japan	Japan	Acquisition of Firestone Rubber Co.
Sullivan and Cromwell	Hatchett Group	France	Purchase of companies
Swaine and Moore	Campeau Group	Canada	Acquire Federated Department Stores
Wilkie, Farr and Gallagher	Bianca Commercial Italiana	Italy	Buy Irving Bank
Coudert Brothers	Sumitomo Life Insurance	Japan	Purchase New York real estate

Source: Labatan (1988)

domestic clients about the legal restrictions they faced abroad, today they often tend to help foreign firms investing in the US securities and real estate markets (Labaton 1988).

Tourism is perhaps the fastest growing part of international services trade (Tuttle 1987). As a centre of entertainment and the arts, New York has long attracted visitors from around the nation and the world; Table 11.10 indicates that foreign visitors have become relatively more important to the generation of tourist dollars. By 1987, foreigners accounted for roughly 20 per cent of the New York region's visitors, spending more than $2.5 billion annually there and sustaining much of the region's hotel and entertainment industries. Japanese tourists now account for the largest single source (more than 250,000) per year (Warf 1988).

IMPACTS ON OFFICE AND LABOUR MARKETS

The reconstruction of the New York metropolitan region in the 1980s has engendered a concomitant restructuring of its economic landscape. The New York region's conglomeration of office spaces – which are not homogeneous but a series of quasi-distinct units with limited substitutability – is easily the world's largest. Roughly one billion square feet – 40 per cent of the US total office space – are found within a fifty mile radius of the CBD (Schwartz 1979). Since 1976, the region has gained over 140 million square feet, the equivalent of the entire city of Houston. Manhattan alone is the site of roughly 300 million square feet, four times the footage of Chicago, seven times that of Washington, DC. With the recovery, Manhattan has experienced a building boom of a magnitude unparalleled in its history (Ponte 1985). Correspondingly, rents and land values have

Table 11.10 Tourists to New York City, 1976–87

	Foreign visitors (millions)	*All visitors (millions)*	*Foreign as % of total*
1976	1.7	16.5	10.6
1977	1.8	16.7	11.0
1978	2.0	17.0	11.8
1979	2.1	17.5	12.1
1980	2.6	17.1	15.2
1981	2.7	17.0	15.8
1982	2.7	16.9	16.0
1983	2.8	17.1	16.4
1984	2.9	17.2	16.8
1985	3.1	17.1	18.1
1986	3.5	17.4	20.1
1987	3.8	17.6	22.1

Source: New York Convention and Visitors' Bureau

risen dramatically: in 1987, median CBD office rents exceeded $50 per square foot (prime spaces approached $500 per square foot), and median land costs exceeded $1,700 per square foot. Notably, the burst of office construction in Manhattan has been exceeded by that in northern New Jersey, which is functionally part of the New York metropolitan economy and one of the fastest growing parts of the United States in the 1980s (Greer 1982).

Foreign investment accounts for much of the recent growth in the New York region's real estate market. Totalling $8.5 billion in 1987, foreign investments account for 60 million square feet, or 20 per cent of Manhattan's total office inventory. While Canada is the leading investor in terms of space, with 50 per cent of the New York region's foreign owned property (most of which is owned by a single firm, Olympia and York), Japanese firms have become the largest foreign owners in terms of value, owning $3.9 billion worth of property in 1987 (Figure 11.5). For Japanese investors, however, real estate in Manhattan is a bargain: prices in downtown Tokyo can be twenty times as high. Finally, the recent flight of capital from Hong Kong in response to fears about the impending return to China has also induced a significant flow of funds into New York's Chinatown community, dramatically raising rents there. Manhattan's role as the home to ever larger numbers of foreign rather than domestic firms well symbolizes its function as a global 'front office'.

While the growth of the CBD is a sign of the health of the regional economy, it creates a serious predicament for the firms caught between rapidly escalating rents and a reliance upon its agglomeration economies. The corporate responses to this changing set of circumstances offer a

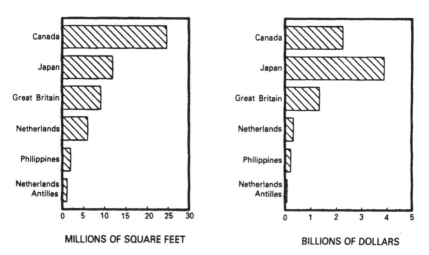

Figure 11.5 Foreign ownership of New York metropolitan region real estate, 1987
Source: Port Authority of New York–New Jersey

fascinating example of the interpenetration of land and labour markets. By driving up rents in the CBD, the internationalization of the New York economy has brought increasing pressures on firms to relocate their 'back offices', which house routinized clerical functions such as data entry, cheques and claims processing, and billing and payroll tasks. Most large finance and business services firms divide their labour forces about equally between skilled administrative, managerial and professional labour, which performs headquarter functions, and less skilled clerical labour, which performs back office functions (see Nelson 1986). Historically such firms have been bound by a need to keep the two sets of workers in close proximity so as to minimize intra-firm transaction times and costs. Increasingly, however, many firms have attained the capacity to uncouple their headquarter and back office functions, and thus permit the latter to flee the high rents of the CBD, a flexibility made possible by the widespread introduction of telecommunications systems (Moss and Dunau 1986). Because back office functions involve little 'face to face' interaction with suppliers and clients, their decentralization is in conformity with the explanations advanced by a substantial corpus of office location literature (e.g. Daniels 1979; Scott 1983).

The decentralization of back offices has allowed some firms to escape the region altogether. Thus American Express moved its back offices to Salt Lake City, Fort Lauderdale and Phoenix; Citibank moved its Mastercard and Visa divisions to Tampa and Sioux Falls, South Dakota; Hertz moved its data entry division to Oklahoma City, while Avis went to Tulsa; and Eastern Airlines shifted its back offices to Miami and North Carolina. Yet other firms have even begun to move back offices to various islands in the Caribbean, Ireland, and China (Office of Technology Assessment 1985).

The explosion of financial and business services has had profound effects upon the demand for labour in the region. Many service sector jobs have undergone a pronounced upgrading in their skill requirements, largely due to office automation systems, and increased their reliance upon professional labour. Large numbers of unskilled jobs – particularly well-paying industrial ones – have been routinized out of existence or have left the CBD for the suburbs and beyond, although retail trade continues to expand as a major source of demand for unskilled workers. As Sassen-Koob (1984) notes, the internationalization of the New York economy has generated a bifurcated labour market and an increasingly bifurcated distribution of incomes. Often the schism between skilled and unskilled occupations has occurred along ethnic lines. Many minorities reliant upon the disastrous New York public education system have been trapped by a lack of adequate skills that excludes them from well-paying jobs, condemning them to a lifetime of poverty and perpetuating the existence of a permanent 'underclass'. The demand for skilled clerical jobs, meanwhile, has increased dramatically, raising fears of a 'mismatch' between the supply and demand for labour in that sector (Bailey and Waldinger 1984; Chall 1985).

INTERNATIONALIZATION AND THE CHANGING RESIDENTIAL LANDSCAPE

Because the workplace and home are intimately interlinked, occupational change frequently induces a reordering of residential spaces; for this reason, the comprehension of neighbourhood and community change can rarely be achieved without reference to the labour market. It is not surprising, therefore, that the new land and labour markets that accompanied the internationalization of the New York region have had tremendous repercussions for the places and ways in which its inhabitants live. Approaching urban growth in this light bridges the artificial and analytically disastrous chasm between the office and industrial location literature on the one hand and issues of housing and social reproduction on the other. In this way, the urban growth is revealed as much more than an abstract object of intellectual discourse; it acquires a deeply human and highly poignant meaning as well.

In restructuring the local labour market and dramatically increasing the demand for skilled professional labour, the internationalization of the New York region has attracted a flood of skilled professionals to the central city. Among these are the bankers and securities traders, many of whom must work on the basis of twenty-four hours per day markets and receive very high incomes. Like many cities in the throes of internationalization (e.g. Los Angeles, Toronto, London), New York has seen an explosion in the cost of housing. A median family home in 1988 cost roughly $179,000, the highest in the United States. Scores of 'yuppies' have taken over several formerly moribund areas such as Manhattan's 'Upper West Side', inducing the construction of innumerable luxury condominium towers and a wave of coop conversions designed to circumvent New York's stringent rent control laws. Schaffer and Smith (1986) note that the tide of gentrification has even begun to infiltrate Harlem, bastion of the black community. Brooklyn – long a lowly step-sister to Manhattan – has also experienced a pronounced 'brownstone revival'. Even many northern New Jersey shore communities, such as Hoboken and Weehauken directly across the Hudson River from Manhattan, have enjoyed a surge in property values, although pockets of poverty remain in places such as Newark, the poorest city in the United States.

The worst aspects of the housing crisis have been felt by the region's low income residents. Caught between rising rents and a labour market which requires ever fewer unskilled clerical workers and pays poverty-level wages for the remaining jobs, many low income residents have been squeezed out of their homes (Marcuse 1986). Consequently, a tide of homelessness has engulfed the city: by 1988, 100,000 residents of the region slept on the streets or in emergency shelters, while some 300,000 more teetered on the brink of losing their domiciles. Their presence has made parts of the New York metropolitan region resemble Calcutta, a tragic testimony to the dark side of internationalization.

It would be erroneous to blame all of the New York region's problems on its increasing linkages to the world economy. None the less, it is likewise evident that during the recent round of internationalization, New York has undergone a significant period of labour market restructuring which has engendered new forms of occupational segregation and an increasingly bifurcated income distribution, both of which are powerful forces increasingly linked to the ongoing realignment of residential areas. When local community change is examined in this light, it is clear that internationalization has brought with it severe and often unexpected costs as well as benefits.

CONCLUSION: THE LESSONS OF NEW YORK

The internationalization of the world economy has reinforced New York's position at the apex of the global financial hierarchy, a position rivalled seriously by London and perhaps surpassed by Tokyo. Service exports as well as imports are an important part of the process of internationalization as global markets have become more important to the region's firms, ranging from debt-equity swaps of Latin American loans to construction contracts in Japan to foreign tourists. Given the changing US macroeconomic position internationally, New York has become a centre of foreign investment and service imports as much as (or more than) a locus of export activity. Increasingly it is foreign banks and securities firms, particularly Japanese, which dictate the location of jobs and incomes in the region.

The region's office market has been profoundly reshaped by these events, not only through the levels of direct foreign investment in real estate, but also through the explosion of rents in the CBD that have induced many firms to relocate their back offices to the suburbs and beyond. As this process has unfolded (accompanied by a wave of office automation), jobs for the unskilled have progressively evaporated, while the demand for skilled professional labour, particularly in the CBD, has risen dramatically.

It is important that these changes be viewed in more than purely economic terms, for the new, internationalized division of labour has ramifications that extend into every corner of the city, restructuring the housing market and local communities. New York's yuppies and low income minorities increasingly have found themselves locked in a fierce competition for scarce housing, and the result has been a wave of displacement and homelessness. Thus, the new international division of labour, suburban office growth, gentrification, and homelessness, far from constituting discrete phenomena, are revealed as intricately interwoven facets of one broader transformation.

What is the applicability of the New York case to other metropolitan areas? Of course, the issues and problems confronting the region are not in and of themselves unique – virtually all cities are confronted with disparities in growth and periodic breakdowns in land, labour and housing markets. But it is also evident that the New York region, in some sense, is different from other urban areas in the United States. The advantages of a Manhattan location are not easily duplicated elsewhere, for they reflect its comparative advantage in the global economy.

This uniqueness detracts from the degree to which the lessons of New York may be applied elsewhere. The region, for example, may occupy a position relative to the United States as a whole similar to that of London *vis-à-vis* Great Britain, that is, a thriving, internationalized core with localized backward linkages situated within a relatively stagnant national economy (see Daniels 1986). In that Los Angles enjoys something of a similar status, the 'bicoastal economy' has emerged on the national political agenda. Nothing guarantees this process will continue, however. The next convulsion in the global economy may well see the emergence of a new set of 'world cities'. Indeed, given the rapid technological change and financial perturbations of the 1980s, the next few decades will likely yield outcomes that few can anticipate at present.

REFERENCES

Bailey, T. and Waldinger, R. (1984) 'A skills mismatch in New York's labor market?', *New York Affairs* 8, 3: 3–18.

Beyers, W. and Alvine, M. (1985) 'Export services in post-industrial society', *Papers of the Regional Science Association* 57: 33–46.

Business Week (1987) 'Japan on Wall Street' (7 September: 82–90).

Chall, D. (1985) 'New York City's "skills mismatch"', *Federal Reserve Board of New York Quarterly Review*, spring: 20–7.

Cline, W. (1983) *International Debt and the Stability of the World Economy*, Washington, DC: Institute for International Economics.

Damanpour, F. (1986) 'A survey of market structure and activities of foreign banking in the US', *Columbia Journal of World Business*, winter: 35–46.

Daniels, P.W. (1979) 'Perspectives on office location research', in P. Daniels (ed.) *Spatial Patterns of Office Growth and Location*, New York: Wiley.

Daniels, P.W. (1986) 'Producer services and the post-industrial space economy', in R. Martin and B. Rowthorn (eds) *The Geography of De-industrialisation*, London: Macmillan.

Danielson, M. and Doig, J. (1982) *New York: The Politics of Urban Regional Development*, Berkeley: University of California Press.

Farnsworth, C. (1987) 'Japan to open its doors to American lawyers', *New York Times* (15 June: D2).

Grava, S. (1985) 'Consequences of the boom: strain at the core', *New York Affairs* 8, 4: 32–47.

Greer, W. (1982) 'Employment growth in New York and New Jersey: the effects of suburbanization', *Federal Reserve Bank of New York Quarterly Review*, 7, 3: 48–52.

Gurwitz, A. (1983) 'New York State's economic turnaround: services or manufacturing', *Federal Reserve Bank of New York Quarterly Review*, 8, 30–4.

Gurwitz, A. and Rappaport, J. (1984) 'Structural change and slower employment growth in the financial services sector', *Federal Reserve Bank of New York Quarterly Review*, winter: 39–45.

Ingo, W. (1988) *Global Competition in Financial Services: Market Structure, Protection, and Trade Liberalization*, Cambridge: Ballinger Books.

Kakabadse, M. (1987) *International Trade in Services: Prospects for Liberalisation in the 1990s*, Paris: The Atlantic Institute for International Affairs.

Labaton, S. (1988) 'US law firms expand to reach global clientele', *New York Times* (12 May: 1).

Marcuse, P. (1986) 'Abandonment, gentrification and displacement: the linkages in New York City', in N. Smith and P. Williams (eds) *Gentrification of the City*, Boston: Allen and Unwin.

Mollenkopf, J. (1984) *Corporate Legal Services Industry in New York City*, unpublished document prepared for New York City Office of Economic Development.

Moss, M. (1987) 'Telecommunications, world cities, and urban policy', *Urban Studies*, 24: 534–46.

Moss, M. and Dunau, A. (1986) 'Offices, information technology, and locational trends', in J. Black, K. Roark and L. Schwartz (eds) *The Changing Office Workplace*, Washington, DC: Urban Land Institute.

Nelson, K. (1986) 'Labor demand, labor supply and the suburbanization of low-wage office work', in A. Scott and M. Storper (eds), *Production, Work, Territory*, Boston: Allen and Unwin.

Office of Technology Assessment (1985) *Automation of America's Offices*, Washington, DC: US Government Printing Office.

Office of Technology Assessment (1986) *Trade in Services: Exports and Foreign Revenues*, Special Report OTA-ITE-316, Washington, DC: US Government Printing Office.

Ponte, R. (1985) 'Manhattan's real estate boom', *New York Affairs*, 8, 4: 18–31.

Quante, W. (1976) *The Exodus of Corporate Headquarters from New York City*, New York: Praeger.

Quinn, J., Baruch, J. and Paquette, P. (1987) 'Technology in services', *Scientific American*, 257, 50–8.

Quint, M. (1989) 'Brazil payments aid Citicorp net', *New York Times* (18 January: D1).

Sassen-Koob, S. (1984) 'The new labor demand in global cities', in M. Smith (ed.) *Cities in Transformation: Class, Capital and the State*, Beverly Hills: Sage Publications.

Schaffer, R. and Smith, N. (1986) 'The gentrification of Harlem?', *Annals of the Association of American Geographers*, 76, 347–65.

Schwartz, G. (1979) 'The office pattern in New York City, 1960–1975' in P.W. Daniels (ed.) *Spatial Patterns of Office Growth and Location*, New York: John Wiley and Sons.

Scott, A. (1983) 'Location and linkage systems: a survey and reassessment', *Annals of Regional Science*, 17: 1–34.

Shefter, M. (1985) *Political Crisis, Fiscal Crisis: The Collapse and Revival of New York City*, New York: Basic Books.

Stanback, T. (1981) 'New York City and the services transformation', in B. Klebaner (ed.) *New York City's Changing Economic Base*, New York: Pica Press.

Starr, R. (1985) *The Rise and Fall of New York City*, New York: Basic Books.

Thrift, N. (1987) 'The fixers: the urban geography of international financial capital', in J. Henderson and M. Castells (eds) *Global Restructuring and Territorial Development*, Beverly Hills: Sage Publications.

Tuttle, D. (1987) 'Whether your business is tourism or not, tourism is your business', *Business America*, 16 February: 3–8.

Vogel, E. (1985) 'Pax Nipponica?', *Foreign Affairs*, 64, 752–62.

Warf, B. (1988) 'Japanese investments in the New York metropolitan region', *Geographical Review*, 78: 257–71.

Wheeler, J. (1986) 'Corporate spatial links with financial institutions: the role of the metropolitan hierarchy', *Annals of the Association of American Geographers*, 76: 262–74.

Young, J. (1979) 'New York's insurance industry: perspectives and prospects', *Federal Reserve Bank of New York Quarterly Review*, spring: 9–19.

12 Foreign banks, telecommunications, and the central city

Mitchell L. Moss and Joanne G. Brion

INTRODUCTION

Much attention is given to the movement of firms out of the central city to suburbs, small towns and overseas. Many theorists argue that telecommunications technologies will eliminate the need for central city locations and that urban land and transportation costs have also contributed to the dispersion of economic activities to the periphery of the metropolitan region (Pascal 1985; Richardson 1985). It is widely assumed that the flow out of the central city to suburban areas is applicable to all business firms. However, considerable variation exists in the vitality of central business districts and the way in which international business services locate in large central cities. All too often, it is assumed that cities are homogeneous, subject to similar economic and technological forces and equal in their capacity to respond to these forces. But the impact of economic and technological change depends on the functions that a city performs. This chapter looks at the way in which changes in international economic activity are affecting the role of selected central cities in the United States.

Despite the loss of manufacturing jobs over the past two decades, several cities are attracting firms that place a high value on proximity to clients, advanced business services and professional contacts of the kind most readily available in the central business district (CBD). Foreign banking is one industry that has expanded its presence in central cities, specifically New York City, Los Angeles, Miami, Chicago and San Francisco. This expansion coincides with the increased levels of foreign investment in US companies, securities, and real estate. Direct foreign investment in the United States in 1987 was about $262 billion, which produced about $10 billion in income (Scholl 1988). Total foreign investment in banking and finance was about $22 billion in 1987 and in real estate that figure was $24 billion (US Department of Commerce 1988). But the majority of this foreign investment is directed at major central cities and is not evenly distributed among US cities. For example, Japanese real estate investors prefer cities such as New York, San Francisco and Los Angeles because they offer stable investments in premier 'trophy' properties (Lindner and Monahan 1986; Kenneth Leventhal 1988). Other medium-sized cities such as Atlanta, Seattle, Boston and Phoenix are receiving attention from Japanese firms, but only after the investment opportunities in the larger urban centres have begun to diminish (Kenneth Leventhal 1988).

Much of the increase in foreign investment in the United States, particularly from the Japanese, can be attributed to the trade imbalance between the United States and other countries. Since 1960, US exports rose from about $35 billion to over $361 billion by 1986; however, foreign imports also have risen steadily since 1960, from about $25 billion to $484 billion by 1986 (Office of Technology Assessment 1987: 59–60). Much of this import activity is what the US Office of Technology Assessment calls 'invisible payments', which by 1985 were $110 billion, $65 billion of which is investment income (Office of Technology Assessment 1987: 64). Services have gradually made up an increasing portion of world trade. From 1978–84, the sum of all countries' exports and imports was $360 billion, of which services, excluding investment income, accounted for about one-fifth (Office of Technology Assessment 1987: 65).

Only a few American cities have fully benefited from this infusion of foreign direct investment, partly because central cities differ in their ability to generate economic growth and to attract international firms. For example, the proximity of Los Angeles and Seattle to the Pacific Rim has helped them capture much of the foreign trade with Asia. New York City is one of the nodes in the financial triangle with Tokyo and London. Miami's connection to Latin America is another example of a city which has grown as a result of international trade, particularly in financial services (Satterfield 1988). The central cities whose economies are participating in these networks have recognized that the platform of success is no longer local but global. New York City, which experienced an enormous loss of manufacturing jobs in the 1960s, has emerged in the past decade as a global business service centre whose economic growth is closely linked to international economic activities (Noyelle 1988a).

The proposition presented here is simple: a few US cities, including New York, Los Angeles, and Chicago are exceptions to the post-Second World War trend in the United States in which economic activity flows out of, not into, central cities. These cities are increasingly hinged to a dynamic set of international business services. The specialized nature of such international services is based on face-to-face transactions combined with a dependence on global communications systems. New York, for example, alone among most American cities, provides the requisite blend of physical proximity and international communications on the island of Manhattan. This chapter examines the role of international activity within the central city and focuses on New York's capacity to adapt to the internationalization of economic activity.

THE CHANGING ROLE OF THE CENTRAL CITY

During the past thirty years, most cities in the United States experienced substantial declines in population and jobs. The movement out of the central city has been attributed to such factors as: the preference of low-density housing, crime and congestion in cities, racial conflict and the rise of the automobile and interstate highway system (Bradbury *et al.* 1982). The widespread growth of decentralized office complexes, such as Tyson's Corner in Virginia, Walnut Creek outside San Francisco, Perimeter Center just north of Atlanta, and Costa Mesa/Irvine in Southern California, highlight the attractiveness of suburban locations for what were once traditional central city office activities (Leinberger and Lockwood 1986).

A report prepared for the US Economic Development Administration pointed to the rapid rise of suburban office centres across the nation.

> With surprising speed in the 1970s and 1980s, suburbs have evolved from a loosely-organized 'bedroom community' into a fully-fledged 'outer city'.... The suburbs have led the way in new job formation in both traditional blue- and white-collar occupations.... Indeed, suburban employment now exceeds central city totals in a large share of major metropolitan area labor markets.
>
> (Hartshorn and Muller 1986: 1)

While this process of geographic dispersal has been occurring, a handful of cities have adapted to the loss of manufacturing and routine office activity by becoming international information capitals. Cities such as New York, Los Angeles and Chicago have undergone a remarkable transformation from 'centers of production and distribution of material goods to centers of administration, information exchange and high-order service provision' (Kasarda 1985: 33). A city such as New York, which once served as the nation's centre for such industries as printing and apparel is now an international information capital in which banks, advertising agencies, law firms and consulting firms, have replaced garment factories and printing plants.

Despite the fact that technology makes it possible to locate office activities at remote sites, near beaches, mountain tops, and in desert resorts, New York City continues to attract and retain business. The attraction of New York lies in the capacity to link telecommunications technologies to the face-to-face activity that has helped sustain the demand for Manhattan's approximate 327 million square feet of office space. The business transactions that occur in New York City are not constrained by the boundaries of the city or of the nation but by the telecommunications systems that link the city to other centres of business activity around the world. Communications technologies allow firms based in New York to convert new information into profit-making services and decisions that result in the production of goods and services around the world.

INTERNATIONAL FINANCIAL CENTRES

The deregulation of financial services and the advent of advanced communications has led to the internationalization of financial services firms. The globalization of the world's economy has created a new role for those cities that are international hubs of business, and those which are linked together by telecommunications technology (Moss 1987b). The emergence of New York, London, and Tokyo as world financial centres has been striking: 'From 1974 to 1986, the amount of world wide capitalization concentrated in the three leading centers grew from 73 percent to 80 percent' (Regional Plan Association 1988: 5). Walter highlights the continuing importance of a physical presence in today's financial markets:

> The reasons for the rapid growth in the activities of financial institutions in various onshore and offshore markets lie primarily in the nature of the services provided. It is often (but not always) imperative for a financial institution to have a presence physically close to the client and an active presence in important markets in order to do business effectively ... the complex nature of financial services and client needs has in many ways enhanced the importance of reliable direct connections.
>
> (Walter 1988: 12)

The increased presence of foreign banks in New York City and other major US cities illustrates how the globalization of the financial services industry affects urban functions. Originally, international banking developed as a complement to international trade; the British banking system was greatly tied to the British empire, with most of its foreign mercantile activities supported by an active international banking system (Kindleberger 1983). Kindleberger notes that the separation of international banking and international trade is a fairly recent event; banks and business firms are now two parts of a world network rather than one. He also states that international banking has long been handicapped by the slowness and uncertainty of communications. This is no longer the case. According to *The Economist*, one-third of the world's foreign-exchange trade is done at the touch of a key (*Economist* 1988). Financial markets operate on a twenty-four hour basis aided by the electronic transfer of billions of pieces of information and funds among a small number of cities around the world.

Although electronic financial services make it possible to disperse the financial services industry, this industry is subject to both centrifugal and centripetal forces. Moreover, certain aspects of the financial services industry benefit from centralization while others benefit from decentralization (Levich and Walter 1989). Kindleberger (1983) and Levich and Walter (1989) note that although centralization is important to the financial industry by offering economies of scale in information gathering and processing, certain diseconomies exist that mandate the need for regional finance centres.

First, costly information about local borrowers, small firms, and local market conditions points to the need for face-to-face contact and decentralized operations ... second, national time zone differences (e.g., New York versus Los Angeles) impose another diseconomy from centralization.

(Levich and Walter 1989: 64)

These centralizing and decentralizing forces have created the need for several international financial centres around the world. As Thrift has wisely observed:

The natural habitat of commercial capital is the 'financial centre'. For three interrelated reasons the organizations of commercial capital tend to group together in these centres ... to be near clients, ... to be in close proximity to relevant markets, ... to tap into information on markets and the operations of banking and industrial corporations and the state rapidly and efficiently.

(Thrift 1987)

Although some US cities are becoming international finance centres, not all cities and regions will capture this type of growth. Chicago has been remarkably successful by expanding the hours of its futures markets and through the use of new technologies to extend the geographic reach of its futures markets. Cost reductions in the processing and transmitting of financial information have created one global financial market (Kindleberger 1978), and 'the emergence of international finance centers has facilitated the emergence of this global market' (Moss 1987a: 77). Further, the headquarters of the independent firms that once thrived in smaller cities are often subsidiaries of large, multinational companies and are located in world financial centres: 'Thus, communications and information technologies are strengthening a small number of world cities while weakening the traditional autonomy of many smaller cities' (Moss 1987a: 77). Of the top 100 US cities, for example, only ten cities have ten or more foreign bank branches. In contrast, forty-five of the top 100 cities have ten or more US commercial banks (Rand McNally 1988).

The internationalization of New York's financial services industry has been critical to the resurgence of the city's economy in the 1980s (Drennan 1988; Noyelle 1988; see also Warf Chapter 11 of this volume). Levich and Walter state that 'Overall, New York has enjoyed preeminence as the financial center of the United States. Tradition and habit are both strong forces favoring New York to continue in this role' (1989: 86). However, they note that this preeminence is not guaranteed because of the changing role of the United States in the world economy.

Foreign bank branch offices located in the United States have increased dramatically since 1970 (Walter 1988; Office of Technology Assessment 1987), with most situated in a limited number of cities. The US Office of Technology Assessment (1987: 90) in *International Competition in*

Services, notes several reasons for the increased presence of foreign banks in the United States, including:

(a) deregulation of the US banking industry;
(b) relatively small involvement in Third World debt (compared with US banks);
(c) more experience providing nationwide services than US banks; and
(d) regulations which favour or make it easier for foreign banks to purchase failing US banks and savings and loans as compared with US banks, which are subject to antitrust laws.

The US Office of Technology Assessment notes that although foreign banks provide competition for US banks, they do not sell services produced abroad; rather they produce services 'with the aid of US workers, the US banking infrastructure, and often, US capital' (1987: 90).

Walter comments that just as American commercial banks followed American multinational corporations overseas during the 1960s and 1970s, as European, Canadian and Japanese investment increased in the United States during the 1970s and 1980s, 'foreign banks followed their respective companies into the American market' (Walter 1988: 27). The objective is to follow the customer; Walter also identifies 'customer leading' (i.e. attracting customers to new locations) and seeking local markets for competitive financial products and services as reasons for the increased presence of foreign banks (Walter 1988). The increased presence of foreign banks is most clearly associated with the need for quick and reliable information for decision-making purposes. In the financial services industry, face-to-face communications, therefore, are still valued as an important way of conducting business.

Walter (1988: 83) identifies three factors that make information important in competitive performance:

(a) that it be used in multiple forms of production or repeatedly for different purposes;
(b) that its half-life tends to be short and decays at a rapid rate; and
(c) due to the increased complexity of the financial environment, clients need assistance in distinguishing between relevant and irrelevant information.

All these factors suggest that a physical presence is necessary if foreign banks are going to compete effectively with American commercial banks and other non-banking financial institutions. Daniels notes that 'in contrast to trends in the location of population and employment (i.e. away from the central city), the internationalization of certain services may be a force for centralization rather than dispersal' (Daniels 1985: 196).

The remainder of this chapter will focus on the location of foreign banks in Manhattan and other US cities. The role of central cities has not diminished with the advent of telecommunications; rather, new industries and economic activities are replacing those that have found it more efficient to locate outside the central city. These new industries and activities are among the driving forces behind metropolitan growth in the 1980s and will continue to shape metropolitan growth for the rest of this century.

FOREIGN BANKS IN MAJOR US CITIES

Studies of the internationalization of the financial services industry have largely focused on trade and regulation rather than business locational patterns. Not only have financial services concentrated in a few global hubs, but within New York, the location of specific types of financial institutions are arranging themselves in identifiable patterns. The locational behaviour of foreign banks in Manhattan supports the above hypothesis, not only on an industry basis as a whole but also in terms of nationality and global regions (i.e. Asia, Latin America, Europe and the Middle East).

Kindleberger notes that the reasons domestic banks establish foreign offices is to find outlets in capital markets and to obtain additional resources. He notes that 'domestic banks may create subsidiaries abroad for separate foreign operations for the same reasons – mainly to find outlets for surplus funds' (Kindleberger 1983). The recent increase in Japanese investments in the United States is related, no doubt, to the current trade imbalance between Japan and the United States. Perhaps most important, Kindleberger (1983) states that:

> a bank may wish to establish a branch in a foreign country ... to have a presence there. This is called 'defensive investment,' investment designed not so much to make a profit in that place as to prevent a loss somewhere else, or in the system as a whole.

The US Office of Technology Assessment supports this statement by noting that 'US operations of foreign banks have seldom been particularly profitable' (1987: 90). Therefore, the focus must generally be on presence first, profit second.

Daniels (1986) states that those countries and certain cities with good access to telecommunications infrastructure and technology have attracted new services specializing in foreign financial transactions. Data from Rand McNally shows that twenty-three of the top 100 US cities (by population) had at least one foreign bank in 1988. New York City has over twice the amount of foreign banks as any other city. Only seven cities have more than twenty-five foreign banks.

Miami has attracted much foreign banking activity in the United States, even though historically it was not considered a business hub. Although thirty-five other US cities have a larger population than Miami, it ranks sixth in terms of number of foreign banks. This fact can largely be attributed to Miami's proximity to the Caribbean and Latin America; Miami serves as a financial gateway to Latin countries, both for outgoing and, frequently, for incoming capital.

As Table 12.1 shows, all of the listed US cities experienced an increase in the number of foreign bank branches from 1980 to 1988. In just eight years, several cities saw anywhere from a 39 per cent to 450 per cent increase. Dallas, Houston, Atlanta, Miami and Washington DC experienced a sharp increase in the number of foreign bank branches, all over 100 per cent. Cities that are traditionally associated with foreign banking such as New York, San Francisco, Los Angeles and Chicago also gained foreign bank branches, although not as many as the south-west and mid-Atlantic cities mentioned in Table 12.1.

The downtown area of the City of Los Angeles – once only identifiable by the intersection of freeways rather than the presence of business activity – has been revived during the 1980s through the influx of foreign banks, financial service firms and business services. Much of this growth can be attributed to the tremendous growth in international trade. In the last seven years, 47 million square feet of office space has been added to the Los

Table 12.1 Number of foreign bank branches in US cities, 1980 and 1988

City	1988	1980	Net increase (%)
New York City, NY	277	210	12
Los Angeles, CA	104	60	73
Chicago, IL	67	48	39
San Francisco, CA	55	32	71
Houston, TX	46	25	84
Miami, FL	39	17	129
Atlanta, GA	28	10	180
Seattle, WA	14	8	75
Washington, D.C.	13	5	160
Dallas, TX	11	2	450
Boston, MA	8	4	100
Portland, OR	5	4	25
Cleveland, OH	3	1	200
Philadelphia, PA	3	2	50
Columbus, OH	2	2	0
Honolulu, HI	2	0	100
Tampa, FL	2	0	100
Charlotte, NC	1	0	100
Denver, CO	1	1	0
Lexington, KY	1	0	100
Long Beach, CA	1	0	100
Tucson, AZ	1	0	100
Total	684	431	47

Source: Rand McNally & Company, *Bankers Directory* (September 1988 and 1980)

Angeles area, making it the fourth largest office market in the country with about 108 million square feet (Shulman *et al.* 1987: 8). According to Shulman, the influx of foreign capital has further increased the value and demand for Los Angeles real estate: 'More than two thirds of all major [downtown] office properties is foreign owned' and 'the Japanese own the largest amount of commercial real estate' (Shulman *et al.* 1987). Today, Los Angeles ranks second only to New York in relation to the number of foreign banks – 104 in 1988. Indeed, Los Angeles has now replaced San Francisco as the dominant financial centre of the West Coast (Lubenow 1987).

Overall, US cities had an overall increase in the number of foreign bank branches of 47 per cent or from 431 in 1980 to 684 in 1988. These figures coincide with the rapid increase in direct foreign investment in US business and real estate over the last decade.

THE LOCATION OF FOREIGN BANKS IN MANHATTAN

According to *The American Banker 1988 Year Book*, New York City has 221 foreign bank branches; Rand McNally calculates the number at 239 (234 of which are in Manhattan and the remaining five in Queens). *The Banker*, in March 1988 published a list of 353 foreign banks, agencies and other US incorporated subsidiaries of foreign banks in Manhattan, which in total have approximately 34,500 employees. Despite the discrepancies in statistics, it is clear that the presence of foreign banks in Manhattan is not insignificant.

Prior to 1970, there were about seventy foreign banks located in Manhattan. Several established a presence in New York City in the nineteenth century; these were typically Canadian or British banks except for one Hong Kong bank, which opened an office in 1880. However, the majority of foreign banks established offices in Manhattan during the 1970s and 1980s (*The Banker* 1988). During the 1970s, 136 foreign banks opened offices in Manhattan and during the 1980s, an additional 148 were opened. These 283 banks represent about 85 per cent of the total number of foreign banks in Manhattan or a growth rate of 411 per cent since 1970.

Japan has by far the most bank branches in Manhattan, that is, fifty-one compared to the next largest presence, the United Kingdom with twenty-five banking offices followed by Italy with twenty-three offices. In the 1920s, there were five Japanese banks in Manhattan; by the 1960s the number increased to thirteen. Clearly, as with foreign banks in general, most of the growth in the number of Japanese banks in Manhattan occurred in the 1970s and 1980s. According to *The American Banker*, Japanese banks took the top seven spots by assets and account for sixteen of the top twenty-five world banks by assets. This is in striking contrast with the fact that only one US bank ranked in the top twenty-five by assets (1988) (the number of US banks ranked among the top 500 world banks (by assets) has continuously declined since 1956 from 295 to 102 in 1986).

Clearly, the above trends coincide with the increased presence of foreign banks in the United States and particularly in Manhattan. Robert Cohen notes that 'In the United States, foreign banks accounted for 16 percent of all commercial and industrial loans in 1987 with the Japanese accounting for nearly 9 percent of these loans' (Cohen 1989: 33). Most economists attribute the increased presence of Japanese banks to the appreciation of the yen; although Cohen also notes the increased competitiveness of foreign banks (Cohen 1989).

Within Manhattan, the actual location of foreign banks is concentrated in two geographic areas: Downtown or Lower Manhattan and Midtown Manhattan. Downtown Manhattan generally refers to the southernmost portion of the island from Canal Street south. Midtown Manhattan generally refers to the area between Fifty-Ninth and Thirty-Fourth Streets bounded by the East and Hudson Rivers. In terms of magnitude, more foreign banks are located in Midtown than Downtown, although for certain foreign banks, Downtown is the preferred location. Most foreign banks have chosen to locate within close proximity to banks of either their own country or region of origin. For example, of the approximately fifty Japanese banks located in Manhattan, thirty-five are located in Downtown and sixteen are located at One World Trade Center. Kindleberger points out that 'the location of banks is generally dictated by the nature of their business. Merchant bankers were originally at ports, court bankers at the capital or seat of power' (Kindleberger 1983: 588). He also comments on the tension between Midtown and Downtown Manhattan as banking centres, with Midtown having the allure of the multinational corporations while Downtown is still the centre of the capital market in New York City.

Asian banks

Japanese banks are located downtown, in close proximity to each other and to the Bank of Japan and the Japanese Finance Ministry. In addition, other Asian banks are situated downtown as well (Figure 12.1). Except for a select few, e.g. one Japanese bank on State Street and one in the World Financial Center, Asian banks in Downtown Manhattan are situated on either Broadway, Wall Street or in the World Trade Center. Asian banks have chosen, for a variety of reasons, to concentrate in a few locations downtown; a similar type of locational pattern is occurring in Midtown Manhattan (Figure 12.2). Most Asian banks are located along Park Avenue, from Grand Central to about Fifty-Eighth Street. A few Asian banks have moved as far as Third Avenue to the east and Madison Avenue to the west. Only one Japanese bank is located in the Rockefeller Center.

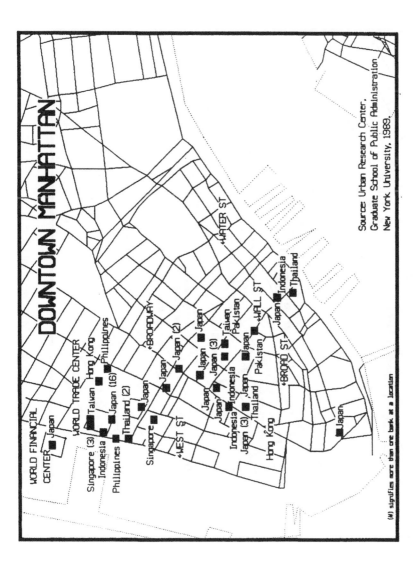

Figure 12.1 Location of Asian banks in Downtown Manhattan

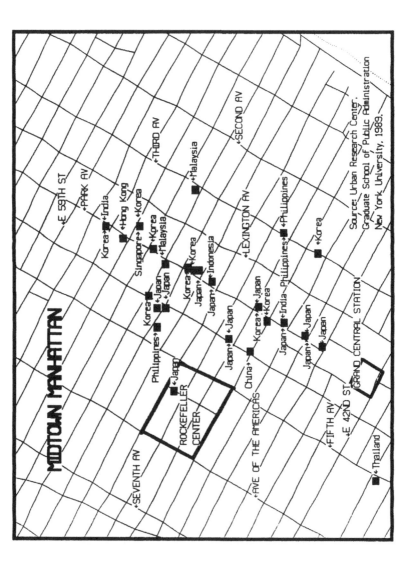

Figure 12.2 Location of Asian banks in Midtown Manhattan

Latin American and other foreign banks

With the exception of Mexican banks and a few other banks, Latin American banks are located in Midtown Manhattan (Figure 12.3). Latin American banks are concentrated along Park Avenue as well as in and around Rockefeller Center. Only a few Latin American banks have chosen to locate east of Park Avenue or south of Forty-Eighth Street. In contrast with the locational patterns of Asian banks, only one Chilean bank has located in the World Trade Center; four Mexican banks on the other hand, are in Downtown, all of which are located within a few blocks of each other between Broadway and Water Street.

The emergence of 'commercial nodes' identified by region or country is also apparent from the location of Middle Eastern banks in New York City. Only one Middle Eastern bank (from Qatar) is located in Downtown Manhattan. The locational patterns of Middle Eastern banks in Midtown Manhattan is similar to the pattern of Asian banks in Midtown although the former are more dispersed and substantially smaller in absolute numbers (twenty-six compared to 100 Asian banks).

European banks

Japanese banks are not the only source of competition for American banks. Cohen notes that in any rigorous competitive analysis of the banking industry, 'a number of European banks are also playing an important role in shaping the global competitive picture in banking', including major banks from West Germany, Switzerland, France and Britain (Cohen 1988: 36). He also states, as our numbers suggest, that 'with the possible exception of British banks, European banks have been placing a great emphasis on establishing a global presence' (Cohen 1988: 36).

European banks, including the United Kingdom, France, West Germany, Italy and Switzerland, have a significant presence in both Downtown and Midtown Manhattan (Figures 12.4 and 12.5). As shown, British banks (labelled UK) have concentrated in Lower Manhattan, predominantly on the eastern side around Wall Street. Unlike Asian banks, only one bank from Switzerland, Swiss Bank Corp., has chosen to locate in the World Trade Center. In Midtown Manhattan, save a few exceptions, all European banks are located between Park Avenue and Fifth Avenue, including several located at Rockefeller Center.

An article in *The New York Times* notes this trend, stating that 'the majority of foreign tenants would rather be Midtown than Downtown ... and rarely stray west of Sixth Avenue and east of Lexington' (McCain 1988). A large West German bank and Canadian bank have occupied much of what was once the E.F. Hutton Building (Fifty-Second Street near Sixth Avenue), which was vacated after Shearson Lehman acquired E.F. Hutton. Deutsche Bank AG has leased approximately 330,000 square feet

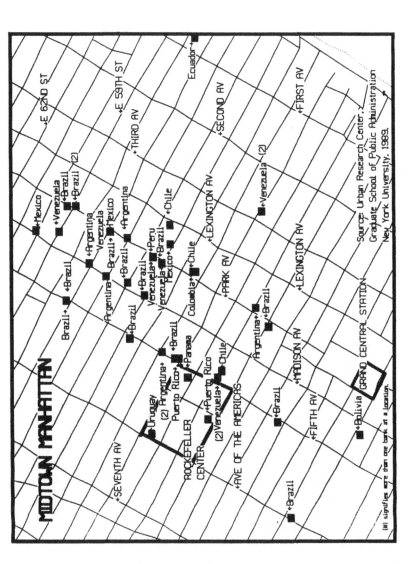

Figure 12.3 Location of Latin American banks in Midtown Manhattan

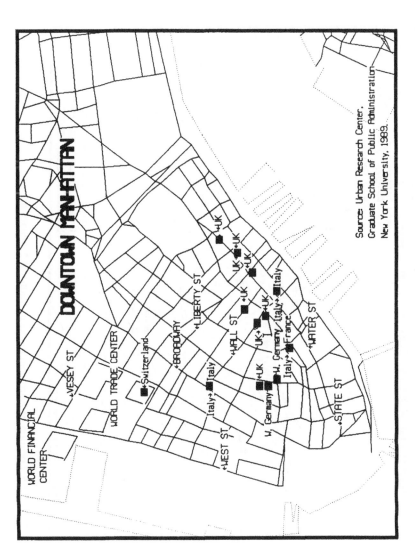

Figure 12.4 Location of European banks in Downtown Manhattan

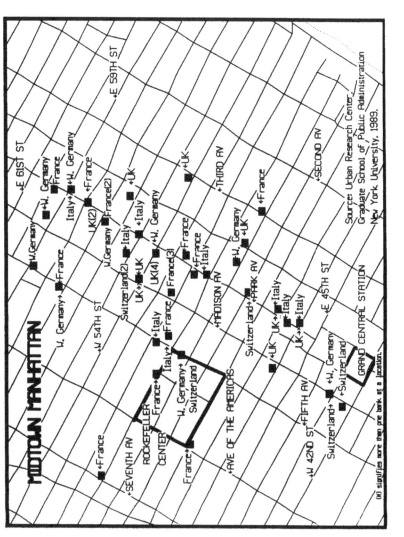

Figure 12.5 Location of European banks in Midtown Manhattan

of this building with options on an additional 200,000 (McCain 1988). The location of foreign banks in New York City indicates – as Thrift (1987) has observed – the high value that international financial firms place on proximity to clients and to information sources on banking and markets.

CONCLUSION

The data presented in this chapter demonstrate the powerful attraction that central business districts, such as Manhattan, possess for international financial service activities. During the post-Second World War period, New York City's role as a corporate headquarters and manufacturing centre has diminished, with corporate headquarters relocating to adjacent suburbs and manufacturing to other metropolitan areas and overseas. However, New York City has adapted to changes in the national and international economy, albeit with a loss in jobs for the unskilled. Once the nation's business centre, New York's economy is increasingly hinged to inter-national financial services. Manhattan may have lost its domestic franchise, but it has emerged as an international hub, closely linked by global telecommunications networks to cities such as London and Tokyo.

Within New York City, there is a remarkable ecology of foreign banking, with specific nodes where banks of similar national or regional origin locate. For example, Water Street, on the eastern edge of the traditional financial district, is the site for major British Banks such as National Westminster, Barclays, Standard Chartered Bank, and Ulster Investment. Further, almost all Asian banks are located in Lower Manhattan on Broadway, Wall Street or in the World Trade Center. By contrast, most Latin American banks are situated in Midtown Manhattan, on Park Avenue or in the Rockefeller Center area. While the location of European banks is bifurcated, split between Midtown and Downtown Manhattan, they are still aggregated within specific sites in both areas.

These locational patterns highlight the importance of proximity in an age of advanced communications technology. Although it is possible for financial services firms to communicate among dispersed locations around the world twenty-four hours a day, the dynamics of international finance still put a high value on the information gained through personal contact and face-to-face communications. New York City may have a diminished national role with the growth of large-scale regional banks in the United States, but its role as an international banking hub has been reinforced with the deregulation of financial institutions and the growth of international telecommunications networks. Furthermore, the widespread use of new, sophisticated financial instruments requires an environment where un-certainty can be reduced by close contact and co-ordination among lenders and borrowers.

The global city – such as New York, London, and Tokyo – provides a setting for banks to express their social and cultural identities through locational and architectural decisions. The appeal of high-prestige locations represents one way to establish an instant identity in a dynamic banking environment, while also attaining geographic parity with one's competitors. Although the flow of money in the international market today transcends geographic boundaries and barriers, individuals and firms engaged in international finance still prefer to be near their key clients and up-to-date sources of information. When considering the role of the central city in a 'global economy' it becomes apparent that technology has not eliminated the need for physical proximity and that major central cities such as New York will continue to dominate and attract international financial service firms.

It is important to recognize that the growth of foreign banking activity in New York City has occurred at the same time that many financial institutions have moved their data processing and back office activities away from Manhattan to other boroughs in New York City as well as to New Jersey and other regions of the United States. Advances in communications make it possible for firms to consider new locations for routine office functions where a skilled labour force is available and where space costs are less than in prime central city sites. In both cases – the immigration of foreign banks and the dispersion of back offices – telecommunications technology has allowed firms to locate activities consistent with their corporate strategy and business interests.

For world financial capitals such as New York, London, and Tokyo, the need for high-value information producers and users to be near each other has created an intense demand for prime office space equipped with advanced communications systems. This has led to the rejuvenation of the Docklands in London, the utilization of new land at Battery Park City (in Downtown Manhattan), and proposed waterfront expansion in Tokyo. Telecommunications technology allows the financial services and products generated by people working in each of these cities to be marketed and distributed to other cities around the world. However, the benefits derived from being physically present at the information source are so powerful that only a handful of cities have been able to thrive as world financial centres in an era of advanced technology.

REFERENCES

American Banker, Inc. (1988) *The American Banker 1988 Year Book,* New York: American Banker.
The Banker (1988) 'New York's foreign banks', New York.
Bradbury, K.L., Downs, A. and Small, K.A. (1982) *Urban Decline and the Future of American Cities,* Washington, DC: The Brookings Institute.
Cohen, R.B. (1989) 'The foreign challenge to US commercial banks', in T. Noyelle (ed.) *New York's Financial Markets: The Challenges of Globalization,* Boulder, Colorado: Westview Press, Inc.
Daniels, P.W. (1985) *Service Industries: A Geographical Appraisal,* London: Methuen.

Drennan, M.P. (1988) 'Information intensive industries in metropolitan areas', forthcoming in *Environment and Planning A*, 1989.

The Economist (1988) 'From foreign desk to foreign exchange', Finance column, 23 July.

Hartshorn, T.A. and Muller, P.O. (1986) *Suburban Business Centers: Employment Implications*, Final Report Prepared for US Department of Commerce, Economic Development Administration, November.

Kasarda, J. (1985) 'Urban change and minority opportunities', in P. Peterson (ed.) *The New Urban Reality*, Washington, DC: The Brookings Institute.

Kenneth Leventhal & Company (1988) *Japanese Investment in US Real Estate, An Interim Report*, Los Angeles, California, August.

Kindleberger, C.P. (1978) *Economic Response: Comparative Studies in Trade, Finance, and Growth*, Cambridge, Mass.: Harvard University Press.

Kindleberger, C.P. (1983) *International Banks as Leaders or Followers of International Business: a Historical Perspective*, Elsevier Science Publishers BV, North Holland.

Leinberger, C.B. and Lockwood, C. (1986) 'How business is reshaping America', *The Atlantic Monthly*, October, 258, 4.

Levich, R.M. and Walter, I. (1988) 'The regulation of global financial markets', in T. Noyelle (ed.) *New York's Financial Markets: The Challenges of Globalization*, Boulder, Colorado: Westview Press, Inc.

Lindner, R.C. and Monahan, E.L. Jr. (1986) *Japanese Investment in US Real Estate: Status, Trends, and Outlook*, Cambridge, Mass.: Center for Real Estate Development Report, Massachusetts Institute of Technology.

Lubenow, G.C. (1987) 'No longer number one', *Newsweek*, National Affairs section, 13 April.

McCain, M. (1988) 'Banks create a bright spot in a sluggish market', *The New York Times*, 21 August.

Moss, M.L. (1987a) 'Telecommunications and international financial centers', in Brotchie *et al.* (eds) *The Spatial Impact of Technological Change*, New York: Croom Helm.

Moss, M.L. (1987b) 'Telecommunications, world cities, and urban policy', *Urban Studies*, 24: 534–46.

Moss, M.L. (1988) 'Telecommunications: shaping the future', in G. Sternlieb and J.W. Hughes (eds) *America's New Market Geography, Nation, Region and Metropolis*, New Brunswick, NJ: Center for Urban Policy Research, Rutgers.

Noyelle, T. (1988a) 'New York's competitiveness and overview, issues for the 1990s', in T. Noyelle (ed.) *New York's Financial Markets: The Challenges of Globalization*, Boulder, Col.: Westview Press Inc.

Noyelle, T. (1988b) 'Smooth landing: a forecast for New York City for 1989 and 1990', Testimony before the New York City Council Committee on Economic Development Hearings on 'Wall Street's crash, one year later', New York City: Conservation of Human Resources, Columbia University, 14 December.

Office of Technology Assessment, US Congress (1987) *International Competition in Services: Banking Building Software Knowhow*, OTA-ITE-328, Washington, DC.

Pascal, A. (1985) *The Vanishing City: How Technology Induces Urban Entropy and What's to be Done About It?*, Chicago: The Rand Corporation.

Poniachek, H.A. (1986) *Direct Foreign Investment in the United States*, Lexington, Mass./Toronto: DC Heath and Company.

Rand McNally & Company, Financial Publishing Division (1988) *Rand McNally Bankers Directory*, September.

Regional Plan Association, Inc. (1988) *The Region in the Global Economy*, New York.

Richardson, H.W. (1985) 'Regional development theories', in *Economic Prospects for the Northeast*, Philadelphia: Temple University Press.

Satterfield, D. (1988) 'Latin troubles, Florida's economic growth boost international bank presence in state', *American Banker*, 14 November.

Scholl, R.B. (1988) 'The international investment position of the United States in 1987', Survey of Current Business, Bureau of Economic Analysis, US Department of Commerce, June: 76–84.

Shulman, D., Canton, M. and Kostin, D.J. (1987) *Los Angeles Real Estate Market*, Salomon Brothers Inc., Bond Market Research – Real Estate, New York.

Thrift, N. (1987) 'The urban dimension of global restructuring', in J. Henderson and M. Castells (eds) *Global Restructuring and Territorial Development*, London: Sage Publications Ltd.

US Department of Commerce, Bureau of Economic Analysis (1988) Survey of Current Business, Table 23, August.

Walter, I. (1988) *Global Competition in Financial Services: Market Structure, Protection, and Trade Liberalization*, Cambridge, Mass.: American Enterprise Institute/Ballinger.

13 Information industries: New York's new export base

Thierry Noyelle and Penny Peace

INTRODUCTION

In the early 1980s, as a new wave of firms began relocating white-collar facilities outside the city, local public and private officials grew increasingly worried that the new computer-communications technology was becoming a potent weapon, one that could rekindle the extensive 'corporate exodus' that had wrecked New York's headquarters sector during the 1970s. As the new trend intensified, the city searched for appropriate policies to help stem this new out-migration, making a strong effort, in particular, to promote the outer boroughs as an alternative office location to Manhattan.

By the mid-1980s, official concern tapered off somewhat as the flow of corporate relocations seemed to moderate. In early 1987 and 1988, however, New York City was struck by a new rash of announcements concerning planned or effective departures of major firms: J.C. Penney to Dallas, TWA to Westchester, NBC to the Meadowlands, and others. Black Monday (19 October, 1987), if anything, aggravated the trend. Suddenly, the city found itself trying to respond to move-out threats from such major firms as Dreyfus Funds, American Insurance Group and Drexel Burnham Lambert. In fairness to local officials, the city, this time, proved to have some effective tools to respond to some of these new pressures – primarily tax abatements for firms willing to relocate to other boroughs – although the largesse with which some of these abatements were at times extended can be criticized.

Despite the corporate move-outs of the 1980s, the city has gained employment every year since 1977, except in 1982 when net employment declined by roughly 13,000 jobs over the previous year. Even Black Monday barely seemed to make a dent in the city's steady record of employment growth. While employment growth slowed during 1988, some 25,000 net new jobs nevertheless were added to the city's economy during that year. By mid-1988, 420,000 net new jobs had been added to the local economy since 1977.[1] Clearly, whatever the negative impact of the job exodus of the 1980s may have been, other, more powerful overriding forces were at work during that same period to sustain continued employment growth within the city.

The fact remains that we do not understand as well as we ought to where this new growth came from during the 1980s, which jobs have stayed and expanded within the city, which ones have moved out, and in which way the job exodus of the 1980s has differed from that of the 1970s.

Urban economists have many theories and models to help them understand how the spatial division of labour in manufacturing can reorganize over time and predict which manufacturing jobs are likely to decentralize and where; they have none to help them explain the evolution of the spatial division of labour in services and white-collar activities under changing market, technological and labour force conditions. What is known is often known piecemeal or is usually not well conceptualized.

In this chapter, we try to take a rigorous look at recent changes that have affected the location of a broad array of white-collar, office-based activities in New York. We group these activities under the rubric 'information industries'. The purpose of this exercise is not simply to review what has happened; it is also to assess what some of the trends portend for the future of New York's economic base.

In the next section, we define the information industries and show their importance to New York's economy, relative to other major US cities, and relative to the 1970s. We then take a detailed look at what is happening in terms of the location and the kinds of jobs that are being created by four major components of the information industries sector: corporate head-quarters; financial services; business, publishing and computing services – with a special focus on accounting and management consulting, and computer software and data processing; and the offices of foreign firms. We conclude with a discussion of the changing nature of the division of labour in white-collar activities and the potential impact of technological, market, and labour force changes on New York's economic development process.

INFORMATION INDUSTRIES: NEW YORK'S NEW EXPORT BASE

Urban economists often describe cities in terms of a distinction between local sector and export base. Output of the local sector is intended to fulfil the daily needs of local firms and the local population; by comparison, output of the export sector is meant to be sold principally to firms or final consumers located outside the city's economic boundaries, be they in the suburbs, in the city's greater hinterland, in other regions of the nation, or even abroad. Urban economists also agree that a city's export base, through the multiplier effect, represents its main engine of new growth.

Analysing the state of New York's economy in the late 1980s requires a minimal understanding of the extent of the transformation in the city's export base over the past two decades or so.

Back in 1969, New York had a varied and complex export base

organized around six principal sectors or groups of industries including: a manufacturing sector, comprising well-developed apparel and printing industries; a corporate headquarters sector, in which large corporations headquartered in the city traded their administrative and managerial services to facilities located elsewhere; a harbour-related sector; a financial and business service sector; a non-profit sector; and a business/tourist hotel/retail sector. While each of those six sectors contributed to the city's exports, at the time the first four sectors were by far the most significant forces behind New York's export base.

The story of the 1970s is, quite simply, that of a frontal attack by various market forces on three of these four major pillars of the New York economy, leading to a major retrenchment in apparel and printing, a massive exodus of jobs in the corporate headquarters sector – mainly to the city's Westchester and lower Connecticut suburbs – and a major shift of harbour-related business services out of the city – mostly to north Jersey and other parts. However healthy the financial and business services may have remained at the time, they clearly did not have the wherewithal to grow fast enough and stem the haemorrhage of jobs from the other three sectors. The outcome is by now well known: between 1969 and 1977, the city experienced a net loss of over 600,000 jobs, at which point the city finally began its recovery.

By the late 1970s, New York's export base looked quite different from eight years earlier. The importance of manufacturing and harbour-related activities to the city's economy had shrunk dramatically, while in the corporate headquarters sector the picture was more mixed. Not surprisingly then, it was increasingly towards its headquarters sector, its financial and business services, and even its non-profit and business/tourist hotel/retail sectors that New York had to turn during the 1980s to redevelop its economic base.

In the remainder of this chapter, we focus our attention on this new economic base, and mostly on New York's headquarters, financial, and business service sectors. Because creating, processing, and transferring information is both central and common to the work carried out by these sectors, we propose to look at them in terms of a single information industry sector and then to focus our attention on how the spatial division of labour is being reorganized around some of those activities.

In their purest definition, the information industries could be looked at as that group of activities that creates, processes, and transfers information – be it data, voice, or image – primarily for use by other parts of the economy. But to be useful, such an abstract definition must be connected to the empirical reality. In particular, we must account for the fact that many firms may serve both other producers and final consumers or that some firms, not primarily involved in the production of information services for others, are nevertheless, heavy producers of information services for their own purposes. The implication is that the concept of information industries must be broad enough to encompass not only a core of industries that produce for others – primarily, telecommunications, data processing, and computer software – but also a periphery of industries involved in one form or another in the creation, processing, or transfer of information either for their own use or for that of others. This periphery might include the publishing industry; the management consulting, accounting and like business service industries; the computer-communications hardware industries, especially in their functions as vendors of services, and software; the broadcasting industry; corporate headquarters; and finally, the financial industries.

Tables 13.1 and 13.2 show the importance of the information industries in New York's economic base during the 1980s in terms of their shares of all jobs and wages and salaries in 1986 (Table 13.1) and in terms of their contribution to net new employment growth between 1979 and 1987 (Table 13.2). The findings are remarkable.

By 1986, the information industries represented nearly 40 per cent (39.3 per cent) of all New York City jobs, compared to 32.1 per cent and 27.1 per cent respectively in Chicago and Los Angeles. In Manhattan only, more than one out of every two workers (52.5 per cent) was employed in the information industries in 1986, compared with 41.9 per cent in 1969. In payroll terms, the contribution of the information industries to both the New York and the Manhattan economies was even higher, with nearly two-thirds of Manhattan's wages and salaries linked to that sector. The growing importance of information industry employment in New York's economy is also underscored by the fact that during the years of recovery – 1979 to 1987 – over two-thirds of the new jobs were created in that sector (Table 13.2).

With the information industries holding such a growing share of economic activity in New York City, we must be prepared to ask whether the dynamics of local employment growth in these industries will remain strong enough to keep the city going during the 1990s or if, on the contrary, forces of decentralization may overwhelm, once again, the local economy.

Here we face two challenges. Conceptually, as indicated earlier, economists have very few theories to help them understand the changing dynamics of job location in the service industries. Empirically, conventional employment and output data developed by government agencies seldom permit analysis of the dynamics of centralization or decentralization within the office and service sectors of the economy. What are called for are models and measures pertinent both to the distinction between front-office and back-office employment and to the changing boundary between these two traditional areas of office work, as the increasing computerization of white-collar work allows the reintegration of a growing portion of back-office tasks into front-office jobs.

We do not portend to develop a full-blown theory of the changing spatial division of labour in the office and service sector in this chapter.

Table 13.1 Employment and payroll in the information industries as a share (%) of total, 1986 and 1969

SIC	Employment	1986				1969
		New York City (5 boroughs)	NYC Manhattan only	Chicago (Cook County)	Los Angeles (LA County)	NYC Manhattan only
48	Communications	2.2	2.6	1.3	1.7	3.0
737	Computer software and data processing	0.5	0.8	1.1	0.8	na
27	Publishing and printing	3.3	4.3	2.7	1.7	4.8
357, 366 & 367	Computing, telecommunications and office equipment	0.3	0.1	2.1	3.0	0.1
60 to 64 & 67	Financial services	14.7	20.7	8.8	6.7	14.3
73, 81 & 89 (−737)	Business and legal services	13.8	18.2	9.4	9.5	11.7
CAO&A	Corporate headquarters	4.5	5.8	6.7	4.4	8.0
	All information industries	39.3	52.5	32.1	27.8	41.9
SIC	Payroll					
48	Communications	2.8	3.3	2.1	2.2	3.7
737	Computer software and data processing	0.7	0.8	0.8	1.0	na
27	Publishing and printing	3.9	4.7	3.2	1.8	5.7
357, 366 & 367	Computing, telecommunications and office equipment	0.3	0.1	5.6	4.1	0.1
60 to 64 & 67	Financial services	23.8	30.6	11.0	7.7	16.2
73, 81 & 89 (−737)	Business and legal services	13.2	16.2	9.7	9.7	11.9
CAO&A	Corporate headquarters	5.1	7.3	7.0	4.0	10.6
	All information industries	49.8	63.0	39.4	30.5	48.2

Source: County Business Patterns (1986, 1969)

Table 13.2 Employment in the information industries in New York City, 1979 and 1987

SIC		1979	1987
48	Communications	79,000	65,400
737	Computer software and data processing	9,986	20,590
27	Publishing and printing	93,500	90,500
357, 366, 367	Computing, telecommunications and office equipment	12,806	8,422
60 to 64 & 67	Financial services	345,600	450,900
SIC 60	Commercial banking	144,000	172,000
SIC 62	Securities industry	76,100	156,600
SIC 63	Insurance carriers	75,700	65,100
SIC 64	Insurance agents and brokers	26,400	29,100
SIC 61, 66, 67	Savings and loan and other financial sectors	23,400	28,100
73, 81 & 89 (−737)	Business and legal services	318,714	430,310
SIC 73	Business services (advertising, consultancy, software, others)	228,800	298,000
SIC 81	Legal services	42,900	68,900
SIC 89	Accounting, architecture, engineering	57,000	63,700
COA&A	Corporate headquarters	na	na
Total non-agricultural employment		3,278,800	3,584,900
Net increases in information industries – 1979-87		206,516 (67.5%)	
Net increases in all sectors – 1979-87		306,100 (100%)	

Source: US Department of Labor, Bureau of Labor Statistics, New York Regional Office

What we try to do, however, is to examine some of the trends and tendencies in the location of white-collar work as they relate to New York City. But because conventional employment and output data do not exist, we can only rely on a patchwork of somewhat non-traditional data relevant to particular subsectors of the information industries.

CORPORATE HEADQUARTERS

In 1959, there were 142 Fortune Industrial 500s headquartered in the city and its suburbs – all but a very few firms in the city.[2] The late 1960s and early 1970s witnessed a sweeping process of change, however, characterized by both an extensive churning within the ranks of the Fortune Industrial 500 (including mergers, acquisitions, declining and rising firms) and a widespread relocation of headquarters, mostly to locations in New York suburbs.

By 1979, there were 132 Fortune Industrial 500s headquartered in New

York and its suburbs combined, representing a relatively minor change over the 1959 total. The difference, of course, is that, by 1979, fifty-four such firms were headquartered in New York suburbs compared with fewer than ten in 1959. The fact that so many Fortune Industrial 500s chose to relocate to New York City's suburbs rather than to other cities during that twenty year period has been analysed by the authors of *The Corporate Headquarter Complex in New York City*.[3] They found that many of those firms chose the suburbs over other locations because they continued to rely on the financial and business service expertise they could obtain from the city's business community.

After 1979, the city continued to lose the headquarters of Fortune Industrial 500 firms, at roughly the same rate as during the previous two decades. But, contrary to what had happened earlier, these losses were no longer balanced by gains in the suburbs. Actually there was even a small drop in the number of Industrial 500 firms headquartered in the suburbs, from fifty-four to fifty between 1979 and 1987. Overall, then, the total number of Fortune Industrial 500s headquartered in the city and its suburbs declined from 132 to 100 during the 1980s. In the city itself, the major cause for losses during the 1980s was the acquisition or dismantlement of New York headquartered firms in the wake of the wave of corporate mergers, rather than the result of relocations as had been the case earlier.

Interestingly, the recent decline in the number of Fortune Industrial 500 firms headquartered in New York City appears to have been partially compensated for by an increase in the number of Fortune Service 500s headquartered in the city. Between 1983 and 1987, these increased from fifty-six to sixty-six. Note, however, that, as in the case of the Industrial 500s, the number of Fortune Service 500s headquartered in New York suburbs declined during the 1980s, from twenty-three to seventeen in 1983.

What can be concluded from this analysis? To begin with, it must be noted that the *Fortune* lists of the largest US Industrial and Service corporations remain quite imperfect in that, with a few exceptions, *Fortune* tracks only publicly held companies. In the wake of the extensive corporate restructurings during the 1980s, including leveraged buy-outs and other forms of privatization of once publicly held companies, how well the *Fortune* lists have been able to keep score of recent changes is somewhat in question.

Also, with 183 headquarters of Fortune Industrial and Service 500 firms located in the greater New York area in 1987 (116 in New York City alone), New York continues to boast the largest concentration of large corporate headquarters in the nation.

Earlier research showed that the headquarters that left the New York area during the 1960s and early 1970s tended to be those of firms with a national market. Typical candidates for move-outs were the food companies, the airlines, the retailers, and so forth.[4] Typical recipients were cities such as Atlanta, Minneapolis, Boston, Dallas and other regional centres that were actively developing an infrastructure of business service firms sufficient to satisfy the needs of national firms. In comparison, firms that either tended to stay in New York or to move to the suburbs, were those with extensive international operations. The pressure for them was to maintain continual close access to New York's business and financial service expertise.

There is little evidence in the recent move-outs to suggest that the aforementioned pattern has changed. What might have changed, however, is that while during the 1960s and 1970s departures tended to be made up in part by new local firms joining the ranks of the 500s, New York may not have been as viable a breeding ground for new large firms during the 1980s. This might be the case, unless the centre of action has indeed moved over either to the realm of private corporations, or the service sectors, or both, where, as noted above, lists such as those developed by *Fortune* are simply not very good in tracking change.

We suspect that the trend lies somewhere in between these two developments: New York's attractiveness as a headquarter location for large industrial firms has continued to decline, its capacity to breed new large firms may have weakened somewhat, at the same time that large service firms have continued to concentrate in the city. Put another way, this analysis does suggest that ever since the late 1960s, the relative attractiveness of other cities as a favoured location not only for clerical workers, but also for professional and executive personnel, has increased and cut into what was once New York's domination of the corporate headquarters office market and of its professional and executive labour markets.

Tracking corporate headquarters move-outs may not be as significant an indicator of change, however, in the spatial division of white-collar work as it once was for other reasons as well. For while firms in the past did indeed move entire headquarters at once, the trend is increasingly towards moving discrete portions of the head offices, independently of each other rather than as a whole: executive offices, computer centres, clerical back-offices, or sales offices. Increasingly also, the locations that might be sought for such facilities may differ based on costs of labour and utilities, rents, access to labour pools, and yet other market factors.

In the next section, we examine some of these developments as they have taken shape among financial firms, a major group of firms within the information industries. When compared to other firms in the information industries, financial firms have often taken a lead in restructuring the location of their head offices along these new, more differentiated patterns.

THE SPLINTERING OF THE HEADQUARTERS: THE CASE OF FINANCIAL SERVICE FIRMS

Three preliminary observations stand out as one begins to assess trends in the geographical restructuring of the offices of financial firms. The first is the diversity of patterns that have emerged; the second is the fact that such geographical restructuring involves much more than simply the decentralization of low-level, clerical processing work; and the third is the fact that head office restructuring among financial firms is a far more mature development than is often assumed, going back nearly two decades. More importantly perhaps, an assessment of the available evidence helps to refine one's understanding of how different market and technological factors are reshaping the geography of office work.

Decentralization in the insurance industry goes back to the late 1950s. It is during this period that Newark-based Prudential, the nation's largest insurance company, initiated a decentralization plan of its clerical processing operations away from Newark, where many of these operations had been located until then. In a process that would become a forerunner of things to come in the rest of the industry, including among many of the New York-based insurance carriers, Prudential created a structure of six regional processing centres located in six major areas (Philadelphia, Jacksonville, Atlanta, Houston, Los Angeles and Minneapolis) each in charge of processing policies for its region. What is interesting is that Prudential initiated the change long before distributed data processing existed, making it clear that, while distributed data processing and advanced telecommunications may have helped firms in their efforts to relocate processing facilities, technology was not the overriding factor. Rather, Prudential's decision to relocate back-office processing was based on a mix of factors including rents, labour costs, and access to pools of clerical labour.

In all truth, Prudential's decentralization strategy was somewhat ahead of its time, and it would take until the late 1960s and early 1970s for the rest of the industry, including many of the New York-based insurance carriers, to follow suit. The process that unfolded then was very similar to that pioneered by Prudential, and has been described for one of the largest New York-based insurance companies in Noyelle's *Beyond Industrial Dualism.*[5]

Because a large share of insurance employment is concentrated in back-offices, the impact of the industry's decentralization on New York was not negligible. As shown in Table 13.2, employment by New York City insurance carriers declined by approximately 10,000 jobs between 1979 and 1987.

Curiously, however, New York's loss of insurance back-office employment to other cities over the past fifteen years or so, may have given the city greater resilience for the future. Indeed, as the industry continues to introduce new, more advanced generations of data processing technology to reorganize and rationalize back-office production further, it is the cities that became the recipients of some of these regional processing centres that may now be most threatened by change. Once again in the lead, Prudential has been engaged over the past few years in a rationalization process which ultimately will result in the closing of its Houston, Atlanta and Los Angeles life insurance processing facilities.

In investment banking, as shown by Moss and Dunau (1986) in their research, back-office relocations by New York City's commercial banks during the 1970s and 1980s focused primarily on the city's immediate vicinity – particularly Long Island and north Jersey – with a few noticeable exceptions however.[6] Chief among the latter are Citibank's credit card operations that were moved to South Dakota, Nevada, and Colorado, Citibank's travellers cheque division relocated to Fort Lauderdale, as well as various facilities of Morgan, Chase, and Chemical decentralized to Delaware.

Contrary to the case of the insurance carriers, it seems that the impact of back-office relocations on New York City commercial banking employment has remained negligible: employment in the sector grew from 144,000 to 172,000 between 1979 and 1987 (Table 13.2). This is partly due to the fact that back-office processing jobs represent a much smaller share of employment in commercial banking than in insurance.

Investment banking demonstrates yet a third pattern of transformation. Here too, it is useful to refer to research by Moss and his colleagues. In their research, Moss and Brion (1988) interviewed eighteen of the city's largest brokerage houses to determine where their processing operations had been or were being located. Their findings are summarized in Table 13.3. Perhaps, their most remarkable finding is that, while the industry went through a major phase of back-office modernization during the mid-1980s, many firms opted to keep the bulk of their operations – including their back-offices – in New York rather than to move them out.

But Moss and Brion's research is interesting in yet another way: namely, because of the distinction they establish between clerical back-offices and data processing centres. To a large extent, the distinction was moot until recently, because firms tended to locate data processing centres either with their headquarters or with their back-office clerical facilities. According to Moss and Brion, however, the 1980s saw nine of the seventeen brokerage houses that they surveyed relocating their data processing centres outside the city, on a free-standing basis (Table 13.3). There is ample evidence from our interviews with New York service firms that the trend is not unique to investment banking but has become widespread in other sectors. While it is true that data processing centres usually employ small numbers of people, this development is nevertheless important in more than one respect.

First, data processing centres typically employ mostly college-educated

Table 13.3 The location of clerical back offices and data processing centres of NYC's largest investment banks

	Clerical back-offices	Data processing centres
1. Merrill Lynch	Manhattan, Sommerset	Princeton, NJ; Stat. Is.
2. Shearson Lehman Hutton	Manhattan	Manhattan
3. Dean Witter Reynolds	Manhattan	Manhattan; Dallas, TX
4. Prudential Bache	Manhattan	Manhattan
5. Paine Webber	Manhattan	Weehawken, NJ
6. Drexel Burnham Lambert	Manhattan, Queens	Manhattan; Weehawken, NJ
7. Kidder, Peabody	Manhattan	Manhattan
8. Smith Barney	Manhattan	Manhattan
9. Thomson McKinnon	Manhattan	Manhattan
10. Bear Stearns	Manhattan	Whippany, NJ
11. Goldman Sachs	Manhattan, Brooklyn	Manhattan
12. Salomon Brothers	Manhattan	Rutherford, NJ
13. Morgan Stanley	Manhattan, Brooklyn	Brooklyn
14. Donaldson, Lufkin & Jenrette	Manhattan	Manhattan; Jersey City, NJ
15. First Boston	Manhattan	Manhattan; Princeton, NJ
16. Oppenheimer	Manhattan	Manhattan
17. Gruntal	Manhattan	Manhattan

Source: Moss and Brion (1988)

technicians, professionals and managers in relatively well-paid jobs. Evidence of their relocation outside New York thus disrupts the conventional notion that the relocation of back-offices is a process that applies mostly to clerical processing jobs, involving only low- or medium-skill personnel, and reinforces our earlier finding, based on the headquarters move-outs, that it is not only the lower echelons of New York's occupational structure that are threatened by decentralization but a wide range of jobs, including high-level jobs. Conversely, Moss and Brion's finding concerning the large number of investment houses that elected to retain their clerical operations in New York suggests that the forces of technology and labour costs need not be sufficient to pressure firms into relocating outside the city. Other forces, such as the need for proximity to customers or, as in the case of the brokerage houses, the need for intense organizational co-ordination between front-office sales personnel and back-office clerks and managers, may prevail over technology and labour considerations.

The theme of growing competition for New York's high-skilled jobs is one that is also illustrated by developments in the business, publishing, and computing service segment of the information sector, to which we now turn our attention.

BUSINESS, PUBLISHING, AND COMPUTING SERVICES

The reasons for the explosive growth of the business, publishing, and computing service segment of New York's information sector during the 1980s, noted earlier, are many and complex. They must be found in an analysis of major structural changes in the US economy during the 1970s and 1980s, which is beyond the scope of this chapter.[7]

More directly related to the present analysis, interviews with business service executives reveal that recent technological and market changes have both reinforced New York's prominence in the area of business services and increased the level of competition between the city and other urban areas. In this section of the chapter, we use two examples to highlight those trends: the accounting/management consulting industry and the computer software industry.

During the first half of the 1980s, the accounting/management consulting industry saw the increasing computerization of the once extensive number-crunching work associated with accounting, auditing, and tax services. In the process, the balance of work in the industry began shifting away from lower level professional tasks associated with book-keeping, auditing, or tax preparation towards higher level expertise: tax expertise, management consultancy, or other specialities. This, in turn, led to a geographical shift.

Whereas extensive geographical decentralization of personnel is required in audit and tax return preparation work, because accountants need to carry out much of their work on the client's premises where accounting records are kept, the new expertise – by nature often difficult to standardize – appears to be best met typically by centralizing high-level professionals in a few large urban centres from which they are able to serve market areas that are large enough to support their expertise. This is why much of the new employment growth in the accounting/management consulting industry has tended to take place in the nation's largest urban centres, including New York, not in smaller cities as had been the case during the 1970s.

In comparison to the accounting/management consulting industry, the software industry offers an example of both the current attractiveness of New York and the growing competition from other locations. Unbeknown to many, computer software and data processing, broadly defined, have become a very large component of New York's information sector. A major difficulty in analysing this sector, however, is that it is largely hidden within other sectors. Electronic hardware manufacturers, management consultants, accountants, publishers, and even large users – banks, airlines, retailers – are all large producers of computer software, information systems, and data processing services, as are independent computer software and data processing firms. Based on this broad definition of the boundaries of the industry, New York enjoyed a software and information system boom during the 1980s, perhaps as significant as that which occurred in better-known centres of the software industry such as Boston or Silicon Valley. This is clearly indicated by the findings from a postal survey of New York-based firms engaged in software, information systems, and other data processing activities that we conducted in 1988 in order to identify some key parameters of this growth. Using NYNEX Yellow Pages and *Data Sources*, an industry directory, questionnaires were mailed to slightly over 1,000 firms. Responses were obtained from nearly 200 firms.[8]

Table 13.4 shows a distribution of firms that responded to our survey by year founded. Of 198 respondents, 142 – almost three-quarters – were firms that had been formed since 1977. In addition, Table 13.5, which shows the distribution of New York software firms by location of their headquarters, suggests that most firms in our sample were New York-headquartered firms – 167 out of 192 – not simply branch offices of firms headquartered elsewhere. Finally, Table 13.6, which shows the distribution of firms by employment size, shows a distribution somewhat skewed towards the small-size employers: 58 per cent of the respondents reported

Table 13.4 Distribution of NYC software firms by year founded

	CHR survey
pre 1969	14
1969–72	17
1973–6	25
1977–80	39
1981–4	74
1985–7	29
	198

Source: *CHR Survey of NYC Software Firms* (1988)

Table 13.5 Distribution of NYC software firms by location of headquarters, 1988

		Multiple locations	
Employment size	*Single location*	*NYC headquartered*	*Non-NYC headquartered*
1–9	67	13	1
10–19	24	8	2
20–49	21	9	3
50–99	2	6	3
100 and more	1	16	16
Total	115	52	25

Source: *CHR Survey of NYC Software Firms* (1988)

Table 13.6 Distribution of NYC software firms by employment size

Number of employees	CHR survey number of firms	%
Less than 10	81 ⎫	
10–19	34 ⎬	58.0
20–49	38 ⎫	
50–99	11 ⎬	42.0
More than 100	35 ⎭	
Sample size	199	100.0

Source: *CHR Survey of New York Software Firms* (1988)

employing fewer than twenty employees. Together these findings underscore the dynamism of the sector and its largely endogenous nature.

To get a sense of the demand for the services of the New York software firms, we asked firms to tell us who their major clients were as a percentage of their annual sales. We had expected to find heavy demand from the financial services. We found that slightly over a third of the industry's 1987 sales, as reported by the responding firms, were sales to the financial industries. However, we were surprised to find the extent of diversification in the client base of the industry since another two-thirds of their sales were geared at other sectors, including, by order of importance, telecommunications and broadcasting, manufacturing, business services, hospitals and educational institutions, retailers and others.

Because several software firms interviewed before conducting our survey had expressed concern for personnel shortages in the New York area – concern shared by firms in several other business service sectors as well – we tried to find out both through our survey and through additional interviews how the software firms were meeting some of their current human resource needs. Our survey showed that one solution to shortages had been for firms to turn increasingly to women and minorities for new recruits, with their share of employment in the responding firms growing from 35 per cent to over 50 per cent between 1978 and 1988. The other solution, revealed through our interviews, not the survey, came through increasing outsourcing, including increasing subcontracting to software firms located in a few key countries trying to develop a niche in the industry: Ireland, Israel, India, Singapore, and even the People's Republic of China.[9] In all cases, what was being subcontracted was not data entry work, but programming and systems analysis work. Important on its own, this latter finding confirms other parts of our analysis in this paper, namely: that New York can no longer take for granted its dominance in areas of expertise that it may have once considered beyond the reach of others. If New York cannot meet the needs for highly skilled personnel from certain industries, they will simply devise ways for moving out.

THE OFFICES OF FOREIGN FIRMS

A last, redeeming feature of the growth of New York's information economy during the 1980s is its accelerating inclusion within the world economy.

One measure of the impact of globalization on New York's economy is foreign trade. Table 13.7 shows the impact of foreign trade (imports and exports) on New York's regional economy. In 1985, all imports and exports of goods and services (exclusive of factor payments) amounted to nearly $150 billion. This was nearly a third of the $425 billion value of the Gross Regional Product of the tri-state metropolitan area as measured by the Regional Plan Association.[10] A similar calculation for the nation for the same year shows a ratio of imports plus exports approximating 20 per cent of GNP.

Our presumption is that a narrowing down of these estimates to New York City's economy only would show an even higher share of the Gross City Product linked to trade than for the region as a whole, a greater dependence of the city's economy on service trade, in particular, and a major portion of the city's service trade linked to the internationalization of its information economy. Still, volume of trade is only one measure of New York's growing internationalization. A more complete picture would involve taking an in-depth look at the employment and revenue impact not only of cross-border trade but also foreign direct investment, both inward and outward.

The measurement issues associated with such an exercise are beyond the

Table 13.7 Trade flows in New York City's regional economy, 1985

Gross Regional Product (GRP)	$ 425 billion
Goods exports	
Ports	$ 10 billion
Airports	$ 13 billion
Services exports	$ 24 billion
All exports	$ 47 billion
Goods imports	
Ports	$ 32 billion
Airports	$ 31 billion
Services imports	$ 19 billion
All imports	$ 82 billion
Exports and imports as percentage of GRP	30.35%

Note: Tri-state metropolitan area as defined by Regional Plan Association
Source: *Regional Plan Association* (GRP figure); *NY & NJ Port Authority* (goods exports and imports); Author's estimates (Service imports and exports)

resources of this chapter. Clearly, however, a look at the impact of foreign direct investment on the New York economy would show that the two consecutive waves of foreign direct investment of the 1980s into the United States (roughly, 1981 to 1983 and 1987 to 1990) have resulted in a growing presence of foreign firms within New York's information economy: North American headquarters of foreign industrial firms establishing a presence in the United States, offices of foreign business service firms, branches or subsidiaries of foreign banks and so forth. Between 1981 and 1987, for example, the number of offices of foreign banks located in New York City increased from 342 to 435.[11]

CONCLUSIONS

A possible conclusion from our analysis would be that the transformation of New York's economy during the 1980s towards a much more specialized export base has resulted in a loss of diversity that will work to the city's detriment in future years. A frequent presumption is that greater diversity adds stability by serving as an additional buffer against the vagaries of economic cycles.[12] As far as New York is concerned, however, this conclusion is a weak one.

As history amply shows, the level of diversity that existed in New York's export base in the late 1960s did not prevent its collapse in the early 1970s. Also, while New York's economic well-being has become far more dependent on a single group of industries, its information industries, the depth and diversity that exist today within that group are clearly far greater than they were in the 1970s.

Still, whatever happens to New York's information industries in the years ahead will undoubtedly influence much of what will happen to New York's economy during the 1990s. In this respect, the research associated with this chapter has helped reveal three major sets of forces that will be likely to continue to reshape New York's information industries: decentralization, globalization, and rejuvenation through the development of new information industries.

Before exploring the implications of some of these factors further, it is useful to restate the obvious. Within the context of a continuously changing US economy, the very nature of New York's current specialization is one of its most potent weapons against the vagaries of the future. As Noyelle and Stanback (1983) have shown in their work on the rise of the service economy, on-going changes in the US economy involve a transformation of the ways in which it produces, with financial, business, and, more broadly speaking, information services becoming increasingly important inputs to the economy's processes of production. One result is that the share of information services consumed by the economy at large has been growing at a steadily, differentially higher rate than the rest of the economy and will most likely continue to do so in the years ahead.[13] Indeed, it is because many of the largest US cities entered the 1980s with some of the strongest concentrations of information industries in the nation, that they were among those that grew the fastest during the 1980s.

Within this context, several developments could either derail or reinforce a continued scenario of growth and development around the information industries. Decentralization is one factor that could derail that scenario. What our analysis has shown, however, is that the forces that continuously create new pressure for decentralization are not as unidimensional as they are sometime meant to be. Market considerations may override other cost-saving benefits that may be obtained by firms if they were to relocate, restricting considerably the use of the relocation option. There is no technological determinism that says that technology will drive out of New York everything that can possibly be driven out. In that respect, the new computer-communications technology is no different from other, earlier technological developments (road transportation, trucking, air transportation, and so forth): it enters in the decentralization equation, but it is not the only variable.

In addition, the city has showed that it can fight back and retain office facilities with its own weapons. There are problems with the city's existing retention policies, however. First, tax abatements tend to be given piecemeal and to leave too much room for firms, not the city, to determine how much is enough. The result is that individual firms asking for incentive to move to the other boroughs are simply trying to outbid each previous deal. Second, current policies are geared primarily to large employers. In effect, they tend to subsidize large firms relative to small ones. In the long run, this may be a major problem, if it is the case that the centre of new growth in the US economy has moved over to the small firms.

In this respect, several points are worth raising. New York, like the rest of the nation, is in the throes of a transformation in which a new premium on market mechanisms and small firms has taken shape. To the extent that market mechanisms have gained renewed importance, this development, in principle at least, has tended to favour large urban economies: the bigger the economy, the wider the opportunities for niches and niche players to emerge, and the better the amount and quality of available information on which markets tend to thrive. Our findings on the computer software industry in New York, and Noyelle's (1989) findings regarding the emergence of a new crop of small financial firms in the city,[14] both suggest that New York has done a good job in fostering this type of new growth. But to succeed in the long run, New York must prove to be not simply a fertile ground for new firms and new activities, but also one that can sustain their development.

New York's tradition is that of a city dominated by large firms and large employers, at least relative to other cities in the nation. Its local infrastructure of services is weak in helping small firms: rents are high; local utilities are often expensive and unresponsive to the needs of small businesses; local banks are used to doing business with large firms and often disdain small customers; and the local bureaucracy is often unresponsive to the needs of the small business enterprise. During our research, we found strong evidence among both the burgeoning software firms and the new crop of financial boutiques that if New York cannot accommodate the growth needs of these new firms, these firms will also seek to move out. In short, New York City's economic development policies will clearly need adjustments.

New York's growing insertion in the international economy should remain one of its biggest assets as long as international trade and investment grow faster than the domestic economy, which has been the case almost every year since the early 1970s. However, if such growth were to slow, perhaps because of a protectionist backlash, then New York would be adversely affected, most likely proportionately more so than other cities in the nation.

But perhaps one of the biggest challenges to New York's future is one that is far more subtle than those highlighted above and revealed by some of the underlying factors that make increasingly possible not only the relocation of clerical back-offices but also that of headquarters, data processing centres, software firms, financial boutiques, and other firms and facilities that thrive on the employment of skilled personnel. As time goes by, New York's once considerable comparative advantage represented by a unique access to large pools of skilled labour will simply continue to erode, as other urban areas and regions strengthen their human capital. More than shortages of lower skilled labour that can be partly compensated for by deepening the use of new technology, New York must worry about the weakening of what was once its unique access to skilled labour. Competition is growing tougher and, as suggested by the findings from our survey of the software industry, is coming not only from other cities or regions in the country but also increasingly from abroad. To retain a competitive advantage, New York will need to pay much greater attention than it ever did to what is needed to maintain an urban environment with amenities that are attractive to skilled labour. Here our conclusions meet part way some of the conclusions from a recent study by the New York and New Jersey Port Authority suggesting that the city is paying an increasingly heavier price for the lack of moderately priced middle-class housing and for its high congestion costs.

NOTES

1. US Department of Labor, Bureau of Labor Statistics, New York Regional Office and GOTHAM, the Drennan Econometric Model and Economic Data Base.
2. All data cited in this section are based on tabulations of the headquarter locations of the 500 largest US industrial and 500 largest service corporations as ranked by *Fortune* magazine. Data for the 500 largest service firms are

available for the 1980s only. Detailed tabulations are available from the authors.

3. Conservation of Human Resources, *The Corporate Headquarter Complex in New York City* (New York, Columbia University Press 1977).
4. Thierry Noyelle and Thomas M. Stanback, Jr, *The Economic Transformation of American Cities* (Totowa, NJ, Rowman & Allanheld 1983).
5. Thierry Noyelle, *Beyond Industrial Dualism* (Boulder, Col.: Westview Press, 1987, Chapter 5).
6. Mitchell Moss and Andrew Dunau, 'The location of backoffices: emerging trends and development patterns' (New York University 1986).
7. For example, Thomas Stanback *et al.*, *Services: The New Economy* (Totowa, NJ, Rowman & Allanheld 1981).
8. For a detailed presentation of our results, see Thierry Noyelle, Penny Peace, and Leo Kahane, 'New York's computer software and data processing industry' (Conservation of Human Resources, Columbia University December 1988).
9. See also 'Indian software exports: taking off!' *C & C* (Bombay, India: March 1988, cover story); 'Northern Ireland: survey' *(Financial Times* 3 December 1987, p. 7).
10. The estimates of service trade presented in Table 13.7 are the authors' estimates and are based on the Office of Technology Assessment of the United States Congress' study of *Trade in Services: Exports and Foreign Revenues* (1986). The estimates of goods trade are from the New York–New Jersey Port Authority.
11. For a more extensive discussion, see Thierry Noyelle, 'Smooth landing: a forecast for New York City for 1989 and 1990' (Testimony before New York City Council, Committee on Economic Development, 14 December 1988) and Thierry Noyelle, (ed.) *New York's Financial Market: The Challenges of Globalization* (Boulder, Co. and London, Westview Press 1989).
12. Samuel Ehrenhalt, 'After the Crash: New York prospects' (Bureau of Labor Statistics, US Department of Labor, New York Regional Office July 1988).
13. Stanback *et al.*, *Services: The New Economy, op. cit.*; Noyelle and Stanback, *The Economic Transformation of American Cities, op. cit.*
14. 'New York's competitiveness', Chapter 5 in *New York's Financial Market, op. cit.*

REFERENCES

C & C (1988) 'Indian software exports: taking off!', March, Bombay, India.

Conservation of Human Resources (1977) *The Corporate Headquarter Complex in New York City*, Columbia University.

Ehrenhalt, S. (1988) 'After the Crash: New York prospects', Bureau of Labor Statistics, US Department of Labor, New York Regional Office, July.

Financial Times (1987) Cover story: 'Northern Ireland: survey', 3 December, p. 7.

GOTHAM, the Drennan Econometric Model and Economic Data Base.

Moss, M. and Brion, J. (1988) *Back Offices and Data Processing Facilities in New York State*, New York: New York University, Urban Research Center, June.

Moss, M. and Dunau, A. (1986) 'The location of backoffices: emerging trends and development patterns', New York University.

Noyelle, T. (1987) *Beyond Industrial Dualism*, Boulder, Co.: Westview Press.

Noyelle, T. (1988) 'Smooth landing: a forecast for New York City for 1989 and 1990', Testimony before New York City Council, Committee on Economic Development, 14 December.

Noyelle, T. (ed.) (1989) *New York's Financial Market: The Challenges of*

Globalization, Boulder, Co. and London: Westview Press.

Noyelle, T., Peace, P. and Kahane, L. (1988) 'New York's computer software and data processing industry', Conservation of Human Resources, Columbia University, December.

Noyelle, T. and Stanback, T.M. Jr. (1983) *The Economic Transformation of American Cities*, Totowa, NJ: Rowman & Allanheld.

Office of Technology Assessment, Congress of the United States (1986) *Trade in Services: Exports and Foreign Revenues*.

Stanback, T. *et al.* (1981) *Services: The New Economy*, Totowa, NJ: Rowman & Allanheld.

US Department of Labor, Bureau of Labor Statistics, New York Regional Office.

14 Networked producer services: local markets and global development

P-Y. Leo and J. Philippe

INTRODUCTION

Producer services in France are concentrated in and around the capital. Over the years, this geographical concentration has been reinforced as it developed: certain service jobs with a low level of qualification have moved towards the provinces whereas highly qualified jobs continue to be concentrated in the Ile de France region. This fact, which is now well known, has been corroborated in other European countries by numerous empirical studies (Illeris 1989). Business services have taken over from industry in the creation of regional and urban inequalities: in all the European countries the head offices of large companies and central administrations have a draining effect and reinforce the hold that the capital city has on services.

For the second rank regions and cities these locational trends are not creating a favourable climate for development. In the European integrated market which is currently being set up, all activities will need specialized services in order to innovate, to get organized and to market their products: those places which are deprived of them will therefore find themselves out of the mainstream of the movement that is transforming productive structures.

This fact and this statement of a problem seem fair to us but appear to lead to an analytical cul-de-sac. Businesses prefer certain locations and choose to operate in certain markets for precise reasons: the global situation of the regions or cities where the business is sited does not necessarily have a decisive influence on their decision. Those studies, including our own (Leo *et al.* 1985) which are the basis for these reflections, are empirical studies founded on global statistics about the localization of services. Those sectors of activity analysed are few in number and in most cases the variables analysed are data connected to employment. The deficiencies in the statistics used are revealed in the quality of the diagnosis and in the relevance of the measures of economic policies which are proposed in order to rectify the imbalances in the siting of services.

Therefore, in this chapter, we would like to tackle the question of the localization of services from another angle, which is that of businesses, using data from the *INSEE Annual Survey of Services*. For companies with more than twenty employees this source of information offers very detailed information. To build an analysis based on data supplied by firms presupposes that two conditions are met: the first concerns the conception of the organization of businesses; the second relates to the way that businesses are grouped and classified in order to be able to generalize from the case of businesses to that of the behaviour of relevant sub-groups.

To satisfy the first condition, it is necessary to determine who does what in the service company. Very often, the distinction is made between activities or jobs carried out in the back office and those in the front office in order to explain the difference in behaviour and constraints when faced with the customer. This distinction is correct, but from certain points of view it is too simplistic. We would prefer to take Eiglier and Langeard's (1987) analysis and distinguish servuction jobs within the company from those in administration and the conception of the service. Servuction includes all the components of the production of a service as well as putting it at the customer's disposal. It is the administration of a service company which makes the management of the company's accounts and logistics possible. Conception activities are responsible for defining new services and for research and consultancy activities, for example, and they constitute the main content of the service.

Space or distance is an important variable in company management, for in order to sell a service there has to be an encounter between the producer and the consumer of that service. The service company is often compact in form because the co-existence of servuction, administration and conception personnel in the same location is the easiest way to deliver and to control the service. To separate these functions in terms of distance is difficult if not impossible for certain services, whereas for industrial or agricultural activities production and distribution can be separated.

The relevance of distinguishing servuction, administration and conception activities becomes evident as soon as one takes an interest in the development of service companies. An increase in turnover suggests the conquest of distant markets. A company which adopts this strategy must set up a particular organizational structure in order to be sure that its different branch establishments deliver identical products/services. The delocalization of servuction jobs ought to be accompanied by a more or less greater number of administration jobs. Network organization is the way this constraint in terms of space/distance is expressed.

The second condition concerns the regrouping of companies in relevant sub-groups. Many lists of service companies exist but very few cover specifically business services. We have therefore chosen to make a fundamental distinction between activities according to the nature of the service. The thirty-four activities selected, taken from the basic classification supplied by the *Annual Survey of Businesses*, can be grouped into four basic sectors: research and consultancy services, services which assist and mediate, services which mainly carry out material activities and services which make means of production (machines and people) available.

The results of the analysis are strongly influenced by the territorial setting: the urban environment is the 'natural' setting for the siting of business services, but the region also represents an ideal size market. One must therefore have the possibility of changing the territorial setting according to the aims of the analysis. In order to do this the business's data pose serious methodological problems. The survey used covers businesses: in order to get to the level of establishments, which on their own allow a complete space/distance analysis, this source of information has to be coupled with another source of statistics: the SIRENE file. However, a certain amount of statistical manipulation is necessary in order to be able to explore the spatial organization of business services.

The chosen field covers all companies with more than twenty salaried workers. There are 3,700 in all in France and they control 12,000 establishments. However, the ground covered still remains only partial for many companies in the service sector have under twenty employees. Nevertheless, we think that it is meaningful to study networked companies since very few small companies actually have a network of establishments.

In this study the geographical networks of service companies are analysed. It is part of an effort to generate a more refined knowledge of the localization of business services in France since the companies' data allow us to better appreciate the influence of head offices. The first part is devoted to an analysis of the localization and the regional markets for business services while the second part deals with an analysis of the forms of networked firms.

REGIONS, MARKETS AND SERVICE ACTIVITIES

The concentration of business services is a phenomenon that has been noticed and measured for a long time. Our analysis confirms this fact while specifying its extent and its limits thanks to the analysis of head offices and of branch establishments and to the observation of the specific behaviour of each sector considered.

Sixty per cent of business services are compact firms and only 24 per cent have two or three branch establishments, often located in the same region. These figures might lead one to think that the analysis of branch establishments is secondary in order to understand business services. But network firms, although few in number, represent 63 per cent of establishments in the field studied. The siting of head offices does not obey the same laws as those governing that of branch establishments. It is appropriate therefore to study first, the siting of head offices and then that of branch establishments.

Head offices location and control of the services markets

The question of the siting of head offices only arises for networked companies. Very often these firms keep their initial establishment as the head office and it then carries out servuction on its market and co-ordinates the different establishments at the same time. The siting of the head office is not innocuous with regard to the possibilities for company development because the search for centrality is important to its prestige. But centrality is expensive: in the end, it is only available to services with a strong non-material content which need to have their 'products' supported by a strong brand image or the prestige derived from being in certain locations.

Head offices and metropolitan activities

The metropolitan siting of head offices is an indisputable fact since over three-quarters of business services' head offices are located in cities of over 300,000 inhabitants, and 54 per cent are in Paris. Historical and commercial reasons explain the dominance of metropolitan locations for business services' head offices. First, there is the effect of decisions taken by large industrial companies to externalize their demand for services (Barcet *et al.* 1983). The considerable expansion of business services during the last twenty years has relied mainly on the decisions taken by large industrial firms to externalize their requirements for services.

This externalization was the result of a desire to control costs (by productivity subcontracting) and a search to improve efficiency (spread between offer and supply) because of the original and creative content of new services (agencies for research, recruiting, advertising, etc.). However, the externalization of tertiary activities has not brought about the geographic dispersal of business services' head offices (Leo *et al.* 1985) which have remained close to those of large industrial companies and to central government administration.

An analysis of the purchasing process for industrial goods and services brings about a good understanding of these proximity phenomena between suppliers and purchasers (Johnston and Bonomy 1981). Selling industrial goods to large firms implies making contact at various hierarchical levels in several departments of the firm and with numerous individuals. As for the sale of services, this requires meeting fewer people but demands much more frequent communication with them because of the intangible nature of services. The siting of business services' head offices near those of large firms is the result therefore of the need to communicate with those who make decisions.

Predominance of Paris

Twenty-four per cent of the French working population is to be found in the Paris region. Most large French companies have their head office there and because of this it has become the top regional market for services. The Parisian head offices of business services exercise control over 61 per cent of all establishments, 64 per cent of those which have more than twenty employees, 67 per cent of all employees and 77 per cent of all turnover. There is well and truly a phenomenon which can be described as the widening of the concentration of the market since the largest service companies have sited their head office in the capital, draining away a large part of the regional market because of their networking.

For some sectors 70 per cent of head offices are concentrated in the region of the capital. This is particularly evident for branches in consultancy: economic, sociological and market surveys (89 per cent), technical control, verification and expertise (79 per cent), data processing engineering (79 per cent), consultancy in organization management (79 per cent), consultancy and creation in advertising (75 per cent), legal and fiscal consultancy (74 per cent), co-ordination and leadership for building operations (73 per cent), rental of various equipment (71 per cent), and rental of data processing facilities (70 per cent). The concentration in Paris of head offices enlarges the geographical concentration of the market because of the ease with which research services can move about (Coffey and Polèse 1984).

The regional head office control index (the relationship between the number of branch establishments commanded by the head offices in a region and the number of branch establishments in the region) demonstrates the difference between Paris and the provinces. The index is 1.8 for Paris and is below 1 for the other regions except for the Pays de la Loire (1.2). The size of the market has a decisive influence at two levels: it favours the setting up of company headquarters in the region and the access to other regional markets. The more a region represents a large market, the more companies which have their headquarters there are large in terms of turnover and staff.

Regional cities of over 300,000 inhabitants have a relatively modest number of head offices, 24 per cent in all, but much less on the basis of turnover (14 per cent). However, some sectors locate their head offices in regional cities at above average levels; the demand for legal and judicial aid (barristers: 38 per cent of head offices) and for staff training (36 per cent), secretarial services (32 per cent), the renting of public works equipment (32 per cent) and architects (30 per cent) is important. Although they are not so numerous, the cleaning enterprises and market research firms located in these cities actually have large turnovers.

Some branches do not reveal a preference for metropolitan locations: in accountancy 48 per cent of head offices are outside metropolises, as they are for land surveyors (47 per cent outside of large towns). These two professions are both regulated and their clientele is geographically dispersed.

The attractiveness of regional markets for branch establishments

As we have seen, servuction imposes on service companies the necessity for proximity to their market. The size of regional markets seems to be the main reason for encouraging firms to set up in the regions. However, there are a number of reasons, linked to specific sectoral contexts, which show that the space/distance constraint of access to the market is managed differently from one service sector to another.

Market size and firms location pattern

An analysis of the number of branch establishments per region gives a preliminary idea of the effect of the size of a market upon the siting of service activities. The larger a regional market, the higher the number of service company branches. This effect is accentuated within and beyond the limits of the size of a market for it is observed that the bigger regions attract a larger number of companies in each sector while the smaller ones are less well served.

The number of different companies in one sector set up in a regional market shows the degree of competition, the level of development of the sector and the attractiveness of the regional market for external companies. Only the Paris region has in its territory a large proportion of the French companies present in each sector (Figure 14.1). For the other regions, the size of the market has a determining influence: in the smaller regions in most sectors the conditions for competition are not met.

The diversity of service sectors present in a regional market shows the development possibilities for firms that are located there. The effect of the size of the market again has a decisive influence; the larger markets are also the ones which offer the most diversified services. Rhône–Alpes and Provence–Côte d'Azur which make up the two largest regional markets after the Paris region benefit from the most complete range of services. This effect of market size is general, but two regions vary slightly from it: the Nord, more specialized, and Alsace, highly diversified despite its small size. Some more modestly sized markets have clear cut specialities. This is the case with Brittany with reference to data processing and industrial engineering, Aquitaine for factoring and Languedoc Rousillon for the realization of advertising campaigns.

Regional markets are always characterized by the presence of a core of business services: accountancy, technical expertise and control, hiring of manpower, advertising supports, cleaning, security services, data processing piecework, warehousing, construction and public works engineering and data processing engineering. Other activities develop around this core, largely due to the effect of market size and very little as a result of local specialization policy for developing a regional market. This is because encouraging service companies to specialize also means that they will be

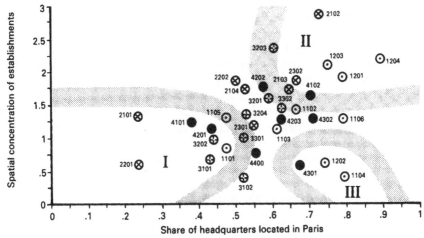

KEY:

⊙ Engineering & consultancy.

⊗ Actually carried out assistance services.

⊕ Actually carried out execution services.

● Made available services.

I Spread spatial pattern.

II Concentrated spatial pattern.

III Concentrated headquarters & spread establishments pattern.

Producer services sectors identification numbers used.

1101 Constr. & public works engineering.
1102 Industrial engineering.
1103 Other engineering.
1104 Technical control, verif. & expertise
1105 Architecture.
1106 Data processing engineering.

1201 Management and organization consultancy.
1202 Legal & fiscal consultancy.
1203 Advertising creation & consultancy.
1204 Marketing & socio/eco. surveys & studies.

2101 Surveyors & land surveyors.
2102 Building operations leadership.
2103 Technical aid.
2104 Staff training.

2201 Accountancy.
2202 Legal & judicial aid.

2301 Factoring & orders management.
2302 Enquiries, market info. & recruiting.

3101 Cleaning.
3102 Security services & cash transport.

3201 Press routeing & mailing.
3202 Data processing piecework.
3203 Secretarial piecework.
3204 Various pieceworks.

3301 Realization of publicity operations.
3302 Publicity items production.

4101 Rental of public works equipment.
4102 Rental of various equipment.

4201 Warehousing.
4202 Rent of professional premises.
4203 Real estate promotion & administration.

4301 Advertising supports management.
4302 Rental of data processing means.

4400 Manpower hiring.

Figure 14.1 Location patterns of producer service activities

obliged to find markets outside their original region. Not all companies are capable of this and cannot easily overcome the friction of distance. They would need to be able to reproduce the organization of the company in another location or else be mobile in order to produce their service.

The nature of the service given does not place all companies in the same position in order to resolve this problem. Research and consultancy activities overcome the distance from their clients with greater ease because their consultants can move about. From the point of view of the constitution of regional markets this can be seen in relation to competition on the basis of distance: the regions close to Paris have fewer consultancy agencies than Rhône–Alpes or the regions in the south and the west.

Sectors and management of the distance problem

The influential role of the size of regional markets can be reinforced or counterbalanced according to the way that companies localize head offices and branch establishments. The problems of distance that they have to manage are particular to each sector of activity. Because of this we can observe the main types of localization behaviour:

1. *Diffused sectors* where the head offices are not very concentrated in Paris and whose establishments are spread out among the regions: accountancy, data processing piecework, construction and public works engineering, architects, land surveyors, rental of public works equipment, cleaning, warehousing.
2. *Polarized sectors* concentrated in Paris both for head offices and branch establishments: consultancy and studies in organization-management, economic, sociological and market studies, creation, consultancy and brokerage in advertising, co-ordination and leadership for building operations, technical assistance, information enquiries, secretarial services and rental of diverse equipment.

We have also noticed that the activity of technical control, verification and expertise, data processing engineering and to a lesser degree, judicial and fiscal counselling, the rental of data processing facilities, advertising supports and studies and advice in organization-management present a different spatial organization: the extreme concentration of head offices is accompanied by a network of establishments in the provinces.

NETWORKED FIRMS

The network is the spatial organization of businesses which need to maintain a close relationship with local markets while hoping to benefit from a global (or inter-regional) development. For a service company which wishes to develop, progress via a network is unavoidable because this is the only method of organization which helps to win slices of the market, allows important spending on publicity and facilitates the promotion and mobility of the labour force. The network is the response by firms which face global economic constraints while obliged to manage local markets. Finally, for many service activities a network is the only way to gain access to substantial economies of scale. Unlike in manufacturing industry it is difficult for services to achieve economies of scale by increasing the size of their establishments: in fact it is often the opposite which happens because of the high degree of human interaction in these activities. It is also rare for services to offer several services/products in their establishments because of the difficulties in reproducing servuctions: access to large scale economies is therefore not easy for them.

The network means that they can reach the critical size in order to set up a powerful marketing policy. Networks are not easy to build because the organizational difficulties posed by reproducing a service at a distance show up in the last resort as elevated running costs. Due to this, we notice that networks are developed incrementally. In some sectors they remain embryonic. The analysis of networks therefore demands, first, an analysis of the service sectors where they are to be found; second, an evaluation of their efficiency, their organizational and geographical forms and, finally, their economic characteristics.

The search for economic size

The first way to examine the forms of service networks is to observe their composition according to the number of establishments. It is immediately apparent that networks comprising two establishments where one is the head office are only a primary form. These firms represent 41 per cent of networked companies and 13 per cent of establishments. Their localization shows their primary status since for 70 per cent of them branch establishment and head office are located in the same region. We can consider that from three establishments upward the network is being developed but that it remains in an embryonic state. Networks with from three to five establishments are more frequently dispersed geographically: 64 per cent of them are sited in several regions. Where there are over five establishments we are dealing with developed networks representing 63 per cent of establishments. They are rarely concentrated in the same region: fewer than 10 per cent of this last type of business fall into this category.

The choice of network is revealing about a service activity's distribution policy. The desire to reach the maximum number of clients and to cover the whole country is natural to all businesses: it reaches its limit in the management constraints of networks. Not all companies are capable of this strategy and not all activities lend themselves to it with the same ease. However, Table 14.1 shows the importance of networks for many numerous service branches. The indicators for the proportion of turnover controlled by networked firms and for the proportion of networked

Table 14.1 Networked firms' main characteristics according to service activity

Service activities	Market share by networked firms[a] (% of sector turnover)	Number of firms with more than two establishments (%)	Number of networked firm. diversified (%)	Networked firms versus compact firms ratios			
				Average turnover per firm	Average turnover per head	Average turnover per establish	Net profits/ turnover ratio
Techn. control, verif. and expertise	87.26	39.47	33.33	10.38	1.22	0.37	0.87
Rental of data processing means	83.27	42.86	88.89	6.22	0.83	0.84	1.28
Staff training	81.88	28.57	50.00	10.64	1.35	1.94	1.47
Manpower hiring	75.94	33.90	0.72	6.69	0.94	0.42	0.89
Enquiries, market info. and recruit.	71.70	27.78	80.00	7.85	1.11	1.03	2.25
Press routeing and mailing	68.35	13.85	66.67	16.48	1.39	1.88	0.48
Data processing engineering	68.09	26.22	81.39	7.78	1.18	0.90	1.02
Industrial engineering	64.19	19.57	51.85	11.68	0.90	1.48	0.66
Data processing piecework	62.48	21.82	78.33	5.86	1.15	0.75	0.96
Constr. and public works engineering	62.09	29.20	60.00	4.10	0.86	0.59	0.82
Security services and cash transport	60.54	23.66	31.82	5.27	1.02	0.50	0.75
Warehousing	59.47	27.45	42.86	3.89	1.18	0.50	1.06
Other engineering	58.81	28.73	36.37	3.51	1.18	0.53	0.84
Accountancy	56.24	31.37	49.00	3.00	0.99	0.32	0.90
Cleaning	52.50	12.96	40.00	7.84	1.00	0.85	0.71
Legal and fiscal consultancy	50.26	22.64	0.00	3.16	0.92	0.23	0.61
Rental of public works equipment	49.37	33.33	35.71	1.91	1.01	0.26	1.11
Advertising creation and consultancy	47.06	14.29	66.67	7.04	1.25	1.01	0.62
Advertising support management	46.67	27.83	81.26	2.10	0.30	0.13	1.52
Rental of various equipment	41.54	38.46	50.00	1.06	0.35	0.17	1.00
Realization of publicity operations	39.99	17.58	50.00	3.39	2.01	0.37	0.85
Promotion and admin. of real estate	38.17	22.39	33.33	2.50	1.42	0.30	0.98
Factoring and orders management	31.65	22.00	54.54	2.29	0.57	0.32	0.66
Leadership in building operations	28.57	33.33	71.43	0.81	0.59	0.19	1.17
Marketing and socio./eco. studies	27.08	15.85	38.46	2.19	1.17	0.30	1.72
Management and organ. consultancy	26.03	10.77	57.15	2.90	1.27	0.34	1.64

Secretarial piecework	23.71	12.24	50.00	2.38	1.14	0.43	1.41
Architecture	23.59	23.08	44.44	0.90	0.93	0.17	0.95
Technical aid	23.43	30.77	25.00	1.55	0.56	0.22	0.40
Rent of professional premises	19.37	17.65	66.67	1.08	0.99	0.05	0.56
Various types of piecework	19.10	15.22	42.86	1.18	0.98	0.23	0.64
Surveyors and land surveyors	14.92	8.33	75.00	1.95	1.31	0.27	0.87
Publicity items production	13.08	17.24	73.33	0.66	0.63	0.10	1.45
Legal and judicial aid	5.39	5.00	50.00	1.08	1.30	0.15	0.20
Mean	55.64	23.54	45.19	4.40	1.03	0.42	0.93

Notes: (a) Networked firms are firms with three or more establishments, compact firms have only one establishment
(b) A value 1.0 means that compact and networked firms have the same average value for the ratios concerned
A value greater than 1 shows an advantage to networked firms and less than 1 to the compact firms

companies over the total number of companies shows the strength of networks for certain activities. The lower part of Table 14.1 groups those activities which as yet are not really affected by this form of development.

The activities where networks are dominant are especially those where a service is directly carried out or made available. There are few research services except for engineering. On the other hand we find many research services as well as facilitating service activities concentrated in firms with poorly developed branch networks. From these observations and taking into account all the activities where the network form has made little progress, we can conclude that there is no sectoral logic for the building up of networks. The capital intensity of production seems to be an important factor since it is strong in networked services and weak in dispersed activities.

Building up a network is the only way to gain access to a larger sized market. This result is clear; on average, networked firms have a turnover that is 4.4. times higher than compact firms. For activities like press routeing, industrial engineering, staff training, technical control, verification and expertise, networked firms have a turnover that is ten times that of compact firms. Activities where the network does not allow the acquisition of significantly larger slices of the market retain an atomistic organization: publicity publishing, leadership on building sites, architecture, equipment rental, hire of professional premises, legal and judicial aid, various piecework.

Network organization permits gains in market share but the profitability of these developments is uncertain. The turnover of each of the branch establishments in a network is often lower than that of a compact firm in the same sector. Networked firms have to put up with the cost of reproducing their services (Philippe and Monnoyer 1989) which are specific to them. The necessity for reproducing an identical service in separate establishments and for controlling production by neutralizing any uncertainty about the result of interaction between customer and contact personnel gives rise to extra management and communication costs. Compact companies do not have to deal with these difficulties. They remain ensconced in the local market where they benefit however from the advantage of proximity.

These elements seem to counterbalance each other, for on average the turnover per head of the two categories of firms are close. Nevertheless, in some activities, networks have a turnover per head which is far superior to that of compact firms (between 10 per cent and 40 per cent).

Economic size brings firms different advantages according to the sectors of activity. Size rarely brings down production costs for services but creates status, which will eventually mean selling at a higher price. For manufacturing industry, size means cheaper production and eventually selling at a lower price if competition so demands. In services, the effects of size are not advantageous for all activities and we note that in a certain number of activities compact firms obtain better turnovers per head than networked firms: building operations leadership, technical assistance factoring, advertising supports management, rental of various equipment, publicity items production.

Firm size (number of employees) confirms the difference between the two types of firms. Eighty per cent of compact firms are small businesses with fewer than fifty employees. Firms made up of a head office and one branch establishment more frequently reach a figure of between fifty and one hundred salaried workers whereas firms having from three to five establishments are present in all the size groups. Highly developed networks are primarily represented amongst other companies with over 100 salaried workers.

Barcet and Bonamy (1988) have noticed differences in added value between branches of business services and explain them as discrepancies in the social valuation of the product/service: some services benefit from social recognition which allows them to demand high sales prices. Here the differences are not between branches but within branches between types of firms. We can invoke the 'famous name' effect in order to explain the differences in turnover per head observed within the activities of advertising creation, realization of advertising campaigns, legal and judicial aid, promotion and administration of real estate. This can be combined with economies of scale for certain services which characteristically travel well and which can concentrate their personnel. Such is the case of surveys in marketing, economics and sociology and management consultancy. In the consultancy sector large size is necessary in order to optimize the labour force according to their qualifications and to the diversity of their markets (Maister 1982). Finally, in the process of capital intensification we can provide an explanation for the greater capacity of production and sales for press routeing and mailing, data processing piecework, data processing engineering, warehousing and to a lesser degree for translation/secretarial work.

Compact firms and networked firms do not differ significantly as far as the economic profitability of companies is concerned. From this perspective there is no consistent pattern. The above-mentioned groups of activity, where network organization allows the simultaneous creation of status and economies of scale are however 20–50 per cent more economically viable than compact firms. On the other hand, compact firms are more economically profitable than networked firms in the activities of press routeing and mailing, legal and judicial aid, advertising creation, industrial engineering, cleaning, security and rental of professional premises.

The statistics used here only cover one year. They must therefore be interpreted with caution. However, general conclusions may be drawn. The choice of a networked form of organization or of development is a business decision and definition by sector has very little to do with it. The advantage of network strategy is clear in terms of economic size but is less so in terms of production results. Space or distance therefore pose serious difficulties for the development of service companies. For a long time this favoured the constitution of local markets for compact firms. It is now apparent that this situation is changing, for networks are developing more and more.

Geographical distribution

In all, 852 service companies have a network organization of more than two establishments. One hundred and forty-four firms (17 per cent) cover over half the country with their branch establishments: sixty-nine have a dense grid and seventy-five companies have adopted a looser grid but which covers more of the country. Total occupation of an area is not a frequent form of behaviour therefore, and on the whole firms prefer to adopt a pattern of development in carefully selected (if geographically dispersed) locations. Three hundred and thirty-nine firms (39 per cent) are represented in two large regions, 314 (37 per cent) are located in just one selected region.

Whether network expansion starts from Paris or from the provinces affects the form of grid development: provincial firms on the whole adopt selective development whereas Parisian firms move away more easily from their original region and tend to cover the whole country. Rather more than in the case of manufacturing industry, the regions seem to have an influence on the building of networks (space forming). Figure 14.2 demonstrates this effect clearly since the development of networks corresponds to the structure of the large regions.

The development of networks starting from head offices situated in the provinces is subject to a law of proximity rather than to regional market size. Intensive distribution in the surrounding area is a characteristic found in all the regions. Proximity is not just geographical: it can be sectoral as is shown by the presence in Upper Normandy of Provençal business networks for services linked to port activities. From Paris, regional markets do not seem to mean the same thing: proximity counts less than the potential market: the Parisian networks cover regions with potential whatever their distance from head offices.

Finally, precise analysis of location shows that networks need an urban environment in order to develop: 90 per cent of the establishments of networked firms are located in towns of more than 50,000 inhabitants.

Economic activities of networks

When a company is organized in a network, how are tasks divided between the branch establishments and the head office? Is there, as in manufacturing industry, specialization of establishments according to the comparative advantages of each place or does the necessity of servuction demand the simultaneous presence of administration and marketing personnel?

A first clue to an answer is supplied by building function. Ninety per cent of establishments are offices whatever the sector of activity. Non-sedentary activities (4 per cent) involve industrial cleaning, a few work-shops (3 per cent), warehouses (2 per cent) and retail shops (1 per cent). These figures confirm the absence of functional specialization of establishments and mark out the distribution of the sale of services to business; the service supplier goes out to his client which is the opposite to services to households. The branch establishment is therefore primarily the place of work for personnel rather than the point of sale of the service.

The constraints of servuction show up in the distribution of personnel between establishments and the head office. If services were as easily transportable as goods, decentralized establishments would only be a relay for orders with reduced personnel. Such sales networks as can be identified by the proportion of establishments with up to two salaried workers over the total number of establishments are in the minority: 13 per cent of companies. The commercial goal of these networks is to be present in the market for, apart from the head office, there are no or few establishments with a strengthened labour force.

On the other hand, servuction networks formed primarily of establishments with more than twenty employees have only a small number of establishments with a small number of staff at their disposal. There is a positive relationship between a firm's staff numbers and the proportion of large sized establishments. This suggests that for certain services having a large staff is a management necessity. Twenty per cent of businesses control a network of servuction establishments with a large staff. Between these two archetypes we can single out four intermediate categories of organizational form.

Sales networks typically involve activities where a service is made available and where intermediation is dominant. Services that are actually carried out (cleaning, maintenance, security, piecework) as well as certain advisory services are characteristic of servuction networks. It is to be noted that group companies are better represented in servuction networks than in other forms of network. This underlines the necessity of having sufficient funds in order to build up such structures. On the other hand servuction networks do ensure that the territory is covered more intensively.

Analysis of dominant activities and the turnover achieved by each type of service provides some information about diversification. It confirms that the diversification of activities among networks is an infrequent strategy. Specialization dominates since 88 per cent of networked services have the same main activity in all their establishments. This behaviour is confirmed at a more detailed level by the analysis of turnover made by different services sold by the same establishment: only 13 per cent of networked businesses offer more than three services. Compact firms are even more

Centre (Orléans, Tours) :
20 firms with
87 branch plants

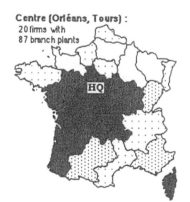

Champagne-Ardennes (Reims, Troyes) :
16 firms with
75 branch plants

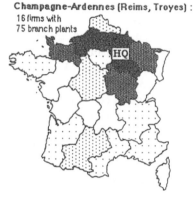

Pays de la Loire (Nantes) :
34 firms with
210 branch plants

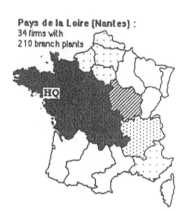

Bretagne (Rennes) :
15 firms with
65 branch plants

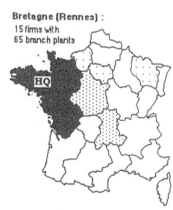

Aquitaine (Bordeaux) :
17 firms with
143 branch plants

Midi-Pyrénées (Toulouse) :
19 firms with
82 branch plants

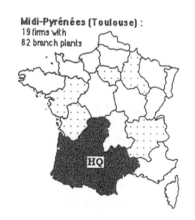

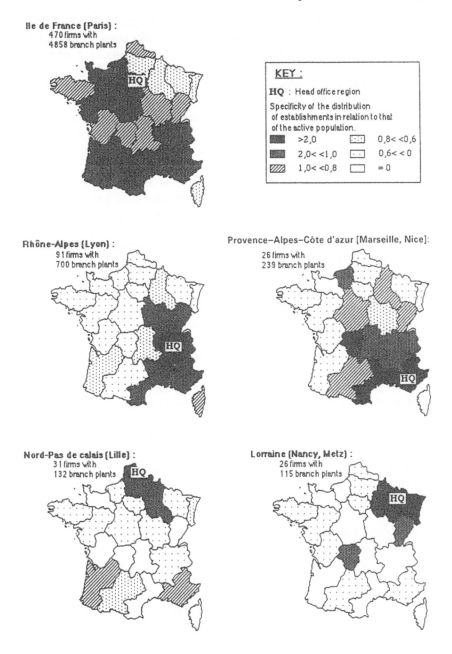

Figure 14.2 Location of non-head office establishments by networked firms, according to the head office region, France

specialized than networked firms. Eiglier and Langeard (1987) rightly point out that the diversified development of networks poses management problems because the heterogeneous nature of the units in a network makes it difficult for the head offices to control them.

Monoactivity which allows the execution of strong service concepts is therefore the attitude adopted by the largest number of firms. The co-existence of different services within one establishment poses problems of quality control of the service and demands greater flexibility on the part of personnel. It is therefore not surprising that firms prefer to manage diversification by creating different specialized establishments, but this tendency only becomes meaningful when three or more services are being marketed. The strategy of multilocalization/multiservices has been inter-preted as a sign of decline for businesses (Eigler and Langeard 1987) and is reached when the geographical market is saturated. This concept is difficult to demonstrate because we can only make a static observation. However, it is apparent that diversified networks may try to cover the national territory in a comprehensive way but their margins of geographical development seem to be reduced as a result.

An inventory of diversified and specialized service sectors strengthens these general impressions. Diversified networks include the hire of data processing facilities (83 per cent of firms are diversified), data processing engineering (81 per cent), enquiries and market information, recruitment (80 per cent), advertising support management (81 per cent), data processing piecework (78 per cent), land surveyors (75 per cent), publicity items production (73 per cent), advertising creation and consultancy (67 per cent), press routeing and mailing (67 per cent), building and public works engineering (60 per cent) and industrial engineering (52 per cent). The diversification of these activities takes place around common elements which are either technological, for data processing, or markets, for advertising and engineering. The high level of competence which is the hallmark of these activities works in favour of the coherence and the quality of the service. These businesses are very highly profitable which goes to show that the effect of the network may be beneficial even with diversified services but only under certain conditions.

Specialized networks bring together legal and fiscal advice (100 per cent of firms are specialized), the hire of manpower (99 per cent), security services (68 per cent), the administration of real estate (67 per cent), the rental of public works machinery (64 per cent), marketing and economic studies (62 per cent), cleaning (60 per cent). For these activities specializ-ation is a necessity either because of corporate organization or because of the importance of the franchise as a form of management of decentralized establishments, or yet again, because of a desire for an image of quality in research services.

It is tempting to reflect upon the influence of the judicial form of the relationship between head offices and branch establishments by comparing networks to group companies with their subsidiaries. Three hundred firms belonging to about one hundred groups have been singled out and examined in a way similar to that used for networked companies. The strongest characteristic of these service groups is their diversification: a large majority offering a range of several services (three or more). Many groups are diversified outside the sector of services to businesses. The group structure is adapted to the diversification in that each unit can have autonomous management and can carry out an efficient local marketing policy. Groups form particular networks because a large proportion of them (25 per cent) consist of embryonic networks (three to five locations), but over 50 per cent of them have more than six establishments. Whatever their composition, group firms are rarely concentrated in a single region but are spread all over the country.

CONCLUSION

Service companies are in direct contact with their clients. Production and distribution remain unified acts, unlike in manufacturing industry where production and distribution of products are carried out in different places and frequently by different firms. In order to overcome this constraint the company has to be able to move its personnel about each time the service is given or else build up distribution networks.

The development of networks requires financial resources but above all requires a clear conception of the service to be given. Firms realize this, but there is a price to pay for reproducing services of the same quality. Although indispensable in order to reach an economic size, networks do not bring any decisive advantages compared to compact firms in relation to profitability. The service sectors have a weak determining role on the constitution of networks: this chapter has clearly demonstrated the heterogeneous nature of business services and the difficulty that this creates when trying to sort them into pertinent groups for analysis.

REFERENCES

Barcet, A., Bonamy, J. and Mayere, A. (1983) 'Economie des services aux entreprises', Lyon: Economie et Humanisme.
Barcet, A. and Bonamy, J. (1988) 'La productivité dans les services, perspective et limite d'un concept', in O. Giarini and J.R. Roulet (eds), *L'Europe face à la nouvelle economie de service*.
Coffey, J. and Polèse, M. (1984) 'Les services qui voyagent le mieux sont ceux qui se concentrent le plus, La localisation des activités de bureau et des services aux enterprises', *Revue d'Economie Regionale et Urbaine*, 5, 717–29.
Eiglier, P. and Langeard, E. (1987) *Servuction, le marketing des services*, New York: McGraw-Hill.
Illeris, S. (1989) *Services and Regions in Europe*, Avebury: Aldershot.
Johnston, J.W. and Bonomy, V.T. (1981) 'Purchase process for capital equipment and services', *Industrial Marketing Management*, 10, 253–64.

Leo, P.Y., Lazzeri, Y., Monnoyer, M.C. and Philippe, J. (1985) 'L'interaction entre les prestataires de services et les PMI et le développement regional', Aix en Provence: CER.

Maister, H.D. (1982) 'Balancing the professional service firm', *Course Development and Research Profile*, 63–179.

Philippe, J. and Monnoyer, M.C. (1989) 'Gestion de l'espace et développement des services aux entreprises', *Revue d'Economie Regionale et Urbaine*, 4.

Index